S0-AJL-401

SCALING RELATIONS IN EXPERIMENTAL ECOLOGY

Complexity in Ecological Systems Series

Complexity in Ecological Systems Series
Timothy F. H. Allen and David W. Roberts, Editors
Robert V. O'Neill, Adviser

Life Itself: A Comprehensive Inquiry into the Nature, Origin, and Fabrication of Life
by Robert Rosen

Toward a Unified Ecology
by Timothy F. H. Allen and Thomas W. Hoekstra

Ecology, the Ascendent Perspective
by Robert E. Ulanowicz

Spatial Optimization for Managed Ecosystems
by John Hof and Michael Bevers

Ecological Scale: Theory and Applications
by David L. Peterson and V. Thomas Parker, Editors

Essays on Life Itself
by Robert Rosen

SCALING RELATIONS IN EXPERIMENTAL ECOLOGY

Robert H. Gardner
W. Michael Kemp
Victor S. Kennedy
John E. Petersen

– editors –

Columbia University Press
New York

Columbia University Press
Publishers Since 1893
New York   Chichester, West Sussex

Copyright © 2001 Columbia University Press
All rights reserved

Library of Congress Cataloging-in-Publication Data

Scaling relations in experimental ecology/
Robert H. Gardner...[et al.], editors
p.   cm.—(Complexity in ecological systems series)
Includes bibliographical references and index.
ISBN 0-231-11498-2 (cloth : alk. paper)—ISBN 0-231-11499-0 (pbk. : alk. paper)
1. Ecology—Methodology.
I. Gardner, R. H. II. Series.
QH541.28 .S32   2001
577'.028—dc21   00-047598

∞

Casebound editions of Columbia University Press books
are printed on permanent and durable acid-free paper.

Printed in the United States of America
c 10 9 8 7 6 5 4 3 2 1
p 10 9 8 7 6 5 4 3 2 1

To Thomas M. Frost

An admired scientist, enthusiastic collaborator, and beloved friend of many ecologists. His work on scale for the design and interpretation of limnological research was an inspiration for the workshop that produced this volume.

CONTENTS

PART I   BACKGROUND

CHAPTER 1   *Scale-Dependence and the Problem of Extrapolation: Implications for Experimental and Natural Coastal Ecosystems* 3

W. Michael Kemp, John E. Petersen, and Robert H. Gardner

PART II   SCALING THEORY

CHAPTER 2   *Understanding the Problem of Scale in Experimental Ecology* 61

John A. Wiens

FIGURES

TABLES

CONTRIBUTORS

Timothy F. H. Allen
Department of Botany
University of Wisconsin-Madison
Madison, Wisconsin, USA

Steve C. Blumenshine
Department of Biological Sciences
Arkansas State University
State University, Arkansas, USA

Walter R. Boynton
Chesapeake Biological Laboratory
University of Maryland
Solomons, Maryland, USA

Denise Breitburg
Estuarine Research Center
Academy of Natural Science
St. Leonard, Maryland, USA

Thomas M. Frost
Trout Lake Station
University of Wisconsin-Madison
Boulder Junction, Wisconsin, USA

Robert H. Gardner
Appalachian Laboratory
University of Maryland
Frostburg, Maryland, USA

James D. Hagy
Chesapeake Biological Laboratory
University of Maryland
Solomons, Maryland, USA

Colleen A. Hatfield
Department of Ecology, Evolution, and Natural Resources
Cook College, Rutgers University
New Brunswick, New Jersey, USA

Mike Heath
Scottish Office Agriculture and Fisheries Department
Marine Laboratory
Aberdeen, Scotland, UK

Edward D. Houde
Chesapeake Biological Laboratory
University of Maryland
Solomons, Maryland, USA

W. Michael Kemp
Horn Point Laboratory
University of Maryland
Cambridge, Maryland, USA

Victor S. Kennedy
Horn Point Laboratory
University of Maryland
Cambridge, Maryland, USA

Anthony W. King
Environmental Sciences Division
Oak Ridge National Laboratory
Oak Ridge, Tennessee, USA

Shahid Naeem
Department of Zoology
University of Washington
Seattle, Washington, USA

Scott Nixon
Graduate School of Oceanography
University of Rhode Island
Narragansett, Rhode Island, USA

Michael L. Pace
Institute of Ecosystem Studies
Cary Arboretum
Millbrook, New York, USA

John E. Petersen
Adam Joseph Lewis Center for Environmental Studies
Oberlin College,
Oberlin, Ohio, USA

John H. Rodgers, Jr.
Research Institute of Pharmaceutical Sciences
University of Mississippi
Oxford, Mississippi, USA

Lawrence P. Sanford
Horn Point Laboratory
University of Maryland
Cambridge, Maryland, USA

David L. Scheurer
Appalachian Laboratory
University of Maryland
Frostburg, Maryland, USA

David C. Schneider
Ocean Sciences Centre
Memorial University of Newfoundland
St. John's, Newfoundland, Canada

Frieda B. Taub
School of Fisheries
University of Washington
Seattle, Washington, USA

Robert E. Ulanowicz
University of Maryland
Chesapeake Biological Laboratory
Solomons, Maryland, USA

John A. Wiens
Department of Biology and Graduate Degree Program in Ecology
Colorado State University
Fort Collins, Colorado, USA

PREFACE

Robert H. Gardner, W. Michael Kemp,
Victor S. Kennedy, and John E. Petersen

A N EXTENSIVE REVIEW OF THE ECOLOGICAL LITERATURE DURING THE preparation of this volume revealed a broad awareness of the problem of understanding scale-dependent relationships in natural and experimental systems. Both the needs for and limitations of a "scaling theory" sufficient to understand and predict relationships have been noted by several authors. John A. Wiens (1992), who has considered many aspects of both theory and methods, has stated that "we must regard scaling not just as a bothersome feature of study design but as a subject meriting study in its own right: a science of ecological scaling." In an oft-cited review, Levin (1992) affirmed the importance of understanding these issues: "The problem of pattern and scale is the central problem in ecology, unifying population biology and ecosystem science, and marrying basic and applied ecology." Of course, theory alone cannot be expected to resolve all issues. On the other hand, Ehrlich (1989) has commented that "[g]ood theory abstracts essential features of a system from the clutter of detail that occurs in the unhappy stochastic real world. It cannot, then, be expected to serve as a tool for flawlessly predicting features of that clutter." Although comments and insights regarding scale-dependent phenomena have been driven more by observation and theory than by experiments, the message has not been lost on those wishing to empirically verify scale-dependent relationships. Perez et al. (1977) have noted that "biotic assemblages and scaling of physical variables within [mesocosm] studies have been simple or arbitrary and usually bear no resemblance to the field system," resulting in most experimental systems failing "to incorporate and, thus, consider the natural levels and/or rates of physical variables such as turbulence and water turnover." However, the advantage of "bottled ecosystems" is clear: "Microcosms make it possible to include much more complexity and biological realism in the modeling effort, including adaptation and self-design properties that are far beyond the state of the art in computer simulation" (Nixon et al. 1979). Nevertheless, issues of enclosure remain: "Since microcosm communities tend to be simplified in comparison with real-world counterparts, natural homeostatic mechanisms of feedbacks and compensatory replacement tend to be reduced" (Kemp et al. 1980), and "large scale, low frequency physical variability can impose a limit on the scale at which biological interactions operate" (Lewis and Platt 1982) in nature and experimental systems. The ultimate usefulness of experimental results often hinges on our ability to extrapolate information across a broad range of temporal and spatial scales from laboratory to nature. However, there are many potential problems in such extrapolations. For example, "the smaller a microcosm, the greater the chance that the values of parameters will be overshadowed by edge effects, exclusion of components, . . . and short-circuiting of transport pathways" (Draggan 1976). The trade-off between size and convenience is obvious. Nevertheless "some of the most important problems facing aquatic microcosm research are those resulting from the small size of laboratory microcosms because not all biological and physical processes present in natural ecosystems can be scaled down to laboratory size" (Dudzik et al. 1979). Potential problems with the reduction in temporal scales of experimental systems have also been noted. "It is critically important that ecologists recognize that short-term experiments mainly give information on transient dynamics, and that transient dynamics can be the opposite of the long-term effects of an experimental manipulation" (Tilman 1989).

In spite of these concerns or perhaps because of them Lawton (1995) has commented that "the criticisms of [experimental ecosystem studies] matter if we blindly extrapolate from the laboratory to the field. They do not matter if we treat the problems as research questions: What differences do size, simplicity, or lack of seasonality make to ecological processes?"

Issues of scale continue to be discussed at meetings and symposia, and there is a multitude of publications dedicated to this subject. Nevertheless, the experimental ecologist is hard-pressed to find specific guidance for the design, execution, and analysis of experiments to produce results that account for scale-dependent effects. Without such guidance the hope that observational and experimental results can be extrapolated across scales is greatly hindered. Exactly what are the fundamental scaling relations that apply to experimental systems? And equally important, how can experimental research benefit from and contribute to the advancement of scaling theory? Will the artifacts inherent in experimental systems affect their realism and, consequently, our ability to extrapolate information across scales? Are scaling relationships, which have been extensively developed for oceanic systems, habitat specific, or can results defined for particular ecosystems be applied across habitat types?

These issues and concerns led to an intense and interactive workshop in December 1997 in St. Michaels, Maryland. A small group of scientists and students, representing diverse backgrounds and specialties, were brought together for two and a half days to discuss a broad spectrum of empirical, theoretical, and practical questions associated with scale. The workshop was organized around an alternating series of presentations and discussions of the issues outlined above. The ideas and insights generated by the workshop have since been written down, refined, reviewed, and are now presented here.

This volume is organized into four parts. The first part, "Background," is composed of a single chapter, "Scale-Dependence and the Problem of Extrapolation: Implications for Experimental and Natural Coastal Ecosystems" by W. Michael Kemp, John E. Petersen, and Robert H. Gardner. This chapter provides an overview of issues of scale and provides the context for the following chapters. This chapter also reviews current theory and identifies both the "rules and tools" required for extrapolation. Examples used throughout illustrate that natural and experimental ecosystems differ in temporal and spatial dimensions and vary both in their normative behavior and responses to perturbation. The conclusion is that existing theoretical and empirical relationships can now be used for improved design of experiments that more realistically represent the dynamics of larger systems and provide a means of extrapolation from mesocosms to nature.

Part II, "Scaling Theory," provides insight into the vigorous dialogue and range of views on the contribution of theory to our understanding of scale-dependent relationships in experimental systems. Chapter 2, "Understanding the Problem of Scale in Experimental Ecology" by John A. Wiens, argues that there are multiple sets of factors that limit extrapolation from experiments (e.g., is the system open or closed, at equilibrium, etc.). Even though these factors may be linked, if properly identified they can reveal whether or not scaling relationships will matter or may be ignored.

Timothy F. H. Allen's chapter, "The Nature of the Scale Issue in Experimentation," further explores the practical relationships between experiments and theory. Of particular note is Allen's observation that experimental failure in the sense of hypothesis rejection can provide unique insight into the qualitative and quantitative effect of scales. Although experiments always require models, the assumptions and limitations of models can only be tested by experimentation. Thus, the judicious use of experimentation provides the critical tests defining the limits and reality of methods of predictions across scales.

Chapter 4, "Spatial Allometry: Theory and Application to Experimental and Natural Aquatic Ecosystems," by David C. Schneider, uses dimensional and power-law relationships to extrapolate information across scales. Following the informative discussion of theory and methods, Schneider uses these techniques to develop scaling relationships for actual experimental systems. The chapter concludes with a discussion of how mesocosms might be used to test and verify the existence of spatial allometric relationships.

Part III, "Scaling Mesocosms to Nature," tackles the central theme of this volume. Chapter 5, "Getting It Right and Wrong: Extrapolations across Experimental Scales," by Michael L. Pace, compares concepts and approaches of experimental results that have succeeded or failed to provide satisfactory extrapolations. For instance, measurements of primary and bacterial productivity in a nutrient-loading experiment were found to be similar to natural systems, while lake enclosure studies were less realistic. Pace notes that the lessons derived from comparison of these cases suggest that it is critical to establish at the outset the precise scale of interest and a clear statement of the context for the study.

Scott Nixon's chapter, "Some Reluctant Ruminations on Scales (and Claws and Teeth) in Marine Mesocosms," reveals the confessions of a true experimentalist. Nixon's extensive experience with the MERL mesocosms at the University of Rhode Island and exhaustive studies of Narragansett Bay have shown that some observations drawn from small samples extrapolate nicely to larger, natural systems. For example, the biogeochemical cycling of N and P (as reflected in mass balances) is similar in Narragansett Bay and the MERL mesocosms. The same is true for relationships among light, chlorophyll, and the ^{14}C uptake rates of phytoplankton. Because such "successes" almost always involve "bottom-up" interactions and small organisms, Nixon concludes that the challenge in designing and interpreting mesocosm experiments is to know when and how the exclusion of larger, longer-lived organisms and large-scale physical processes will modify the resulting behavior of experimental systems.

Michael R. Heath and Edward D. Houde collaborated on chapter 7, "Evaluating and Modeling Foraging Performance of Planktivorous and Piscivorous Fish: Effects of Containment and Issues of Scale," which investigates top-down trophic effects. Although the role of large, mobile predators in aquatic communities is important, sampling of these predators is rarely sufficient due to logistic costs and constraints. Enclosing marine predators (e.g., fish or carnivorous zooplankton) in experimental mesocosms creates special problems. Heath and Houde used an individual-based model of fish foraging behavior within mesocosms to explore the possibility that general rules might exist to allow results of experimental systems to be extrapolated. Model results predict changes in behavior and growth dynamics that scale with enclosure size, and provide appropriate dimensions for experimental research on fish consumption and growth.

Shahid Naeem outlines three classes of experiments in chapter 8, "Experimental Validity and Ecological Scale as Criteria for Evaluating Research Programs." These classes of experiments—field, model ecosystems (e.g., macro-, meso-, or microcosm), and simulation studies—each provide powerful but unique insights into nature. Naeem argues that the use of all three is ultimately required for sufficient understanding to predict across scales. A comparison of all three classes of experiments to investigate biodiversity and ecosystem functioning is used to illustrate the benefits of synthesis across multiple approaches.

Part IV, "Scale and Experiment in Different Ecosystems," provides an overview of the four discussion groups that met throughout the workshop. Records of these discussions were made during the workshop and subsequently documented and refined by the participants. The purpose of each group was to consider a series of questions revolving around the observations that: (1) experimental systems, by their very nature, are simplified versions of natural systems; (2) the artifacts introduced by size, shape, and reduction in biological complexity make extrapolation to other experimental or natural systems difficult; and (3) because different ecosystem types may be more or less amenable to experimentation, our ability to extrapolate across scales may be critically dependent on the type of system being studied.

The first of these chapters, "Scaling Issues in Experimental Ecology: Freshwater Ecosystems," was organized by Thomas M. Frost, Robert E. Ulanowicz, Steve C. Blumenshine, Timothy F. H. Allen, Frieda B. Taub, and John H. Rodgers Jr. This chapter provides an overview of the varied design, and equally varied responses, of experimental freshwater systems. The second discussion chapter, "Terrestrial Perspectives on Issues of Scale in Experimental Ecology" by Anthony W. King, Robert H. Gardner, Colleen A. Hatfield, Shahid Naeem, John E. Petersen, and John A. Wiens, notes that scaling theory has had insufficient impact on experiments within terrestrial systems. A discontinuity between theory and experimentation is evident that must be bridged to adequately resolve this deficiency. The discussion "Issues of Scale in Land-Margin Ecosystems," by Walter R. Boynton, James D. Hagy, and Denise L. Breitburg, provides interesting insights into land-margin ecosystems, which, by their very nature, are ecosystems that integrate physical and biological interactions across terrestrial and aquatic ecosystems. The final chapter, "Scaling Issues in Marine Experimental Ecosystems: The Role of Patchiness," by David L. Scheurer, David C. Schneider, and Lawrence P. Sanford, reviews the classic observations of scale-dependence within oceanic systems and notes the special difficulties associated with ocean experimentation due to the wide range of physical factors that drive ocean dynamics.

We believe that the discussions within this volume will shed new light on the problems of understanding and identifying scale-dependent behavior in natural and experimental ecosystems. Multiple examples are presented throughout the text that demonstrate the rationale and use of scaling theory to design and interpret experimental ecosystems and to extrapolate this information across spatial and temporal scales. We also hope that this volume illustrates the critical role that experimental ecology can play in advancing as well as supporting scaling theory. Knowledge of the differences between natural and experimental ecosystems is ultimately required if we are ever to predict the responses of natural systems to the multitude of factors that may modify dynamics and induce measurable change.

ACKNOWLEDGMENTS

Special thanks are due to Fran Younger for the preparation of the figures throughout the book, to Paulette Orndorff for manuscript preparation, and to Sandi Gardner for assistance in proofreading and organization of the final copy submitted to the publisher. Preparation of this volume was supported by the EPA Star program as part of the Multiscale Experimental Ecosystem Research Center (MEERC) at the University of Maryland Center for Environmental Science.

LITERATURE CITED

Draggan, S. 1976. The microcosm as a tool for estimation of environmental transport of toxic materials. *International Journal of Environmental Studies* 10:65–70.

Dudzik, M., J. Harte, A. Jassby, E. Lapan, D. Levy, and J. Rees. 1979. Some considerations in the design of aquatic microcosms for plankton research. *International Journal of Environmental Studies* 13:125–130.

Ehrlich, P. R. 1989. Discussion: Ecology and resource management—Is ecological theory any good in practice? In J. Roughgarden, R. M. May, and S. A. Levin, eds., *Perspectives in Ecological Theory*, pp. 306–318. Princeton: Princeton University Press.

Kemp, W. M., M. R. Lewis, F. F. Cunningham, J. C. Stevenson, and W. R. Boynton. 1980. Microcosms, macrophytes, and hierarchies: Environmental research in the Chesapeake Bay. In J. P. Giesy Jr., ed., *Microcosms in Ecological Research*, pp. 911–936. Springfield, Va.: National Technical Information Service.

Lawton, J. H. 1995. Ecological experiments with model systems. *Science* 269:328–331.

Levin, S. A. 1992. The problem of pattern and scale in ecology. *Ecology* 73:1943–1967.

Lewis, M. R., and T. Platt. 1982. Scales of variability in estuarine ecosystems. In V. S. Kennedy, ed., *Estuarine Comparisons*, pp. 3–20. New York: Academic Press.

Nixon, S. W., C. A. Oviatt, J. N. Kremer, and K. Perez. 1979. The use of numerical models and laboratory microcosms in estuarine ecosystem analysis— Simulations of winter phytoplankton bloom. In R. F. Dame, ed., *Marsh-Estuarine Systems Simulation*, pp. 165–188. Columbia: University of South Carolina Press.

Perez, K. T., G. M. Morrison, N. F. Lackie, C. A. Oviatt, S. W. Nixon, B. A. Buckley, and J. F. Heltshe. 1977. The importance of physical and biotic scaling to the experimental simulation of a coastal marine ecosystem. *Helgoländer wissenschaftliche Meeresuntersuchungen* 30:144–162.

Tilman, D. 1989. Ecological experimentation: Strengths and conceptual problems. In G. E. Likens, ed., *Long-term Studies in Ecology*, pp. 136–157. New York: Springer-Verlag.

Wiens, J. A. 1992. Ecology 2000: An essay on future directions in ecology. *Bulletin of the Ecological Society of America* 73:165–170.

PART I

BACKGROUND

CHAPTER 1

Scale-Dependence and the Problem of Extrapolation
Implications for Experimental and Natural Coastal Ecosystems

W. Michael Kemp, John E. Petersen, and Robert H. Gardner

EXPERIMENTS DESIGNED TO ELUCIDATE CAUSE-AND-EFFECT RELATIONSHIPS underlying the workings of natural ecosystems are fundamental to the advancement of ecological science (Lawton 1995). During the last two decades there have been two parallel trends reflected in the ecological literature that are relevant to the goal of improving the quality of ecosystem-level research. The first is an increased recognition of the importance of temporal and spatial scale as determinants of ecological pattern and dynamics in nature (figure 1.1a). The second is a growing reliance on controlled, manipulative experiments, both in the field and in enclosed experimental ecosystems, as a means of testing ecological theory (figure 1.1b; Ives et al. 1996). These parallel emphases on scale and experimentation have occurred somewhat independently of each other, creating a unique opportunity for cross-fertilization. On one hand there is a need to apply advances in scaling theory toward improving the design and interpretation of ecological experiments so that results can be more systematically extrapolated across scales to nature. On the other hand, there is a clear need for ecological experiments designed to explicitly test and advance our understanding of how scale governs ecological dynamics in nature. Although this chapter is primarily intended to provide researchers with practical insights for addressing scale in experimental design, the goals presented in the preceding two sentences are inextricably linked. . It is germane to this discussion to consider the three essential steps of experimentation, which include (1) a clear statement of hypotheses, (2) experimental design that allows for statistically rigorous and repeatable hypothesis testing, and (3) analysis of results to accept or reject the stated hypotheses. Step 2 entails manipulation of the independent variable(s) of interest with adequate control, replication,

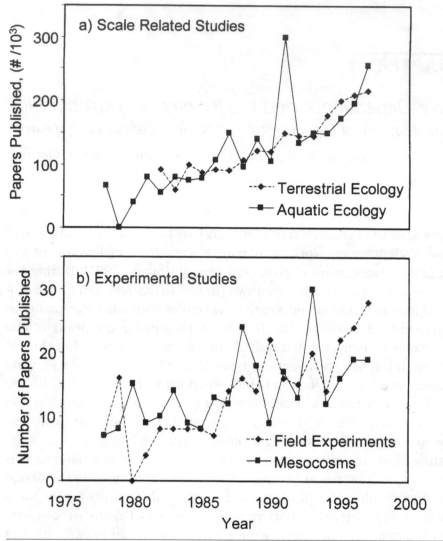

FIGURE 1•1 *Recent Trends in Experimental and Scale-related Studies in Ecology*
Trends in use of experimental approaches and scaling concepts in ecological studies
between 1978 and 1997: (a) Separate searches were conducted by year for the term "scale" in
keywords and abstracts of journals emphasizing terrestrial research (*Ecology, Oikos, Oecologia*)
and journals publishing only aquatic research (*Limnology and Oceanography, Marine Ecology
Progress Series*). The number of papers identified in each year was then standardized to the
total number of papers published for that year. (b) A similar search was conducted with all
five of these journals to identify field studies (operationally defined as those responding to
keywords "field experiment" or "field study"), and mesocosm experiments (keywords =
"mesocosm," "microcosm," "enclosure," "limnocorral"). Given these operational
definitions, it is likely that there was overlap (e.g., mesocosm studies conducted in the field)
and that many field and mesocosm studies were excluded because they did not use these
keywords. As a result, the absolute number of papers and the actual balance between field
and mesocosm studies may be in error; however, we are confident that the temporal trends
are representative.

and randomization of sampling procedure. The need for control in experiments poses particular challenges for ecosystem-level research because the key independent variables driving dynamics in natural ecosystems (e.g., light, temperature, chemical composition) are often highly variable and strongly correlated in both space and time. The increased use of enclosed experimental ecosystems ("microcosms" and "mesocosms") can be largely attributed to the perceived need for control, replication, and repeatability (Kemp et al. 1980). Steps 1 and 3 are also uniquely challenging for ecological research because they must include an assessment of the scope over which the stated hypotheses and experimental approach are valid. A number of researchers have argued that the inherently reduced scale of mesocosms restricts the degree to which hypotheses that are either confirmed or rejected through simplified small-scale experiments can be extrapolated to natural ecosystems (e.g., Roush 1995; Carpenter 1996; Resetarits and Fauth 1998; Schindler 1998).

Control, Realism, and Scale in Ecological Experiments

The problems raised in the preceding paragraph are frequently framed in terms of a balance between control and realism (e.g., Lundgren 1985; Crossland and La Point 1992; Kareiva 1994). Realism, or the extent to which an experimental system accurately represents the dynamics of natural ecosystems, is posited to be positively related to experimental scale, whereas control is thought to exhibit an inverse relationship (Kemp et al. 1980). It has been argued that an adequate degree of realism can only be obtained through in situ manipulation of whole ecosystems in nature (Schindler 1987; Carpenter et al. 1995). These researchers emphasize experiments on small aquatic systems with clearly defined physical boundaries (e.g., ponds, coves, and small lakes). In addition to the problem of obtaining adequate control and replication for such systems, however, there is no a priori reason to assume that results from these experiments can be directly extrapolated to the larger, more open, and more heterogeneous natural ecosystems that are the implicit focus of inference (Fee and Hecky 1992; Schindler 1998). Thus, in situ experiments on whole ecosystems are also subject to the same set of scaling constraints affecting "bottle experiments" (Petersen et al. 1999).

Debate regarding the relative value of mesocosm and whole ecosystem manipulation has been heated (e.g., Carpenter 1996, 1999; Drenner and

Mazumder 1999) and perhaps misdirected (Petersen et al. 1997). There are clearly numerous examples of microcosm and mesocosm studies that have provided insights in both basic and applied science (Huffaker 1958; Kimball and Levin 1985; Drake et al. 1996). The debate does, however, focus attention on the important problem of scale in experimental ecology. A crucial challenge remains to develop a satisfactory theory of scaling that allows the reliable extrapolation of results from experiments (Frost et al. 1988). In this chapter we attempt to develop an approach for generating and applying theoretical and empirical scaling relationships so as to extrapolate results from experiments conducted at inherently reduced scales to the broader scales of natural systems.

Scales in Nature and Observation

Organisms and ecological processes operate at a range of temporal and spatial scales in natural ecosystems, and several relationships between scale and properties have been well established. For example, at the ecosystem level, biotic diversity is often directly related to the horizontal scale of the habitat area (e.g., Diamond and May 1976). Vertical dimension also controls ecological pattern and process. For instance, the structure of a forest ecosystem can be related to canopy height (Oliver and Larson 1990). Similarly, the abundance and structure of marine benthic faunal communities are directly proportional to the height of the overlying water column (Suess 1980; Parsons et al. 1984). Scale is equally important at the organism level. For instance, organism size is correlated with a wide range of ecological attributes, including home range and trophic position (Sheldon et al. 1972; Peters 1983; Steele 1985; Cohen 1994).

Ecological properties are, thus, strongly influenced by the dimensions of the physical boundaries that define organisms and ecosystems. It is also clear that the patterns detected by ecologists are strongly influenced by the scale at which the observation is made. Specifically, the spatial patterns that a researcher detects have been shown to vary both with the size of the observation window (observational grain) and with total spatial and temporal extent over which the observations are made (Wiens 1989). These effects of observational scale have been well described for terrestrial (Krummel et al. 1987; Turner 1989) and aquatic ecosystems (Platt and Denman 1975; Haury et al. 1978; Lewis and Platt 1982; Hall et al. 1994; Horne and Schneider 1997; Legendre et al. 1997). The introduction of new sampling technologies at both micro (e.g., Duarte and Vaqué 1992) and macro scales (e.g., García-Molinar et al. 1993) have contributed to a growing number of quantitative descriptions of scale-dependent patterns. However, our understanding of the basic processes responsible for generating these patterns remains limited (Hutchinson 1953; Fasham 1978; Deutschman et al. 1993). It is increasingly clear that the problem of scale is not just a statistical nuisance; advancing the "science of scale" (sensu Meentemeyer and Box 1987) is a necessary prerequisite to developing a more complete understanding of ecosystem dynamics (Wiens 1989; Levin 1992).

Scale and Experimentation in Coastal Ecosystems

Although problems of scale are inherent to all experimental research, the study of estuaries and other coastal ecosystems poses unique challenges. For instance, the inherent variability in factors driving coastal ecosystems (e.g., light, temperature, tides, winds, precipitation, riverflow) reflect a complex mix of the distinct signatures of fluctuating forces imposed at terrestrial and oceanic ends of the land-sea gradient. The variability of driving forces in terrestrial habitats tends to be relatively independent of the frequency at which they are delivered, whereas in marine environments there tends to be an inverse relationship between variance and frequency of physical factors affecting ocean biota (Steele 1985). In addition, the temporal and spatial scales that characterize ecological processes and organism behavior (e.g., life-span, patch sizes, migration distances) are markedly different in marine and terrestrial ecosystems (Scheffer et al. 1993; Cohen 1994; Steele and Henderson 1994).

Estuaries are relatively unbounded open ecosystems, with bidirectional fluxes from both landward and seaward ends. The dendritic connections of estuaries to upland watersheds and the strong tidal exchange at the seaward end make it difficult to mimic physical transport in experimental estuarine ecosystems, and virtually impossible to conduct controlled experiments in situ. In addition, strong gradients of important environmental properties (e.g., salt, nutrients, and water clarity) along the land-sea gradient of estuaries (e.g., Day et al. 1989) tend to magnify effects of variable exchange with surrounding habitats. As a consequence of the inherent difficulties in conducting controlled in situ experiments, coastal ecologists have relied extensively on the use of diverse enclosed experimental ecosystems (e.g., Strickland 1967; Oviatt 1994; Petersen et al. 1999), some of which attempt explicitly to simulate physical conditions that drive the systems (Sanford 1997). Thus, the unique complexity of coastal ecosystems presents a two-edged sword. On one hand, it necessitates the use of enclosed experimental ecosystems for manipulative studies. On the other hand, the act of enclosing an estuarine community in a small container and then reducing the high degree of environmental variability that it experiences in its natural environment creates enormous difficulties for extrapolation from these studies to natural ecosystems.

Clearly we need to develop systematic methods for extrapolating information from experiments to nature, and it seems reasonable to look toward lessons derived from the increasing number of studies that have examined effects of scale (figure 1.1a). The objective of this chapter is to establish a framework for incorporating theoretical and empirical scaling relations into the design and interpretation of ecological experiments, with particular focus on coastal ecosystems. Toward this end, we consider how both means and variances for measured ecological properties change with spatial and temporal scales in natural and experimental ecosystems. We review scale-dependence in nature and consider its relevance for design of enclosed experimental ecosystems. We then discuss the prospects for developing "scaling relationships" that might allow rigorous extrapolation of results from reduced-scale experiments to full-scale conditions in nature. We have examples from recent research to describe how existing quantitative methods can be applied toward this end.

EXPERIMENTS AND SCALE: KEY CONCEPTS

Coastal Ecological Experiments

In general, the term "experiment" refers to the controlled, deductive scientific processes in which scientists are engaged, as opposed to their exploratory inductive research activities (Popper 1962). Four classes of ecological experiment can be distinguished based on complexity, scale, and degree of exchange: (1) enclosure studies of populations and ecological community, where there is little attempt to simulate biological or physical complexities of the natural habitats; (2) studies in enclosed experimental ecosystems containing artificial boundaries that restrict exchange of matter and energy; (3) whole-system manipulations of naturally bounded ecosystems in nature; (4) open, marked-plot experiments in nature. The last of these approaches is distinct from all others in that the experimental unit is completely open to exchanges with external (unmarked) environments. Although these plot-type studies can be used to address certain questions in benthic and coastal wetland habitats, rapid rates of advection and diffusion render this approach difficult to use in open pelagic environments. The high degree of openness that characterizes all types of coastal habitats also makes manipulation of whole natural ecosystems (experiment type 3, above) difficult. Thus, as we have already argued, mesocosms are the principal tool available for ecosystem-level manipulative research in the coastal environment.

Enclosed experimental ecosystems can be characterized by a number of related criteria, including complexity, initiation, location, and scale. For instance, mesocosms range in complexity and specificity from relatively simple, tightly controlled models of generic ecosystems (e.g., Nixon 1969; Taub 1969) to highly complex models of specific natural ecosystems (e.g., Oviatt et al. 1981). The former are often initiated piecemeal from constituent components (e.g., sediment material, chemical media, individual populations of organisms), whereas the latter are typically constructed using water and intact pieces of sediment/soil taken from natural ecosystems. This latter category may be initiated in situ with installation of enclosing structures (e.g., rigid walls, bags, cages) directly within a larger ecosystem or by removing sections of nature and installing them in a remote enclosure. Obviously, the degree of experimental control varies with experimental design and tends to be greater in relatively simple generic systems than in very complex in situ systems.

The scale of a mesocosm study is defined by a number of attributes. Spatial scale is defined by dimensions of the containment structure (width, depth, volume), by the size and shape of internal physical structure (e.g., bottom substrate, coral, or macrophyte surfaces), and by the size and ambit of the organisms contained. The temporal components of scale include the duration of study, the frequency of sampling, the rate of water exchange, the life-span and generation times of the organisms involved, and the rates of the various biogeochemical processes of interest. Some of these scales (viz., container depth and volume) tend to be reduced in comparison with the natural environment that is being simulated. However, other scales that characterize an experimental ecosystem can be controlled within limits by the researcher (e.g., organism size, experimental duration).

One of the principal ways that mesocosm systems differ from natural ecosystems is in the presence of walls. These physical structures are typically designed to confine and/or exclude specific mobile organisms, to define and retain the experimental volume (sediment, water, air) for measurement over time, and to limit (by flow rate or filtration) exchange of fluid media and associated materials to and from the experimental space. The degree to which these physical structures (e.g., fences, walls, domes) precisely regulate exchange of material and/or energy is one measure of the degree of experimental control. Although these containment structures enhance controllability, they also tend to create experimental artifacts. These artifacts derive from two sources: (1) the physical surfaces of containment structures provide habitat for an undesirable community of attached organisms that can alter the ecology and biogeochemistry of the system; and (2) the structures restrict exchange of material and physical energy.

Scaling Concepts

First, for our purposes a useful definition of scale is "the spatial or temporal dimension of an object or process, characterized by both grain and extent" (Turner and Gardner 1991). Grain is the spatial or temporal resolution chosen to analyze a given data set, whereas extent is the size of the study and the total duration of over which measurements are made (Wiens 1989; Allen and Hoekstra 1991). It is helpful to distinguish three distinct contexts for the terms "grain" and "extent" that vary depending on whether the data of interest are observed in nature, collected through experimental manipulations, or measured as intrinsic scales of the natural system. These are explained below.

Observational grain and extent refer to the scaling characteristics of data collected in spatial or temporal series. The observational grain is the selected level of resolution, and observational extent is the total area or duration over which observations are made. These definitions are solely dependent on the nature of the data collection method, and they say nothing about the underlying structure of the ecological system. For instance, satellite imagery has characteristic grain and extent defined by the instrument measuring the spectral characteristics of the earth's surfaces. Most of the literature of landscape ecology uses the terms "grain" and "extent" in this context.

Second, *experimental grain and extent* are similar, but refer more specifically to the spatial (length, area, or volume) and temporal (frequency, duration) scales of an experimental system and study design (MacNally and Quinn 1998). For example, the spatial and temporal grain of a particular experiment might be 1 L of water sampled at an interval of once per day. Experimental extent, on the other hand, would refer to the size of the system being sampled (for instance, the volume of an experimental ecosystem) and the total duration of the study. For both observation and experimentation it is not possible to make inferences about spatial dynamics that operate at scales finer than the grain size or broader than the total extent of the experiment (Wiens 1989). For both observation and experimentation, a particular ecological property is said to be scale-dependent if the magnitude (or variability) of that property changes with a change in either the grain or extent of the measurements (e.g., Schneider 1994).

Third, *natural or characteristic grain and extent* refer to spatial and temporal scales associated with boundaries that characterize natural phenomena. For instance, in a school of fish characteristic grain might be the size or generation time of an individual fish, whereas characteristic extent might be the size and longevity of the school itself. These characteristic scales often differ from observational and experimental scales in that the temporal and spatial dimensions are always defined by objectively identifiable natural boundaries. Characteristic scales are somewhat analogous to the "levels" described in hierarchy theory (O'Neill et al. 1986), and defining them may provide important insights into system dynamics because processes with similar grain or extent are likely to interact most strongly with each other. These natural scales can be identified by systematically varying observational grain and extent; rapid changes and discontinuities in the measured process that occur over small changes in observational grain or extent indicate boundaries that define the characteristic scales.

It is important to recognize that in all three contexts discussed above, grain and extent are dependent either on the frame of reference and sampling technology used by the investigator or on the definition of processes of interest (e.g., Wiens 1989; Allen and Hoekstra 1991). For instance, from the perspective of a population ecologist working with insects, a small (e.g., 100 m) plot and the specific season (e.g., summer) may define the experimental extent scale. From the perspective of an ecosystem scientist, the same plot may represent experimental grain, and the size of the entire watershed might appropriately define both experimental and characteristic extent. From the perspective of an ecologist interested in problems of global change, that same watershed may represent experimental and characteristic grain size (i.e., single pixel) in a model that defines extent as the regional landscape or even the whole biosphere.

A first principle of designing "scale-sensitive" studies (sensu Bissonette 1997; Petersen et al. 1999) is to maximize coherence among observational, experimental, and characteristic scales. For instance, in a given study it might be advantageous to set experimental duration (i.e., experimental extent) as an even multiple of the generation time (i.e., characteristic extent) of the dominant consumer. Likewise, it is important to consider the home range (i.e., characteristic extent) of a dominant organism in selecting mesocosm size (i.e., experimental extent). Experimental designs that fail to match the characteristic scales of key organisms and processes frequently result in erroneous conclusions (Tilman 1989). Because it is part of the study design, experimental scale can also be treated as an independent variable. We will discuss later the substantial implications of tracing changes in system dynamics as a function of scale.

There are other contexts in which the term "scale" is commonly used in the literature. For instance, ecologists may use scale to refer to the levels of organization under investigation (organism, population, community, etc.), the number or diversity of different types of organisms, the number of biogeochemical pathways included, and the number of different habitats or subsystems (e.g., Frost et al. 1988; Steele 1989). It is tempting to suggest that these attributes of ecological complexity represent a third scaling axis equivalent to time or space. Complexity is, however, distinct from time and space in that its meaning is highly context-dependent and can never be reduced to relationships among fundamental units; measures of species diversity and of biogeochemical complexity will always be apples and oranges.

THEORY OF SCALING RELATIONS

Many ecological properties change quantitatively with changes in scale, and the scale-dependence of these properties can be considered in terms of both mean and variance. On one hand, we may directly observe how mean values for the property change with scale. For example, small schools of fish behave differently (e.g., in movements or effect on prey populations) than do larger schools. Changes in mean values with scale can be measured by direct observations on schools of different size. On the other hand, continuous space- or time-series data can be used to reveal changes in variability with scale. For instance, continuous changes in observational grain might be used to examine how the relative variances of fish species and their prey change with scale. These two approaches have produced the bulk of existing information on scale-dependence of ecological properties, and both provide potentially useful insights of relevance to the design and interpretation of experiments. To make these concepts useful to experimentalists we must go beyond these superficial observations and clearly understand how scale-dependence effects change the system being investigated.

General Scaling Relationships

SCALING MEANS WITH EXTENT. Mean values of ecological properties often exhibit continuous, monotonic changes with extent (e.g., area, height, duration, age) of the system containing them. For example, interactions between pelagic and benthic habitats vary with mean water depth of lakes and coastal bays (e.g., Hargrave 1973; Kemp et al. 1992), whereas seagrass production and biomass can vary directly with depth of organic sediment (Zieman et al. 1989). In addition, diversity and biomass of respective plant communities tend to increase with ecosystem age along successional series for both terrestrial and aquatic ecosystems; however, time-scales are 100- to 1,000-fold longer for terrestrial habitats (e.g., Odum 1971). A great number of physiological rates and behavioral traits vary as a power function of organism size throughout the plant and animal kingdoms (Peters 1983; Calder 1984). It is also well established that biotic diversity varies as a logarithmic function of area within a given habitat (e.g., Odum 1971) and among islands within an archipelago (Diamond and May 1976).

In some cases, variations in ecological properties with extent scale are discontinuous, with evidence of thresholds at transitional scales. For instance, zooplankton population abundance in coastal ecosystems increases gradually with decreasing water residence-time (increased nutrient delivery rate) until residence-time approaches the zooplankton reproductive time-scale, beyond which animals are flushed from the system more rapidly than they can reproduce and abundance declines abruptly (Ketchum 1954). Changes in water retention time for natural and experimental aquatic ecosystems can also cause a selection for fast- or slow-growing planktonic species, resulting in sharp shifts in relative species abundance (Margalef 1967).

Even when mean ecological properties exhibit smooth monotonic relationships with extent, ability to extrapolate across scales may be limited. This is illustrated in a conceptual diagram with two hypothetical scaling relationships that follow different trajectories (figure 1.2). At small extents the upper curve follows a first-order function of extent, grading to a zero-order relationship at broader scales. Conversely, the lower curve grows from an initial scale-independence at small extents to a first-order function at greater extents. The crosshatched area of the figure at small extents indicates the inherent limited range for controlled experiments. Thus, without quantitative description of the full relationships across a broad range of scales, extrapolations will tend to be biased. For the upper curve, extrapolations from observations at small extents will tend to overestimate conditions at greater extents, whereas such extrapolations from small to large extents along the trajectory of the lower curve would lead to underestimates. These examples emphasize that rigorous extrapolation of information from one scale to another requires an identification of factors affecting changes in processes with changes in scale.

SCALE-DEPENDENT VARIANCE. Based on continuous series of observations, the relative variance (e.g., variance/mean) for most ecological properties tends to follow a consistent general pattern with scale. As the grain of observations is increased (with extent held constant), relative variance between grains (or samples) tends to decline (Wiens 1989; figure 1.3a). Conversely, as the resolution of observation window decreases to finer and finer scales, the relative variance among observations tends to increase (figure 1.3b). When these scaling relationships are replotted as the log of the variance versus the log of the scale, the resulting slope reveals information about the scaling pattern. A slope of -1 indicates that the property is randomly distributed in time or space; the degree of deviation from slope $= -1$ provides a measure of spatial patchiness or temporal pulsing (e.g., Gardner 1998). In hierarchically structured systems, we would expect that the pattern involves a staircase sequence of discontinuities abruptly punctuating relatively scale-independent regions (dashed line in figure 1.3a), which when linked together give the

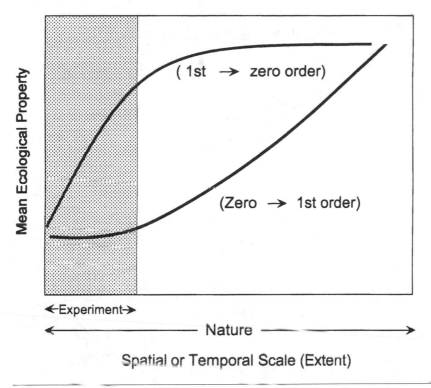

FIGURE 1•2 *Hypothetical Scaling Trajectories from Experiment to Nature*
Conceptual diagram illustrating hypothetical variations in mean ecological properties with changes in spatial or temporal scale (extent) in nature. The crosshatched area at smaller scales indicates the possible limited range for controlled experiments. Shown are two hypothetical trajectories along which extrapolation of observations at one scale to conditions at another scale is complicated by the nonlinear scaling relations.

appearance of a relatively smooth scale-dependent relationship (e.g., O'Neill et al. 1991). There is a series of boundaries or thresholds that separate hierarchical levels, with the height of these jumps being a measure of the distinctiveness of adjacent levels. Although relative variance tends to decrease with increasing grain, the inverse tendency holds with increasing extent. As one increases the extent, or full range, of observations (with observation grain held constant), the heterogeneity of physical parameters and biological resources are also likely to increase, causing the relative variance to be positively related to the extent of the area sampled. As was the case for grain-scaling relationships, the overall pattern of increasing variance with extent will tend to exhibit discontinuities and plateaus that reveal hierarchical levels and the boundaries separating them (Wiens 1989).

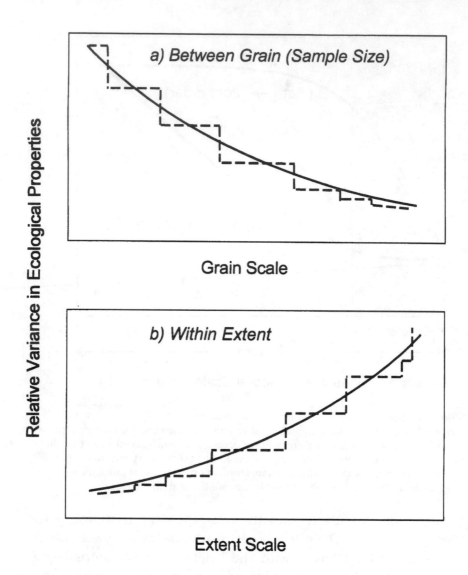

FIGURE 1•3 *Conceptual Relationships Between Scale of Observation and Ecological Variance*
Generalized relationship between relative variance (variance/mean) in ecological properties
and scales of observation, including both (a) grain scale (size of sample or unit of
observation) and (b) extent scale (overall size of study area). Dashed lines indicate that there
are important discontinuities in the trajectory from small to large scales, representing
boundaries and thresholds encountered. (Adapted from Wiens 1989)

There is an analogous set of hypothetical relationships that might
generally characterize how relative variances for ecological properties
scale with size of experimental ecosystems (figure 1.4a, b). In this case, it

is convenient to think of the size of the experimental ecosystem as equivalent to the grain scale of observations made in natural ecosystems. We hypothesize that relative variance among replicate systems tends to decrease as the size of experimental ecosystems increases. Conversely, we anticipate that the spatial heterogeneity possible within larger systems will cause relative variance among replicate samples taken within a single experimental ecosystem to increase with system size. In this latter case, size of the experimental system is a measure of extent. Obviously, the strength of this pattern will depend on the degree of internal mixing within the experimental system. Physical mixing is of primary importance in aquatic systems, whereas terrestrial systems may depend more on biotic mechanisms (seed dispersal, organism motility) for internal homogenization.

A major source of variance among replicate experimental ecosystems is the sum of differences in initial conditions, which become amplified with longer duration of study, but which tend to be buffered by internal feedback effects (e.g., predation, competition, nutrient cycling), particularly in larger systems. Sensitivity to initial conditions is more pronounced in closed systems with limited external exchange (Beyers and Odum 1993). Temporal scaling relations for experimental ecosystems are less well defined than spatial scaling, but they also derive from "founder effects" (initial conditions) and subsequent community dynamics. Our experience suggests that variance among replicate systems (figure 1.4c) tends to increase with experiment duration, as small differences in initial conditions are enhanced over time. Once internal resources are depleted, however, variance declines rapidly, and systems become dominated by the most efficient components. We hypothesize that variance within an experimental system (figure 1.4d) tends to decline with study duration due to internal homeostatic selection processes (e.g., Summers 1988), following an incipient rapid increase as diverse habitats are initially occupied.

SCALE DEPENDENCE OF EXPERIMENTAL ARTIFACTS. There may also be a set of general relationships whereby the artifacts of experimental ecosystems change systematically with spatial and temporal scales of the systems. By artifact we mean the departure of a specific ecological property measured in an experimental system from that observed in nature. Container effects, such as those related to wall growth and reduced exchange, tend to decrease with increasing container size or spatial scale (figure 1.5a). This general relation occurs because of the geometric reality that "edge

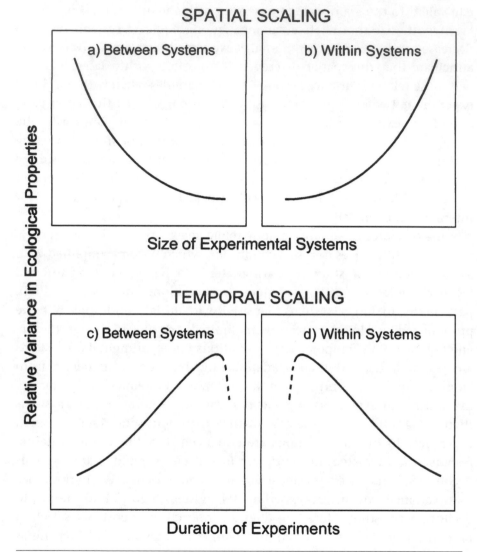

FIGURE 1·4 *Generalized Relationships Between Experimental Scales and Ecological Variance*
Generalized relationships between relative variance (variance/mean) in ecological properties and scale in experimental systems. Spatial scaling relations derive directly from figure 1.3, with (a) relative variance between replicate systems decreasing with system size (grain), and (b) relative variance within experimental systems increasing with size (extent). Hypothetical temporal scaling relations suggest that variance among replicate systems (c) tends to increase with experiment duration, as small differences in initial conditions are enhanced until a point where internal resources are depleted; (d) variance within experimental systems tends to decline with time (after an initial rapid increase) due to internal selection processes.

effects" (volume: area or area: perimeter) decrease with system size, particularly diameter (Chen et al. 1997). On the other hand, artifactual effects of containment tend to increase with experimental duration, as the changing experimental system has more time to adapt to its confined conditions (figure 1.5b). Here, small initial differences are magnified by reduced exchange with the external world that would otherwise serve to renew and stabilize the experimental system (Beyers and Odum 1993).

SCALING RELATIONS IN NATURAL AND EXPERIMENTAL ECOSYSTEMS

In previous sections we have discussed idealized scaling relationships that appear to be broad and general in their application. Here, we discuss examples of scaling relationships observed in natural and enclosed experimental ecosystems and suggest possible relevance of these relationships to the design and interpretation of ecosystem experiments.

Scaling Ecological Properties with Ecosystem Extent

Mean and integral values of ecological properties often vary with the size or age of an ecosystem or habitat. Such scaling relationships derive from a variety of mechanisms, including physical or chemical gradients created by forces or inputs applied at one end of the system, rates of fluid exchange, frequency of habitat destruction events, and reproduction times or ambits for specific organisms.

PELAGIC-BENTHIC INTERACTIONS SCALED TO WATER DEPTH. Two potent physical forces dominating most ecosystems are sunlight and gravity. In aquatic environments sunlight incident at the water surface is absorbed and reflected as it passes down through the water column, creating an exponential gradient of light-diminution along the ecosystem's depth dimension. Particulate organic material that is formed photosynthetically in the well-lighted upper layers of aquatic ecosystems tends to sink through the water column depth toward the sediment surface. Here the sinking is driven by the force of gravity mediated by fluid viscosity and the size, shape, and relative density of the suspended particles. Gravity also acts to allow water masses of different density to be separated vertically, creating a stratified water column that limits the rate of exchange of dissolved solutes along

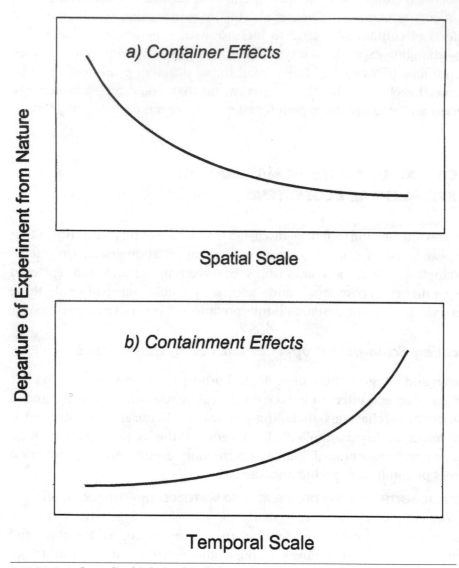

Departure of Experiment from Nature

a) Container Effects

Spatial Scale

b) Containment Effects

Temporal Scale

FIGURE 1•5 *Generalized Relationships Between Experimental Scales and Artifacts*
General trends by which experimental artifacts tend to vary with scale of experiments:
(a) container effects (e.g., wall growth) tend to decrease with increasing spatial scale
(container size), whereas (b) containment effects tend to increase with experimental
duration.

vertical concentration gradients (e.g., retarding fluxes of oxygen and
nutrients from surface to bottom and bottom to surface layers, respectively).
In aquatic environments, quantitative and qualitative aspects of these

"pelagic-benthic interactions" are all related to the depth of the water column. Although analogous interactions between ecological processes in the forest canopy and soil duff exist, these coupling processes along the vertical dimension tend to depend more on physical constraints in aquatic ecosystems (Kemp and Boynton 1992).

Comparative analyses of data from a variety of aquatic ecosystems have demonstrated significant relationships between processes associated with pelagic-benthic interactions and water column depth. For instance, the relative fractions of phytoplankton production that sink through the upper mixed layers of marine ecosystems are inversely related to depth of the water column, considering systems with depth differences ranging over 1000-fold (figure 1.6a; Suess 1980). Other investigators have also described significant relationships between water depth and rates of particulate organic matter deposition in lakes and coastal marine ecosystems (Hargrave 1973, 1979; Baines et al. 1994). Obviously, water depth may also be important in determining the fraction of primary production sinking to the sediments in experimental ecosystems (e.g., Oviatt et al. 1993), and such information could be used in relating experimental conditions to those in nature.

Rates of benthic community respiration, associated nutrient recycling, and other sediment biogeochemical transformations also vary with water column depth. As was the case for particulate organic deposition, comparative analysis has revealed that sediment respiration rates (Jørgensen 1983) and nutrient recycling rates (Harrison 1980) tend to decrease across large decreases in water depth (e.g., 1 to 10,000 m) in marine environments. Similar inverse relationships between benthic respiration and water depth have been reported over a smaller depth range more relevant to coastal ecosystems and their experimental replicas (figure 1.6b; Kemp et al. 1992). Given the diverse effects of other factors on benthic community processes, it is somewhat surprising that such relationships can be discerned within the smaller range of length scales. Recent scaling experiments involving mesocosms ranging in water depth from 0.4 to 2.1 m also revealed an inverse relationship between benthic community recycling of ammonium and depth (B. Bebout and J. Cornwell, unpublished data).

Scaling patterns are not always identical in mesocosms and nature. For instance, relative rates of benthic N-recycling/N-input were found to follow water column depth by an inverse exponential relationship for both coastal ecosystems (figure 1.7a) and mesocosms representing those

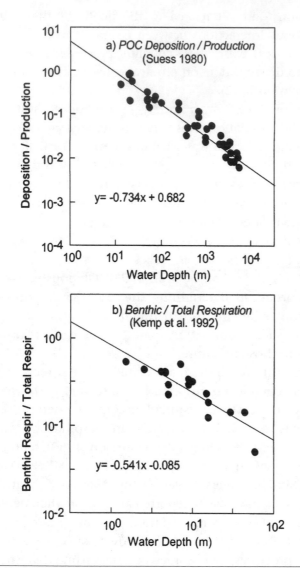

FIGURE 1•6 *Scaling of Pelagic-Benthic Processes to Water Column Depth*
Examples of pelagic-benthic processes (mean values) scaling to water column depth:
(a) proportion of phytoplankton primary production that is deposited as particulate organic
carbon (POC) through the thermocline of the world's oceans (Suess 1980); (b) proportion of
total ecosystem respiration (Respir) associated with benthic community in coastal marine
environments (Kemp et al. 1992).

environments (figure 1.7b). Although the shape of the scaling
relationship was similar for the natural estuarine ecosystems and their
experimental replicas, relative recycling rates were lower in mesocosms

than would have been predicted from the relationship for natural ecosystems based on depth alone. This is because most of the nutrient pools and recycling rates had accumulated in periphytic wall communities in the experimental ecosystems (Chen 1998). Hence, when experimental artifacts such as wall growth are allowed to dominate mesocosm systems, scaling relationships may be altered.

SCALING PLANKTONIC SYSTEM FUNCTION TO DEPTH AND RADIUS. The preceding discussion implies that two separate classes of scale effects can be distinguished in mesocosm research (Petersen et al. 1997). The first class can be termed "fundamental effects" of scale. These include differences in the behavior of ecosystems that can be directly attributed to those dimensions of scale, such as ecosystem depth, that have a common effect on all ecosystems of a given type. The second class of scaling effects can be termed "scaling artifacts" associated with enclosure. Scaling artifacts are unique characteristics of experimental ecosystem structure and function that separate them from natural ecosystems, purely as a result of enclosure. Artifacts of enclosure in aquatic systems include differences that can be attributed to factors such as periphyton growth on mesocosm walls, alteration in material exchange rates, and distortions in the mixing and light regimes. Developing an improved understanding of and ability to distinguish between fundamental effects and scaling artifacts is essential for comparing results among mesocosm experiments and for extrapolating information from enclosed experiments to natural ecosystems.

Experiments conducted at the University of Maryland's Multiscale Experimental Ecosystem Research Center suggest that these two classes of scaling effects can be explored through multiscale studies in enclosed experimental ecosystems. Because depth varies in natural as well as mesocosm ecosystems, depth-effects tend to be fundamental scaling factors. Radius (r), on the other hand, controls the wall area per unit volume (wall area/volume = $2/r$), such that effects of varying experimental radius tend to be attributable to artifacts of scale. To elucidate potential scaling effects, we conducted a series of experiments in planktonic mesocosms subjected to systematic variation in both depth (from 0.5 to 2.1 m) and radius (from 0.2 to 1.8 m).

We found that ecosystem gross primary productivity (GPP) was strongly related to depth scale, but that the specific form of the scaling relationship was different under nutrient and light-limited conditions (Petersen et al. 1997). Under light-limited conditions, total system GPP was similar in tanks of different depth when expressed per unit area

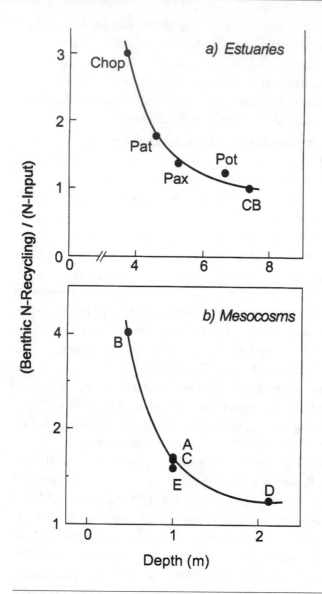

FIGURE 1•7 *Scaling of Relative Nutrient Recycling to Water Column Depth*
Example of pelagic-benthic process (mean benthic nitrogen recycling per nitrogen input) scaling differently to water column depth in: (a) natural estuarine ecosystems (adapted from Boynton and Kemp 2000) [Chop = Choptank; Pat = Patapsco; Pax = Patuxent; Pot = Potomac; CB = upper mainstem Chesapeake Bay] and (b) experimental estuarine mesocosms of differing size and shape (adapted from Cornwell and Bebout, unpublished); mesocosms designated as A-E, by increasing radius (Petersen et al. 1997).

(GPP$_a$), and therefore GPP expressed per unit volume (GPP$_v$) decreased with increasing depth (by definition GPP$_v$ = GPP$_a$/ z). In contrast, under nutrient-limited conditions, total system GPP was similar in the different tanks when expressed per unit volume, and therefore GPP_a increased with increasing depth (z, by definition $GPP_a = GPP_v * z$). Thus, primary productivity was proportional to horizontal surface area under light-limited conditions and to volume under nutrient-limited conditions (figure 1.8a). Although the theoretical and empirical work of others also suggests these scaling patterns (Wofsy 1983; Sand-Jensen 1989), a systematic quantitative test would have been difficult (at best) in the natural environment. Other studies have revealed depth-scaling patterns in nutrient uptake and zooplankton biomass that are also consistent with observations made in nature (figure 1.8b, c).

In contrast, an associated series of experiments indicated that growth of periphyton on experimental walls (and associated artifacts) was inversely related to mesocosm radius (Chen et al. 1997; Chen 1998). Experimental observations (figure 1.9a, b) revealed that periphyton biomass, nutrient uptake, and GPP per water volume were indeed related to container radius (= 2 * (water volume/wall area)). When these periphyton attributes were expressed per unit wall area, they increased with mesocosm radius, indicating the importance of ecological interactions beyond simple container geometry (Chen et al. 1997; Chen 1998). There is a suggestion that total algal biomass and nutrient uptake for natural and experimental ecosystems followed a consistent pattern with system width (figure 1.9). Large variability around the trend lines makes it unlikely, however, that we could yet predict with any precision natural levels of these properties from those observed experimentally. Clearly, considerable theoretical development and empirical confirmation are needed before systematic rules can be established to correct for effects of these artifacts when extrapolating experimental results to nature.

TROPHIC POSITION SCALED TO SYSTEM SIZE. Maximum length of food chains appears to be regulated by a variety of factors, including total production (Hutchinson 1959), system size and habitat variability (Briand and Cohen 1989), organism size and physiology (Peters 1983), as well as other features of the physical environment (e.g., Cohen 1994). In some environments, top predators may be small, with limited direct requirements of space. In terrestrial habitats, for example, predatory spiders can feed high on the food chain but occupy very limited space (e.g., Elton 1927), whereas small planktonic carnivores such as chaetognaths and ctenophores can also act as top predators in relatively small water volumes of the pelagic ocean (e.g., Landry 1977). However, the predominant situation in both terrestrial and

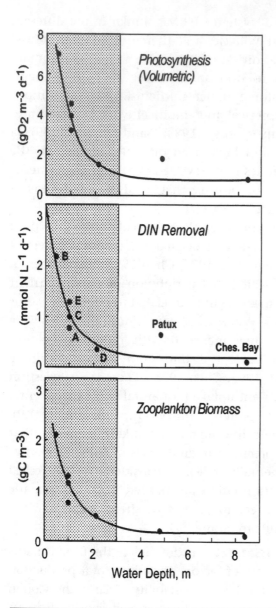

FIGURE 1•8 *Relating Experimental Depth-scaling Patterns to Conditions in Natural Ecosystems*

Example of how key properties of experimental (shaded) and natural (clear) estuarine ecosystems scale to depth of water column: in spring, mean values for total ecosystem photosynthesis, rates of dissolved inorganic nitrogen (DIN) uptake, and zooplankton biomass all decreased with water depth in experimental and natural estuarine ecosystems. Mesocosm data (A to E) are derived from Chen et al. (1997); Petersen et al. (1997); Chen (1998), whereas data from the Patuxent River estuary (Patux) and the mainstem mesohaline Chesapeake Bay (Ches. Bay) were obtained from Kemp and Boynton (1984) and Smith and Kemp (1995), respectively.

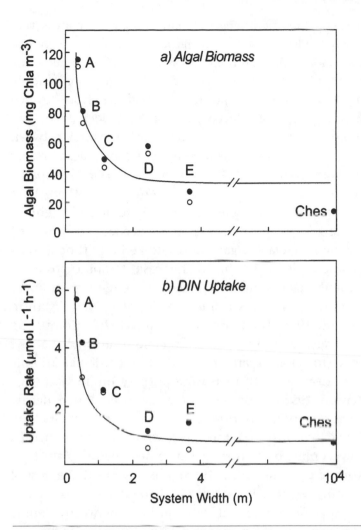

FIGURE 1•9 *Relating Experimental Width-scaling Patterns to Conditions in Natural Ecosystems*
Example of how key properties of experimental and natural estuarine ecosystems scale to ecosystem width in summer, where relationships appear to be related to artifacts of container walls: (a) algal biomass for the entire ecosystem (closed circles) and the wall periphyton communities (open circles) decreases consistently with width for experimental ecosystems (designated A–E in order of increasing radius, Chen et al. 1997); (b) uptake of dissolved inorganic nitrogen (DIN) by entire ecosystems and wall periphyton communities both decline with diameter of experimental containers (Chen 1998). Values for mainstem mesohaline Chesapeake Bay (Ches) are given as a reference for algal biomass (Smith and Kemp 1995) and DIN uptake (Boynton and Kemp 2000).

aquatic ecosystems involves food chains in which prey are smaller than predators, with relatively large organisms occupying the highest trophic positions (Elton 1927). Thus, panthers and hawks roam vast areas that define the Florida Everglades landscape, while large sharks are forced to cover great distances stalking prey on the Great Barrier Reef. In general, the home range or ambit of consumer organisms is, in fact, allometrically related to the animal's size and trophic position, with greater space covered by larger organisms at higher trophic levels (Peters 1983; Calder 1984).

For both herbivorous and carnivorous animals, relationships between the size of their home range and organism body-mass can be described by simple power functions (Calder 1984; West et al. 1997). The spatial requirements for carnivores are, however, consistently larger than those of herbivores of comparable size (figure 1.10a). Therefore, it might be postulated that changes in the relative area needed to provided nutritional support for an average organism would be proportional to its average trophic position (figure 1.10b). In this case, "trophic position" indicates the weighted mean number of feeding steps (from primary producers, whose trophic position is 1.0) in the diet of a given organism (e.g., Odum and Heald 1975; Ulanowicz and Kemp 1979). Comparing among different coastal ecosystems, it is reasonable to assume that the relative area needed to support an organism at a particular trophic position will tend to decrease with increasing primary production (e.g., Pauly and Christensen 1995). Although there has been much debate about the postulated relationship between primary productivity and food-chain length (e.g., Pimm 1982), the ability of organisms to succeed at higher trophic levels must be constrained at some scales by plant food production and space (e.g., Odum 1971, 1983). It is obvious that a host of other factors, including feeding behavior and habitat complexity, will contribute to the specific functional relationships between trophic position and spatial scale (e.g., With and Crist 1996; Chesson 1998).

These fundamental relationships between spatial scale and trophic position set real constraints on the use of experimental ecosystems for studies involving top predators. This limitation has contributed to recent strong critiques of mesocosm studies (e.g., Carpenter 1996). A substantial number of studies have, however, been conducted successfully with fish in experimental aquatic ecosystems (e.g., Threlkeld 1988, 1994). Although there is indirect evidence suggesting that the outcome of studies examining effects of fish predation on aquatic food-web structure may be influenced by the size of experimental systems used, interesting and provocative results have been generated from many such studies regardless of their experimental scales (DeMelo and France 1992). The

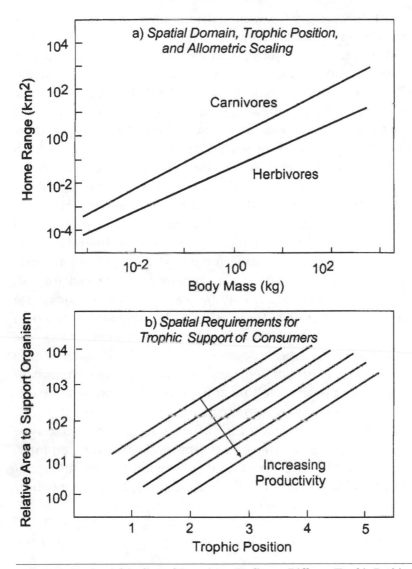

FIGURE 1•10 *Spatial Scaling of Organisms Feeding at Different Trophic Positions*
Spatial scaling of average organisms at different trophic positions in natural ecosystems:
(a) variation in mean home range (km^2) with body mass for herbivorous and carnivorous
animals (after Peters 1983); (b) postulated changes in relative area needed to provide
nutritional support for organisms feeding at a given trophic position for ecosystems of
differing primary productivity (families of curves) assuming 10 percent efficiency at each
trophic level.

surprising thing about this concern about use of experimental ecosystems
is the fact that it comes post hoc to the experiments themselves. Fish tend
to require relatively large spaces for both bioenergetic and behavioral
reasons that are probably linked. Because they feed relatively high on the

food chain, and because of fundamental limitations due to respiratory losses at each step in the food chain, the rates of fish production tend to be several orders of magnitude lower than the rates of primary production on which they depend. This is the basis for the classic trophic pyramid (Odum 1972) and one reason why large fish tend to forage over large areas. Presumably, the constraints set by these ecological and bioenergetic forces are entrained into the behavior of these organisms, such that there would tend to be convergence of spatial requirements to support both food demands and normal behavior. Of course the presence of container walls may have disruptive effects on fish behavior even if the size of the enclosed domain is adequate (e.g., Heath and Houde, this volume).

Although there has been considerable discussion about problems of maintaining fish in experimental tanks (e.g., Harte et al. 1980; Threlkeld 1994), it is surprising that few have attempted to apply bioenergetic concepts for calculating the minimum container size needed for fish experiments. A simple calculation may be made as follows. Consider a hypothetical estuary in which there is a single dominant food-chain leading from primary producers to the planktivorous fish, the bioenergetics of which has been described with standard relationships. Assume that a constant trophic efficiency (ξ, ratio of consumption or production at one trophic level to that at the preceding trophic level) can be used to describe food webs in this ecosystem (e.g., Kitchell et al. 1977). Assuming you wanted to conduct controlled experiments involving these trophic interactions, it can be shown (see Appendix) that the minimum water volume (V) needed to support an experimental fish assemblage of N_n fish with an average weight of W_n feeding at trophic level n (where $n = 1$ indicates primary producers) is

$$V = [\, (N_n)\, (a_n\, W_n^{b_n}) \,]\, [\, \xi^{(n-1)}\, (C_1) \,]^{-1} \qquad \textbf{(EQ 1•1)}$$

Here we define C_1 as primary production, in units of carbon or energy flow per water volume per time, and a_n and b_n are the intercept and slope, respectively, of a log-log allometric relationship describing fish consumption. To illustrate the application of equation 1.1, consider an experiment involving the small ($W_n = 100$ mg C) zooplanktivorous ($n = 3$) fish in this estuary. For this example we assume that "normal" behavior requires the fish to move in schools of at least 5 animals ($N_n = 5$), and that phytoplankton are the dominant autotrophs, with rates of gross primary production (C_1) of 1000 mg C m^{-3} d^{-1}. We, furthermore, assume that food consumption by this fish is related to body size by the allometric formulation, C_n (g C $^{d-1}$) = 0.08 $W_n^{0.7}$, and that ξ = 0.1. Using these values in equation 1 gives the result that a sustainable experimental ecosystem equal to 1.0 m^3 or greater is needed for this hypothetical fish experiment.

TEMPORAL SCALING OF ECOLOGICAL PROPERTIES. Examples of scaling functions that relate ecological properties to the temporal extent of an ecosystem are more difficult to find. In part this is because of the difficulty in identifying the age of an ecosystem or habitat, and in part it is because the characteristic time-scales of natural ecosystems tend to be considerably longer than those in experimental ecosystems. Differences between time-scales of nature and experiment tend to be smaller for aquatic than for terrestrial ecosystems. The concepts of ecological succession developed near the turn of the nineteenth century imply systematic changes in the structure, function, and community composition of ecosystems as they age (Cowles 1899; Shelford 1911). Changes in biomass, production, and biotic diversity that occur on scales of years on land take place within days in pelagic environments, a difference in time-scales that follows that of the dominant primary producers in the respective ecosystems (Odum 1971). Changes in many of these ecological properties follow relatively smooth patterns, reaching steady-state plateaus within decades on land and weeks in pelagic habitats. In principle, therefore, duration of experiments with plankton communities can markedly affect their outcomes.

In aquatic systems, the "ecological age" of a given water mass in a particular geographic zone tends to be inversely related to its residence time in that zone. Water residence-time (Tr, the ratio of habitat water volume to mean flow) can vary from days to years to decades among lakes and estuaries. It constrains phytoplankton (e.g., Malone 1984) and zooplankton (Ketchum 1954) growth and population maintenance within pelagic estuarine habitats. Concepts of chemostat research (e.g., Margalef 1967) reveal that when nutritional resources are not limiting to production, plankton growth rates (T_g^{-1}, production/biomass) can be regulated by rates of water dilution (inverse of residence time, T_r). Similarly, in a natural pelagic ecosystem when T_r approaches the range of values for T_g that are biologically realizable, plankton rates will be inversely related to water residence-time. However, where $T_r \ll T_g$, planktonic organisms are unable to grow and maintain a population, and where $T_g \ll T_r$, growth rates will tend to be unrelated to T_r. Even for biogeochemical processes that are concentrated in benthic communities, water residence time appears to be an important controlling variable. For example, the relative fractions of nitrogen and phosphorus inputs from watersheds to estuarine ecosystems that are exported to the coastal ocean tend to be inversely related to water residence time (Nixon et al. 1996). Conversely, the fraction of N inputs from land and atmosphere that is not exported (i.e., either buried or denitrified in sediments) is inversely related to the ratio of water depth to residence time, and the fraction of N inputs lost via denitrification is directly related to residence time (Nixon et al. 1996).

Water residence time is easily regulated in experimental aquatic ecosystems (e.g., Petersen et al. 1999), and it is thus a fundamental property that must be considered in the design and interpretation of experiments. It is likely that many properties and processes of natural and experimental pelagic and benthic communities are affected by water residence time. Clearly, careful attention must be given to water residence time when using experimental ecosystems to investigate any processes that involve biological-physical interactions; failure to do so may lead to misinterpretation of experimental results (e.g., Nowicki and Oviatt 1990).

Scaling Ecological Variance with Grain of Observation

Continuous observations made as time-series at geographically fixed points or as synoptic space-series from moving measurement platforms (e.g., ships, airplanes, satellites) provide records of changes in properties over wide ranges of time and space scales. Recent technological developments have allowed a rapid expansion in the amount of continuous series data that can be conveniently collected. Consequently, there is an expanding literature providing descriptions of changes in variance with changes in grain. There remains, however, a very limited understanding of the processes that either generate or regulate scale-dependent changes in variance. Furthermore, although it is widely accepted that patterns of scale-dependent changes in variance also lead to changes in mean ecological processes, there is a paucity of quantitative descriptions of such relationships and/or insights into possible underlying mechanisms (e.g., Turner 1989). We illustrate the ubiquity and diversity of patterns in which variance changes with grain by providing representative examples for aquatic ecosystems.

SPECTRAL PATTERNS OF PLANKTON. Techniques of spectral analysis have been used to describe the scale-dependent distributions of plankton biomass with changes in temporal and spatial resolution (Platt and

Denman 1975). Spectral plots provide a quantitative description of how relative variance of a particular property changes with grain of observations (cycle frequency). In general the log-log power spectra of relative variance versus cycle frequency tend to exhibit negative slopes ranging from 0 to –2. Hydrodynamic properties tend to follow slopes of –5/3 arising from laws of turbulent power dissipation (e.g., Platt and Denman 1975). As the slope approaches zero, the power spectrum is said to be "white" (analogous to "white noise"), and as it approaches –2, it is describe as becoming "red" (Steele 1985). Although power spectra (e.g., figure 1.11) and the "scale-variance plots" of landscape ecology (e.g., figure 1.3) both depict changes in relative variance with scale (or scale^{-1}), they are quite different. The latter represents how variance among samples changes with sample window size, whereas the former represents how total variance in serial-observation data can be approximately partitioned over a spectrum of cycles (recurrence intervals^{-1}) measured in length or time units.

It has been generally reported that spectral distributions of phytoplankton biomass (chlorophyll-a) follow those of temperature (e.g., slope –2, figure 1.11a), suggesting that physical forces regulate the distribution of phytoplankton (e.g., Lewis and Platt 1982; Powell et al. 1986). Spectral distributions of herbivorous zooplankton (e.g., krill), however, tend to exhibit patterns that depart from those of the physical environment at fine scales (figure 1.11a), indicating effects of swimming and reproductive behavior in regulating spatial and temporal distributions (e.g., Mackas et al. 1985; Weber et al. 1986; Piontkovski and Williams 1995). This divergence of spectral distributions for zooplankton and phytoplankton at fine scales also suggests that the integrated (over large spatial and temporal domains) rate of individual feeding interactions among all planktonic algae and their grazers will differ substantially from the total feeding rate estimated assuming homogeneous distributions (e.g., using average values for plankton size and abundance). Hence, controlled experiments examining zooplankton feeding on phytoplankton in homogeneous mesocosms will yield substantially different results from those with variable encounters between grazers and food resources (e.g., Dagg 1977).

DISTRIBUTION OF FISH. Similar spectral techniques have been used to characterize spatial distributions of fish (e.g., Horwood and Cushing 1978) and to explore the spatial predator-prey interactions among fish and between fish and birds (e.g., Rose and Leggett 1990; Schneider 1994;

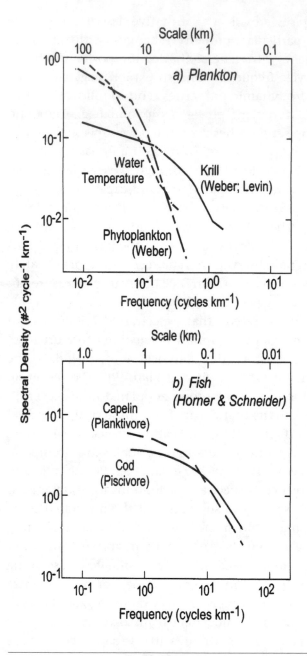

FIGURE 1•11 *Variance Spectra for Marine Organisms and Their Physical Habitat*
Comparison of typical variance spectra for organisms at four trophic levels and their
physical habitat (temperature) in pelagic marine ecosystems: (a) relative variance of
phytoplankton biomass, and herbivorous krill across a range of frequencies in Antarctic
Ocean waters (adapted from Weber et al. 1986; Levin et al. 1989); (b) changes in relative
variance of water temperature, capelin, and Atlantic cod with frequency along transects in
the North Atlantic Ocean. (Adapted from Horne and Schneider 1997)

Logerwell and Hargreaves 1996). An example of these relations can be seen for the North Atlantic Ocean, where spectral distributions for planktivorous fish (capelin, *Mallotus villosus*) and their piscivorous predators (cod, *Gadus morhua*) appear to follow patterns widely differing from those of temperature (index of physical habitat) and of plankton at lower trophic levels (figure 1.11b, Horne and Schneider 1997). Both fish species, however, exhibited flat spectra at high to intermediate spatial scales indicating formation of schools. Both species also had sharp declines in spectral density at finer scales, suggesting fairly even spacing of schooling animals (Horne and Schneider 1997). In this case, a surprising lack of coherence between the variance spectra of capelin and cod was explained by the fact that spatial coupling occurs at time scales outside those of the data set. In relation to ecological experiments, these observations imply that animal behavior dominates trophic interactions at the fine scales relevant to experimental manipulations. A challenge to experimental ecology is to devise protocols that effectively simulate grazer and predator behaviors and their effects on trophic dynamics and related processes (e.g., Harass and Taub 1985). Indeed, experiments can be devised to examine differences in effects of fish feeding under the relatively constant predation pressure common in most controlled experiments versus under the pulsing patterns of predation that occur in nature with highly mobile schools of fish moving in and out of adjacent habitats (e.g., DeMelo and France 1992).

BENTHIC ORGANISM DISTRIBUTIONS. Seagrasses and other submersed vascular plants, which occupy shallow habitats of many lake and coastal marine environments, are typically distributed in heterogeneous spatial patterns. A number of studies have described the dynamics of changing patchiness for seagrass communities (e.g., Duarte and Sand-Jensen 1990; Olesen and Sand-Jensen 1994). Spatial distributions of submersed plants are regulated by a number of factors, such as seed and propagule dispersal (Orth et al. 1994), waterfowl grazing (Kiørboe 1980), fish excavations (Orth 1975), disturbances from boats (Walker et al. 1989), and sediment movements (Marba and Duarte 1995). Although it might be interesting to describe and analyze these patterns using methods of spatial statistics or series observations (Legendre and Legendre 1983; Turner 1989; Turner and Gardner 1991), we are not aware of any such published application of these methods. Similarly, although numerical models have been developed to examine aspects of seagrass dynamics in relation to nutrient loading (e.g., Bach 1993; Fong and Harwell 1994; Kemp et al. 1995;

Madden and Kemp 1996; Buzzelli et al. 1998), few if any have explicitly considered spatial distributions of plant beds.

A number of important ecological feedback interactions may be affected by spatial patterns of submersed plant distributions. For example, many mobile invertebrates and small fish depend on submersed plant beds as refugia from predation, but also require nonvegetated habitats for food during nocturnal forays. The production and well being of such animal populations might be enhanced in littoral habitats with patchy distributions of plants (e.g., Lodge et al. 1988), where the ratio of bed perimeter to area is relatively high. On the other hand, submersed plant distributions characterized by high perimeter/area ratios might be more susceptible to inhibitory effects of nutrient enrichment and associated algal shading (e.g., Kemp et al. 1983; Orth and Moore 1983; Twilley et al. 1985; Shepard et al. 1989). This is because, when seagrass beds are relatively large, the plants are able to reduce local nutrient concentrations and thereby out-compete attached and planktonic algae (e.g., Bulthuis et al. 1984; Kemp et al. 1984; Murray et al. 2000). Although submersed plant interactions and responses to perturbations can be studied reasonably in experimental ecosystems, effects of spatial distributions in modulating observed relationships must be inferred from field and modeling studies (e.g., Murray et al. 2000).

In recent decades, there has been considerable interest in spatial heterogeneity of benthic macrofaunal distributions. Spatial patterns of animal abundances have been described, at scales ranging from 101 to 103 m, with autocorrelation techniques and correlograms from which patch sizes and frequencies can be inferred (e.g., Lodge et al. 1988; Hall et al. 1994). Much work has been done to investigate effects of disturbance in generating patchiness and how spatial patterns modulate responses to perturbations (Hall et al. 1994). Numerical models have been used to explore how spatial heterogeneity interacts with organism life cycles and community dynamics at various scales (e.g., Paine and Levin 1981; Barry and Dayton 1991; McArdle et al. 1997). Much of the experimental work in benthic ecology has used in situ enclosures, exclosures, and manipulations applied to open (uncaged) plots without walls. In a recent series of papers, detailed sampling of natural habitats at multiple scales was combined with manipulative experiments and spatial models to consider questions of extrapolation from small to larger scales (e.g., Legendre et al. 1997). Conclusions from several studies have revealed that spatial heterogeneity, which is not typically represented in experimental studies, alters the strength of ecological feedback at organism, population, and community levels (Legendre et al. 1997; Schneider et al. 1997; Thrush et al. 1997). How to include these effects of spatial variance in the design and interpretation of benthic experiments, however, remains to be demonstrated.

CONSIDERING VARIANCE SCALING IN DESIGN OF EXPERIMENTS. The minimization of variance between and within replicate experimental units is a major concern for experimental ecologists. This is particularly true for experimentalists working at the ecosystem level of organization, where complex interactions sometimes result in divergence among replicates over time. This concern for minimizing variance derives from the ultimate goal of maximizing statistical power for detecting treatment effects in the face of inherent fluctuations. Consistent with earlier theoretical discussions (figure 1.4a), we have observed that variance in phytoplankton community properties among replicate estuarine mesocosms tends to decrease with increasing size of experimental ecosystems (figure 1.12a). There are surprisingly few of these kind of data available to test the generality of this relationship (e.g., Petersen et al. 1999). A similar pattern appears when relating the number of replicates to size of experimental systems (figure 1.12b). This relationship has been attributed to logistic limitations associated with use of large experimental units (Kareiva and Andersen 1988). The pattern may, however, reflect an intuitive understanding by experimentalists that the relatively large variances associated with smaller scale experiments require increased replication to maintain adequate statistical power. But what about the variance that occurs in nature—how would this affect experimental outcomes?

In fact, it is with some irony that we suggest here that changes in relative variance that occur with scale in nature need to be considered in design and interpretation of experiments, which tend to be inherently reduced in scale. The focus for most experimental studies is on mean or integral behaviors (e.g., responses to perturbations) of ecosystems and organisms. However, the so-called fallacy of the means suggests that, for nonlinear systems, the calculated interaction between mean values for ecosystem properties (averaged across all time and space scales) is not necessarily equal to the mean of all the individual interactions distributed in time and space (e.g., Rastetter et al. 1992). Thus, for example, because grazing and predation interactions tend to be highly nonlinear in nature, using mean grazing rates in controlled experiments will not necessarily provide a reliable estimate of the mean effects of grazing in a variable natural environment.

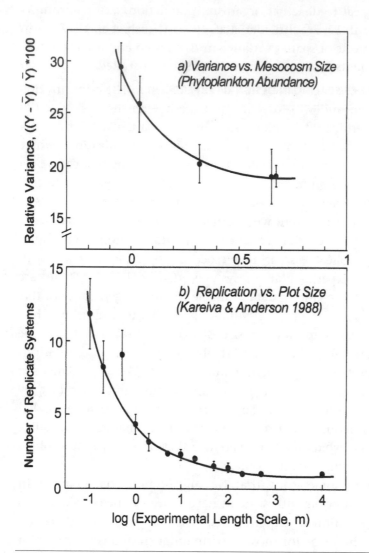

FIGURE 1•12 *Scaling Replication and Variance to Experimental System Size*
Apparent similarity of scale-dependent variance among replicate experimental ecosystems
and scaling of experimental ecologists' replication of study units: (a) changes in relative
variance among replicate systems for composite phytoplankton community properties
(chlorophyll-*a*, phaeophyton, particulate carbon, particulate nitrogen) versus size of
experimental ecosystems (Kemp et al., unpublished); (b) variation in number of replicates
used in experimental community ecology studies with increasing size of experimental plots
(Kareiva and Andersen 1988).

A challenge to experimentalists is to simulate the variability of key
processes, which occur in nature over many scales, within the confines

and limited scales of experimental systems (MacNally and Quinn 1998). For example, at relatively fine scales (1 to 100 m) in estuarine ecosystems, the effects of schooling predatory fish appear as large intermittent pulses as schools move through a given water parcel. At broad scales (103 to 105 m), we might be tempted to consider the predation pressure associated with these fish based on their average abundance, average size, and average physiological state of prey and predators. However, it is not difficult to imagine that the effect of "pulsed" predation would be radically different from that associated with continuous constant predation ("press"). Most studies of top-down effects of fish predation in enclosed experimental ecosystems have attempted to create conditions of relatively constant predation to minimize variance (e.g., Frost et al. 1988; DeMelo and France 1992). It is possible, however, to design experiments to explicitly examine effects of variable conditions, including nutrient loading (e.g., Turpin and Harrison 1979; Sturgis and Murray 1997) or fish predation (e.g., Harass and Taub 1985). Such experiments need to be designed in relation to natural scale-dependent variability of these ecosystem attributes (nutrients, predation) to improve extrapolation of results to conditions in nature.

MODELS TO INTEGRATE SCALE-DEPENDENT VARIANCE INTO EXPERIMENT. Numerical ecological models simulated in explicit spatial grids offer a tool to understand how scale-dependent variance affects ecosystem dynamics and to apply that understanding to the design and interpretation of controlled experiments (e.g., With and Crist 1996). For example, a spatial model of Yellowstone Park ecosystem was developed to quantify how changes in spatial variability (heterogeneity in distribution of snow) and temporal fluctuations in input functions (snowfall patterns) altered ungulate population dynamics and associated processes (Turner and O'Neill 1995). To be useful for the design of experiments, such models need to be constructed and analyzed in parallel with controlled empirical experiments (Nixon et al. 1979; Hill and Wiegert 1980; Brinkman et al. 1994). Here the model would be used to explore how changes in spatial resolution (and associated changes in heterogeneity) alter ecosystem dynamics and responses to perturbations. Such information could then be transferred to interpret results from small-scale experiments conducted in relatively homogeneous containers. As with landscapes, coastal ecosystems typically contain mosaics of different habitats including relatively deep pelagic and benthic regions, flanked by shallow areas with benthic primary producers (e.g., seagrasses) and fringes of emergent but tidally inundated marsh systems (figure 1.13).

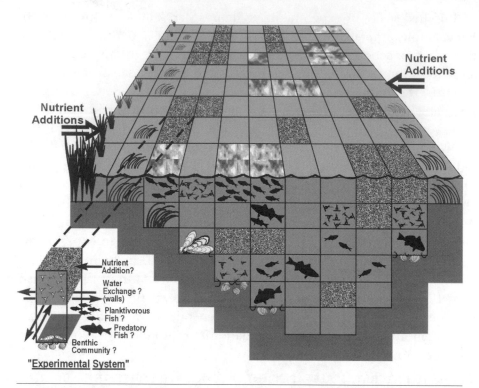

FIGURE 1•13 *Conceptual Model of How Ecological Interactions Depend on Spatial Variability*
Schematic conceptual diagram depicting a three-dimensional (3-D) view of an estuarine
ecosystem, with heterogeneous distributions of phytoplankton (grainy shaded area);
zooplankton (fat "tees"); small planktivorous fish; benthic clams, oysters, and polychaetes;
benthivorous fish; large predatory fish; submersed seagrasses; and emergent marsh plants.
Different mosaics of patchiness can be seen from cross-section versus plan views of the
system. A 3-D grid represents a hypothetical spatial grain used in developing a spatial model
of this ecosystem; the grid could conceivably be set at smaller or larger scales, but the
resulting perception of the ecosystem would change. Water advects and fish swim between
adjacent spatial cells. If the grid volume is isolated from its surroundings by walls limiting
exchange of water and organisms, the volume becomes a numerical version of an
experimental ecosystem.

One approach would be to structure ecosystem processes in these
models using conventional finite-difference equations (e.g., Nixon et al.
1979), while simulating movements of motile organisms such as fish
using individual-based methods (e.g., DeAngelis 1992). In this case,
model behavior could be calibrated by comparing mean dynamics and
spatial patterns to a natural estuarine ecosystem. Although the exact
spatial configurations of natural habitats would be virtually impossible to
reproduce, coherent patterns of variability (e.g., power spectra) might be

achieved. For both model and nature, such patterns could be described using spectral analysis and spatial statistics (e.g., Platt and Denman 1975; Turner 1989). It would be important to make the model code sufficiently flexible to allow for grid dimensions (spatial grain) to be adjusted. A series of simulation experiments could compare how responses to perturbations change with reductions in grid scales, including a version of the model with a single well-mixed volume to simulate an experimental mesocosm. These "numerical mesocosm" simulations could examine how responses vary with changes in water volume, depth, wall growth, water exchange rates, and fish abundance. There are many possible applications of such numerical experiments; however, the main focus here would be to generate guidelines or extrapolation from small to large scales by comparing simulations of isolated experimental ecosystems against equivalent simulations of larger, spatially variable ecosystems.

SCALING EXPERIMENTAL ECOSYSTEMS: APPROACHES AND EXAMPLES

The goal of this chapter is to consider how the results of mesocosm studies are affected by temporal and spatial scales of the experiment. Quantifying these effects potentially allows extrapolation of experimental results to the scales of nature. Three critical steps are necessary to achieve this goal.

1. *Defining the scales of natural systems.* The first step in the design of experimental ecosystems is to identify the range of scales (both grain and extent) at which the ecological processes of interest operate in nature. Because processes operate at a variety of temporal and spatial scales, it is probably impossible to define specific scales for each and every ecosystem process. Therefore, experiments must focus on a subset of processes and compare the experimental scales for these variables against those of the natural system. The temporal extent of processes in natural systems may be measured by the time required to complete a cycle or produce a measurable signal, whereas the spatial extent might be the depth of the water or the distance between interacting components. Feedback pathways by which processes might be modified, and the identification of the scales over which these feedbacks occur, also should be quantified—a much more difficult task. An example of an important feedback is the spatial variation of seagrass uptake in reducing nutrient concentration (and thereby reducing uptake rate). Defining the grain of natural systems might be

accomplished by observing the frequency of measurements needed to detect changes in variability with scale and/or the spatial dimensions over which individuals interact (e.g., the size of a fish school).

2. *Experimental alterations of scale.* Most ecological interactions in aquatic environments depend on the coupling between physical transport and ecological (or biogeochemical) processes. Therefore, in designing ecological experiments, the key physical processes must be identified and changes in their scales induced by the experimental system must be defined. The list of important physical properties to consider includes water depth and volume, areas of sediments and hard substrates (e.g., walls), vertical mixing times, boundary layer thickness, and water-residence times. In many cases, once the ecological and physical scales are identified a set of dimensionless ratios of biological-to-physical processes can be defined as scaling devices. Some examples of useful dimensionless ratios are (plankton growth rate)/(water exchange rate); (seagrass uptake rate)/(water exchange rate); (benthic suspension feeding rate)/(vertical mixing rate); and (vertical mixing depth)/(Secchi depth). In these cases, experimental systems can be designed such that the key ratio(s) are unchanged between experimental and natural ecosystems. When this is done, results are directly applicable to conditions in nature.

It may not always be obvious, however, how to adjust the physical scales of the experimental system design to match those of the counterpart in nature. If the appropriate dimensionless ratios cannot be readily defined, but there is reason to believe that experimental outcomes depend on temporal and spatial dimensions, it may be appropriate to run a series of scaling experiments. Such experiments, which involve parallel controlled manipulations in systems of different length and time scales, have been used to address a range of scaling topics, including variation in predator behavior and feeding rates with container size (de Lafontaine and Leggett 1987; Sarnelle 1997), alteration of plankton responses to nutrients with chamber volume (Gerhart and Likens 1975), changes in ecosystem productivity with water depth (Petersen et al. 1997), or variations in wall periphyton growth with container diameter (Chen et al. 1997). Changes in variability of ecological properties with grain can be investigated by examining how ecosystem responses to perturbation differ depending on the variance of the perturbation (e.g., "pulse" versus "press" perturbation, Frost et al. 1988). For example, experiments have been designed to examine how variability of input alters nutrient responses for phytoplankton (Turpin and Harrison 1979) and for submersed vascular plants (Sturgis and Murray 1997) or to consider how zooplankton responses to fish predation differ when predation is applied in pulsed rather than continuous rates (e.g., Harass and Taub 1985). The point is we can empirically examine effects of experimental and natural scales by conducting parallel experiments at different scales and relating these scales to those of nature using the approach of dimensional analysis.

3. *Extrapolation of experimental results.* The denouement of all of these considerations lies in solving "the problem of extrapolation" from inherently reduced experimental to the larger scales of nature. At this stage, we are far from having a unified theory from which simple "scaling rules" might derive, and we can only offer two approaches that have been proven to be effective. When experimental designs can be based on dimensional analysis of coupled biological-physical interactions, concepts of similitude suggest that the extrapolation could be direct. In all other cases, extrapolation from experiment to nature must involve use of scaling models that may be numerical or statistical in structure and that may emphasize ecological processes or organism behavior.

Dimensional analysis is a technique that is based on the proposition that universal relationships should apply regardless of the specific dimensions of a particular system (e.g., Legendre 1983). The method allows us: (1) to derive insights from differential equations without solving them; (2) to make quantitative predictions about complex process; and (3) to infer dynamic relationships of natural systems based on experiments with small-scale replicas (Platt et al. 1981). In each case, the method allows us to derive a set of dimensionless variables that represent the balance between processes or forces governing the dynamics of a particular system; however, in the context of this chapter, the third of these uses is most relevant. Indeed, a number of ecologists have advocated use of this technique as a tool for systematically adjusting for scaling distortions in the design and interpretation of mesocosm studies (e.g., Kemp et al. 1980; Platt et al. 1981). In this case, the central idea is that distortions in one dimension or variable can be counterbalanced by distortions in others in order to conserve key relationships. Examples of formal application of this approach to extrapolate from experiment to nature or to design "properly scaled" mesocosms include studies of phytoplankton growth (e.g., Platt et al. 1981) and light adaptation (Lewis et al. 1984), zooplankton grazing (McClatchie and Lewis 1986), benthic macrofaunal feeding (Miller et al. 1984), soil community dynamics (Shirazi et al. 1984), decomposition of dissolved organic matter in ponds (Uhlmann 1985), and mixing regimes in experimental water columns (Sanford 1997). A number of researchers have taken a more intuitive approach to counterbalance scaling distortions, without formal application of dimensional analysis (e.g., Luckinbill 1973).

There are important caveats to the application of dimensional analysis in design of and extrapolation from ecological experiments (Platt et al. 1981). One is that biological (as opposed to physical) variables tend to be much more constrained in the range over which their scales can be modified between experiment and nature. Whereas depth, width, turbulence, and water velocities can vary over large ranges in both natural and experimental systems, the potential range of variations for biological factors (e.g., organism size, growth rates, and swimming speeds) is much more limited, as is the experimentalist's ability to prescribe them. Also, the strong interdependence among factors in ecosystems may make it difficult to experimentally change any one (e.g., depth) without affecting others (e.g., photosynthesis). There are typically many dimensionless numbers governing ecosystem dynamics, and it may be difficult to conserve all simultaneously. The art of applying dimensional analysis to mesocosm design lies in identifying dimensionless numbers, which use key variables that pertain to the research question being experimentally addressed. The choice of design criteria is, thus, context dependent.

Numerical simulation models provide another potentially useful tool for extrapolation of experimental results to conditions in nature. Although mesocosms and ecological models are both abstractions and simplifications of nature, time and space scales are far easier to manipulate on a computer than in a living ecological model. On the other hand, mesocosms are genuine living systems and therefore represent a higher degree of realism, including properties of adaptation and self-organization not often represented in the computer code of simulation models. To some extent, the advantages and disadvantages of the two approaches are counteracting and therefore potentially complementary (Nixon et al. 1979). This is evident in a number of the instances where mesocosm studies and simulation models have been used in parallel (e.g., Brockmann 1990; Parsons 1990; Baretta-Bekker et al. 1994). Simulation models have also frequently been developed primarily as tools for post facto analysis and interpretation of mesocosm experimental results (e.g., Brockway et al. 1979; Gard 1981; Laake et al. 1983; Lassiter 1983; Liepmann and Stephanopoulos 1985; Parsons et al. 1986; Andersen et al. 1987; Swartzman et al. 1990; Brinkman et al. 1994).

One recent example may serve to illustrate how numerical modeling can be used to interpret and extrapolate mesocosm results. A model developed to simulate dynamic interactions among planktonic, benthic, and periphytic communities in coastal ecosystems was calibrated for conditions in experimental ecosystems of different sizes and shapes (Petersen et al. 1997; Chen 1998). Simulation studies revealed that artifacts associated with periphyton growth on mesocosm walls could be reduced to acceptable levels with twice-weekly wall cleaning and by using tanks with a radius of 2 m or greater. The model was also used to extrapolate observed dynamic responses of experimental ecosystems to nutrient enrichment under conditions when container walls were numerically removed (Chen 1998). A future challenge will be to develop systematic numerical methods to incorporate effects of scale-dependent variance in modifying results from controlled experiments.

COMMENTS

Mesocosms are likely to continue to gain importance as tools for controlled ecological experimentation. This is particularly true in coastal aquatic habitats, where options for controlled manipulations in the field are severely constrained by the characteristically strong spatial gradients of solute concentrations and high rates of tidal-exchange. At this point, comparison of results among experimental ecosystems and extrapolation to nature remain largely qualitative endeavors. In this chapter we have reviewed numerous concepts of scale that are relevant to experimental ecosystem research and have provided examples of how theoretical and empirical insights can be applied. Our review suggests that existing theory and empiricism can be effectively used to design experiments that more realistically represent the dynamics of larger-scale natural systems and extrapolate results from mesocosms to nature. Systematic application of the scaling concepts outlined in this chapter and elsewhere in this book would go a long way toward creating "scale sensitive" experiments (sensu Petersen et al. 1999).

The "science of scale" (Meentemeyer and Box 1987) is increasingly recognized as a legitimate and important field of inquiry. In addition to passively benefiting from advances in our understanding of scale derived from other types of studies, we suggest that mesocosms can be explicitly designed to test and advance our understanding of scaling theory. Indeed mesocosm experiments have already been successfully used to elucidate the effect of physical scale on a variety of ecological relationships including predator-prey relations (Luckinbill 1974), species diversity (Dickerson and Robinson 1986), food web structure (Spencer and Warren 1996), ecological impact of toxins (Morris et al. 1994), and patterns of primary productivity (Petersen et al. 1997). These kinds of multi-scale experiments represent a logical approach for identifying mechanisms that link pattern and process across scale. Ultimately the results of scaling experiments hold the promise of advancing our understanding of scale in all ecological systems as well as improving our practical ability to extrapolate information among natural ecosystems and from mesocosm studies to whole ecosystems in nature.

APPENDIX

Estimating Minimum Enclosure Size for Fish Experiments

Consider the situation in which an enclosure study is being designed to involve fish in experimental aquatic ecosystems. Assume that there is a single dominant food chain leading from primary producers to the experimental fish. The food ration (C_n) required to sustain this fish in an experimental ecosystem can be calculated by a standard allometric relation (Kitchell et al. 1977)

$$C_n = (a_n W_n^{b_n}) (N_n) (V)^{-1} \qquad \text{(EQ 1•2)}$$

where C_n is in units of carbon or energy flow per water volume per time, V is the water volume of the experimental system, a_n and b_n are the intercept and slope, respectively, of the log-log allometric relationship, and W_n and N_n are the weight (e.g., g carbon per volume) and number (animals per volume), respectively, of fish in the experiment. Trophic levels in this simple food chain are identified by i, where i = 1 indicates the autotroph level, and i = n indicates the top consumer, which is the experimental fish in this system. Heterotrophi9c consumption at any trophic level, i, is indicated by C_i, while C_1 indicates gross primary production (e.g., mg carbon per cubic meter per day). W_i and N_i indicate mean weight and number of organisms at trophic level, i. The total consumption at any trophic level, n, can be calculated as follows

$$C_n = [\prod_{i=1}^{n-1} (\xi_i)] C_1 \qquad \text{(EQ 1·3)}$$

where ξ indicates "trophic efficiency" defined as $(C_i)/(C_{i-1})$, and $\prod$ is the serial product operator. By combining equations 1.2 and 1.3, the minimum size (water volume) of experimental ecosystem needed to support a total of (N) fish of size (W) at trophic level (n) can be calculated as follows

$$V = [(N_n) (a_n W_n^{bn})] [(\prod (\xi_i)) (C_1)]^{-1} \qquad \text{(EQ 1·4)}$$

In the special case where ξ_i can be assumed to be constant (ξ), the equation simplifies as follows

$$V = [(N_n) (a_n W_n^{bn})] [\xi^{(n-1)} (C_1)]^{-1} \qquad \text{(EQ 1·5)}$$

If V exceeds logistic constraints of an experimental facility, then one can consider adjustments to W and N.

ACKNOWLEDGMENTS

This work was supported by U.S. EPA STAR Program as part of the Multiscale Experimental Ecosystem Research Center at the University of Maryland Center for Environmental Science (R819640). We are indebted to our many colleagues, students, and friends for numerous stimulating discussions that contributed to the ideas presented in this chapter. In particular, we want to thank Walt Boynton, Chung-Chi Chen, Jeff Cornwell, Ed Houde, Vic Kennedy, Tom Malone, Laura Murray, Larry Sanford, and Bob Ulanowicz for exchanges of ideas and data that laid the groundwork for this effort. Contribution No. 99-3353 of the University of Maryland Center for Environmental Science.

LITERATURE CITED

Allen, T. F. H., and T. W. Hoekstra. 1991. Role of heterogeneity in scaling of ecological systems under analysis. In J. Kolasa and S. T. A. Pickett, eds., *Ecological Heterogeneity*, pp. 47–68. New York: Springer-Verlag.

Andersen, V., P. Nival, and R. Harris. 1987. Modelling of a planktonic ecosystem in an enclosed water column. *Journal of the Marine Biological Association of the United Kingdom* 67:407–430.

Bach, H. 1993. A dynamic model describing the seasonal variations in growth and the distribution of eelgrass. *Ecological Modelling* 65:31–50.

Baines, S. B., M. L. Pace, and D. M. Karl. 1994. Why does the relationship between sinking flux and planktonic primary production differ between lakes and oceans? *Limnology and Oceanography* 39:213–226.

Baretta-Bekker, J. G., B. Riemann, J. W. Baretta, and E. Koch Rasmussen. 1994. Testing the microbial loop concept by comparing mesocosm data with results from a dynamical simulation model. *Marine Ecology Progress Series* 106:187–198.

Barry, J. P., and P. K. Dayton. 1991. Physical heterogeneity and the organization of marine communities. In J. Kolasa and S. T. A. Pickett, eds., *Ecological Heterogeneity*, pp. 270–320. New York: Springer-Verlag.

Beyers, R. J., and H. T. Odum. 1993. *Ecological Microcosms*. New York: Springer-Verlag.

Bissonette, J. A. 1997. Scale-sensitive ecological properties: Historical context, current meaning. In J. A. Bissonette, ed., *Wildlife and Landscape Ecology: Effects of Pattern and Scale*, pp. 4–31. New York: Springer-Verlag.

Boynton, W. R., and W. M. Kemp. 2000. Influence of river flow and nutrient loads on selected ecosystem processes and properties in Chesapeake Bay. In J. Hobbie, ed., *Estuarine Science: A Synthetic Approach to Research and Practice*, pp. 296-298. Washington, D.C.: Island Press.

Briand, F., and J. Cohen. 1989. Habitat compartmentation and environmental correlates of food-chain length. *Science* 243:238–240.

Brinkman, A. G., C. J. M. Philippart, and G. Holtrop. 1994. Mesocosms and ecosystem modelling. *Vie et Milieu* 44:29–37.

Brockmann, U. 1990. Pelagic mesocosms: II. Process studies. In C. M. Lalli, ed., *Enclosed Experimental Marine Ecosystems: A Review and Recommendations*, pp. 81–108. New York: Springer-Verlag.

Brockway, D. L., J. I. Hill, J. R. Maudsley, and R. R. Lassiter. 1979. Development, replicability, and modeling of naturally derived microcosms. *International Journal of Environmental Studies* 13:149–158.

Bulthuis, D. G., G. Brand, and M. Mobley. 1984. Suspended sediments and nutrients in water ebbing from seagrass-covered and denuded tidal mudflats in a southern Australian embayment. *Aquatic Botany* 20:257–266.

Buzzelli, C. P., R. Wetzel, and M. B. Meyers. 1998. Dynamic simulation of littoral zone habitats in lower Chesapeake Bay. II. Seagrass habitat primary production and water quality relationships. *Estuaries* 21:673–689.

Calder, W. A. 1984. *Size, Function, and Life-History*. Cambridge, Mass.: Harvard University Press.

Carpenter, S. R. 1996. Microcosm experiments have limited relevance for community and ecosystem ecology. *Ecology* 77:667–680.

Carpenter, S. R. 1999. Microcosm experiments have limited relevance for community and ecosystem ecology: Reply. *Ecology* 80:1085–1088.

Carpenter, S. R., S. W. Chisholm, C. J. Krebs, D. W. Schindler, and R. F. Wright. 1995. Ecosystem experiments. *Science* 269:324–327.

Chen, C.-C. 1998. Wall effects in estuarine mesocosms: Scaling experiments and simulation model. Ph.D. diss., University of Maryland.

Chen, C.-C., J. E. Petersen, and W. M. Kemp. 1997. Spatial and temporal scaling of periphyton growth on walls of estuarine mesocosms. *Marine Ecology Progress Series* 155:1–15.

Chesson, P. 1998. Spatial scales in the study of reef fishes: A theoretical perspective. *Australian Journal of Ecology* 23:209–215.

Cohen, J. E. 1994. Marine continental food webs: Three paradoxes? *Philosophical Transactions of the Royal Society of London, Series B* 343:57–69.

Cowles, H. C. 1899. The ecological relations of the vegetation on the sand dunes of Lake Michigan. *Botanical Gazette* 27:95–117.

Crossland, N. O., and T. W. La Point. 1992. The design of mesocosm experiments. *Environmental Toxicology and Chemistry* 11:1–4.

Dagg, M. 1977. Some effects of patchy food environments on copepods. *Limnology and Oceanography* 22:99–107.

Day, J. W., A. S. Hall, W. M. Kemp, and A. Yanez-Arancibia. 1989. *Estuarine Ecology*. New York: Wiley.

DeAngelis, D. L. 1992. *Dynamics of Nutrient Cycling and Food Webs*. New York: Chapman and Hall.

de Lafontaine, Y., and W. C. Leggett. 1987. Effect of container size on estimates of mortality and predation rates in experiments with macrozooplankton and larval fish. *Canadian Journal of Fisheries and Aquatic Sciences* 44:1534–1543.

DeMelo, R., and R. France. 1992. Biomanipulation: Hit or myth. *Limnology and Oceanography* 37:192–207.

Deutschman, D., G. Bradshaw, W. M. Childress, K. Daly, D. Grunbaum, M. Pascual, N. Schumaker, and J. Wu. 1993. Mechanisms of patch formation. In S. Levin, T. Powell, and J. Steele, eds., *Patch Dynamics*, pp. 184–209. New York: Springer-Verlag.

Diamond, J. M., and R. M. May. 1976. Island biogeography and the design of natural reserves. In R. M. May, ed., *Theoretical Ecology: Principles and Applications*, pp. 163–186. Philadelphia: Saunders.

Dickerson, J. E. J., and J. V. Robinson. 1986. The controlled assembly of microcosmic communities: The selective extinction hypothesis. *Oecologia* 71:12–17.

Drake, J. A., G. H. Huxel, and C. L. Hewitt. 1996. Microcosms as models for generating and testing community theory. *Ecology* 77:670–677.

Drenner, R. W., and A. Mazumder. 1999. Microcosm experiments have limited relevance for community and ecosystem ecology: Comment. *Ecology* 80:1081–1085.

Duarte, C. M., and K. Sand-Jensen. 1990. Seagrass colonization: Patch formation and patch growth in *Cymodocea nodosa*. *Marine Ecology Progress Series* 65:193–200.

Duarte, C. M., and D. Vaqué. 1992. Scale dependence of bacterioplankton patchiness. *Marine Ecology Progress Series* 84:95–100.

Elton, C. 1927. *Animal Ecology*. London: Sidgwick and Jackson (reprinted in 1966 by Methuen).

Fasham, M. J. R. 1978. The statistical and mathematical analysis of plankton patchiness. *Oceanography and Marine Biology: An Annual Review* 16:43–79.

Fee, E. J., and R. E. Hecky. 1992. Introduction to the northwest Ontario lake size series (NOLSS). *Canadian Journal of Fisheries and Aquatic Sciences* 49:2434–2444.

Fong, P., and M. Harwell. 1994. Modeling seagrass communities in tropical and subtropical bays and estuaries: A mathematical model synthesis of current hypotheses. *Bulletin of Marine Science* 54:757–781.

Frost, T. M., S. M. DeAngelis, D. J. Bartell, D. J. Hall, and S. H. Hurlbert. 1988. Scale in the design and interpretation of aquatic community research. In S. R. Carpenter, ed., *Complex Interactions in Lake Communities*, pp. 229–258. New York: Springer-Verlag.

García-Molinar, G., E. Mason, C. Green, A. Lobo, B.-I. Li, J. Wu, and G. B. Bradshaw. 1993. Description and analysis of spatial patterns. In S. Levin, T. Powell, and J. Steele, eds., *Patch Dynamics*, pp. 70–89. New York: Springer-Verlag.

Gard, T. C. 1981. Persistence for ecosystem microcosm models. *Ecological Modeling* 12:221–230.

Gardner, R. H. 1998. Pattern, process, and the analysis of spatial scales. In D. L. Petersen and V. T. Parker, eds., *Ecological Scale: Theory and Applications*, pp. 17–34. New York: Columbia University Press.

Gerhart, D. Z., and G. E. Likens. 1975. Enrichment experiments for determining nutrient limitation: Four methods compared. *Limnology and Oceanography* 20:649–653.

Hall, S. J., D. G. Raffaelli, and S. F. Thrush. 1994. Patchiness and disturbance in shallow water benthic assemblages. In P. S. Giller, A. G. Hildrew, and D. G. Raffaelli, eds., *Aquatic Ecology: Scale, Pattern, and Process*, pp. 333–375. Oxford: Blackwell Science.

Harass, M. C., and F. B. Taub. 1985. Effects of small fish predation on microcosm community bioassays. In R. D. Cardwell, R. Purdy, and R. C. Bahner, eds., *Aquatic Toxicology and Hazard Assessment*, pp. 117–133, ASTM STP 854. Philadelphia: America Society for Testing Materials.

Hargrave, B. T. 1973. Coupling carbon flow through some pelagic and benthic communities. *Journal of the Fisheries Research Board of Canada* 30:1317–1326.

Hargrave, B. T. 1979. Factors affecting the flux of organic matter to sediments in a marine bay. In K. R. Tenore and B. C. Coull, eds., *Marine Benthic Dynamics*, pp. 243–263. Columbia: University of South Carolina Press.

Harrison, W. G. 1980. Nutrient regeneration and primary production in the sea. In P. G. Falkowski, ed., *Primary Productivity in the Sea*, pp. 433–470. New York: Plenum.

Harte, J., D. Levy, J. Rees, and E. Saegebarth. 1980. Making microcosms an effective assessment tool. In J. P. Geisy Jr., ed., *Microcosms in Ecological Research*, pp. 105–137. Springfield, Va.: National Technical Information Service.

Haury, L. R., J. A. McGowan, and P. H. Wiebe. 1978. Patterns and processes in the time-space scales of plankton distributions. In J. H. Steele, ed., *Spatial Pattern in Plankton Communities*, pp. 277–327. New York: Plenum.

Hill, J. I., and R. G. Wiegert. 1980. Microcosms in ecological modeling. In J. P. Geisy Jr., ed., *Microcosms in Ecological Research*, pp. 138–163. Springfield, Va.: National Technical Information Service.

Horne, J. K., and D. C. Schneider. 1997. Spatial variance of mobile aquatic organisms: Capelin and cod in Newfoundland coastal waters. *Philosophical Transactions of the Royal Society of London, Series B* 352:633–642.

Horwood, J. W., and D. H. Cushing. 1978. Spatial distributions and ecology of pelagic fish. In J. H. Steele, ed., *Spatial Pattern in Plankton Communities*, pp. 355–383. New York: Plenum.

Huffaker, C. B. 1958. Experimental studies on predation: Dispersion factors and predator-prey oscillations. *Hilgardia* 27:343–383.

Hutchinson, G. E. 1953. The concept of pattern in ecology. *Proceedings of the National Academy of Sciences of the United States of America* 105:1–12.

Hutchinson, G. E. 1959. Homage to Santa Rosalia, or why or there so may kinds of animals? *American Naturalist* 93:145–159.

Ives, A. R., J. Foufopoulos, E. D. Klopfer, J. L. Klug, and T. M. Palmer. 1996. Bottle or big-scale studies: How do we do ecology? *Ecology* 77:681–685.

Jørgensen, B. B. 1983. Processes at the sediment-water interface. In B. Bolin and R. Cook, eds., *The Major Biogeochemical Cycles and Their Interactions*, pp. 477–509. Scientific Communications on Problems of the Environment Publication I. Stockholm: SCOPE.

Kareiva, P. 1994. Diversity begets productivity. *Nature* 368:686–687.

Kareiva, P., and M. Andersen. 1988. Spatial aspects of species interactions: The wedding of models and experiments. In A. Hastings, ed., *Community Ecology*, pp. 38–54. New York: Springer-Verlag.

Kemp, W. M., and W. R. Boynton. 1984. Spatial and temporal coupling of nutrient inputs to estuarine primary production: The role of particulate transport & decomposition. *Bulletin of Marine Science* 35:522–535.

Kemp, W. M., and W. R. Boynton. 1992. Benthic-pelagic interactions: Nutrient and oxygen dynamics. In D. E. Smith, M. Leffler, and G. Mackiernan, eds., *Oxygen Dynamics in the Chesapeake Bay: A Synthesis of Recent Research*, pp. 149–221. Maryland Sea Grant Publication. College Park: University of Maryland.

Kemp, W. M., W. R. Boynton, and A. J. Herman. 1995. Simulation models of an estuarine macrophyte ecosystem. In B. Patten and S. E. Jørgensen, eds., *Complex Ecology*, pp. 262–278. Englewood Cliffs, N.J.: Prentice-Hall.

Kemp, W. M., W. Boynton, R. Twilley, J. C. Stevenson, and L. Ward. 1984. Influences of submersed vascular plants on ecological processes in upper Chesapeake Bay. In V. S. Kennedy, ed., *The Estuary as a Filter*, pp. 367–394. New York: Academic Press.

Kemp, W. M., M. R. Lewis, F. F. Cunningham, J. C. Stevenson, and W. R. Boynton. 1980. Microcosms, macrophytes, and hierarchies: Environmental research in the Chesapeake Bay. In J. P. Giesy Jr., ed., *Microcosms in Ecological Research*, pp. 911–936. Springfield, VA: National Technical Information Service.

Kemp, W. M., P. A. Sampou, J. Garber, J. Tuttle, and W. R. Boynton 1992. Relative roles of benthic versus planktonic respiration in the seasonal depletion of oxygen from bottom waters of Chesapeake Bay. *Marine Ecology Progress Series* 85:137–152.

Kemp, W. M., R. R. Twilley, J. C. Stevenson, W. R. Boynton, and J. C. Means. 1983. The decline of submerged vascular plants in upper Chesapeake Bay: Summary of results concerning possible causes. *Marine Technology Society Journal* 17:78–89.

Ketchum, B. H. 1954. Relation between circulation and planktonic populations in estuaries. *Ecology* 35:191–200.

Kimball, K. D., and S. A. Levin. 1985. Limitations of laboratory bioassays: The need for ecosystem-level testing. *BioScience* 35:165–171.

Kiørboe, T. 1980. Distribution and production of submerged macrophytes in Tipper Grund (Ringkøbing Fjord, Denmark), and the impact of waterfowl grazing. *Journal of Applied Ecology* 17:675–687.

Kitchell, J. F., D. J. Stewart, and D. Weininger. 1977. Applications of a bioenergetics model to yellow perch (*Perca flavescens*) and walleye (*Stizostedion vitreum vitreum*). *Journal of the Fisheries Research Board of Canada* 34:1922–1935.

Krummel, J. R., R. H. Gardner, G. Sugihara, and R. V. O'Neill. 1987. Landscape patterns in a disturbed environment. *Oikos* 48:321–324.

Laake, M., A. B. Dahle, K. Eberlein, and K. Rein. 1983. A modeling approach to the interplay of carbohydrates, bacteria, and non-pigmented flagellates in a controlled ecosystem experiment with *Skeletonema costatum*. *Marine Ecology Progress Series* 14:71–79.

Landry, M. 1977. A review of important concepts in the trophic organization of pelagic ecosystems. *Helgoländer wissenschaftliche. Meeresuntersuchungen* 30:8–17.

Lassiter, R. R. 1983. Microcosms as ecosystems for testing ecological models. In S. E. Jørgensen, ed., *State-of-the-Art in Ecological Modeling*, pp. 127–161. New York: Springer-Verlag.

Lawton, J. H. 1995. Ecological experiments with model systems. *Science* 269:328–331.

Legendre, L. 1983. Dimensional analysis in ecology. In L. Legendre and P. Legendre, eds., *Numerical Ecology*, pp. 53–80. Amsterdam: Elsevier Science.

Legendre, L., and P. Legendre. 1983. *Numerical Ecology*. Amsterdam: Elsevier Science.

Legendre, P., S. F. Thrush, V. J. Cummings, P. K. Dayton, J. Grant, J. E. Hewitt, A. H. Hines, B. H. McArdle, R. D. Pridmore, D. C. Schneider, S. J. Turner, R. B. Whitlatch, and M. R. Wilkinson. 1997. Spatial structure of bivalves in a sandflat: Scale and generating processes. *Journal of Experimental Marine Biology and Ecology* 216:99–128.

Levin, S. A. 1992. The problem of pattern and scale in ecology. *Ecology* 73:1943–1967.

Levin, S. A., A. Morin, and T. H. Powell. 1989. Patterns and processes in the distribution and dynamics of Antarctic krill. In Scientific Committee for the Conservation of Antarctic Marine Living Resources, *Selected Scientific Papers Part 1*, pp. 281–299. SC-CAMLR-SSP/5. Hobart: SC-CAMLR.

Lewis, M. R., J. J. Cullen, and T. Platt. 1984. Relationships between vertical mixing and photoadaptation of phytoplankton: Similarity criteria. *Marine Ecology Progress Series* 15:141–149.

Lewis, M. R., and T. Platt. 1982. Scales of variability in estuarine ecosystems. In V. S. Kennedy, ed., *Estuarine Comparisons*, pp. 3–20. New York: Academic Press.

Liepmann, D., and G. Stephanopoulos. 1985. Development and global sensitivity analysis of a closed ecosystem model. *Ecological Modelling* 30:13–47.

Lodge, D. M., J. W. Barko, D. Strayer, J. M. Melack, G. G. Mittelback, R. W. Howarth, B. Menge, and J. E. Titus. 1988. Spatial heterogeneity and habitat interactions in lake communities. In S. R. Carpenter, ed., *Complex Interactions in Lake Ecosystems*, pp. 181–208. New York: Springer-Verlag.

Logerwell, E., and N. B. Hargreaves. 1996. The distribution of sea birds relative to their fish prey off Vancouver Island: Opposing results at large and small spatial scales. *Fisheries Oceanography* 5:1–13.

Luckinbill, L. S. 1973. Coexistence in laboratory populations of *Paramecium aurelia* and its predator *Didinium nasutum*. *Ecology* 54:1320–1327.

Luckinbill, L. S. 1974. The effects of space and enrichment on a predator-prey system. *Ecology* 55:1142–1147.

Lundgren, A. 1985. Model ecosystems as a tool in freshwater and marine research. *Archiv für Hydrobiologie* 70:157–197.

Mackas, D. L., K. L. Denman, and M. R. Abbott. 1985. Plankton patchiness: Biology in the physical vernacular. *Bulletin of Marine Science* 37:652–674.

MacNally, R., and G. P. Quinn. 1998. Symposium introduction: The importance of scale in ecology. *Australian Journal of Ecology* 23:1–7.

Madden, C. J., and W. M. Kemp. 1996. Ecosystem model of an estuarine submersed plant community: Calibration and simulation of eutrophication responses. *Estuaries* 19:457–474.

Malone, T. C. 1984. Anthropogenic nitrogen loading and assimilation capacity of the Hudson River estuarine system, USA. In V. S. Kennedy, ed., *The Estuary as a Filter*, pp. 291–311. New York: Academic Press.

Marba, N., and C. M. Duarte. 1995. Coupling of seagrass (*Cymodocea nodosa*) patch dynamics to subaqueous dune migration. *Journal of Ecology* 83:381–389.

Margalef, R. 1967. Laboratory analogues of estuarine plankton systems. In G. Lauff, ed., *Estuaries*, pp. 515–521. Washington, DC: American Association for the Advancement of Science.

McArdle, B. H., J. E. Hewitt, and S. F. Thrush. 1997. Pattern from process: It is not as easy as it looks. *Journal of Experimental Marine Biology and Ecology* 216:229–242.

McClatchie, S., and M. R. Lewis. 1986. Limitations of grazing rate equations: The case for time-series measurements. *Marine Biology* 92:135–140.

Meentemeyer, V., and E. O. Box. 1987. Scale effects in landscape studies. In M. G. Turner, ed., *Landscape Heterogeneity and Disturbance*, pp. 15–34. New York: Springer-Verlag.

Miller, D. C., P. A. Jumars, and A. R. M. Nowell. 1984. Effects of sediment transport on deposit feeding: Scaling arguments. *Limnology and Oceanography* 29:1202–1217.

Morris, R. G., J. H. Kennedy, P. C. Johnson, and F. E. Hambleton. 1994. Pyrethroid insecticide effects on bluegill sunfish in microcosms and mesocosms and bluegill impact on microcosm fauna. In R. L. Graney, J. H. Kennedy, and J. H. Rodgers Jr., eds., *Aquatic Mesocosm Studies in Ecological Risk Assessment*, pp. 373–395. Boca Raton, Fla.: CRC Press.

Murray, L., R. B. Sturgis, R. Bartleson, W. Severn, and W. M. Kemp. 2000. Scaling submersed plant community responses to experimental nutrient enrichment. In S. Bortone, ed., *Seagrasses: Monitoring, Ecology, Physiology, and Management*, pp. 241–258. Boca Raton, Fla: CRC Press.

Nixon, S. W. 1969. A synthetic microcosm. *Limnology and Oceanography* 14:142–145.

Nixon, S. W. et al. 1996. The fate of nitrogen and phosphorus at the land-sea margin of the North Atlantic Ocean. *Biogeochemistry* 35:141–180.

Nixon, S. W., C. A. Oviatt, J. N. Kremer, and K. Perez. 1979. The use of numerical models and laboratory microcosms in estuarine ecosystem analysis-simulations of winter phytoplankton bloom. In R. F. Dame, ed., *Marsh-Estuarine Systems Simulation*, pp. 165–188. Columbia: University of South Carolina Press.

Nowicki, B. L., and C. A. Oviatt. 1990. Are estuaries traps for anthropogenic nutrients? Evidence from estuarine mesocosms. *Marine Ecology Progress Series* 66:131–146.

Odum, E. P. 1971. *Fundamentals of Ecology.* 3rd ed. Philadelphia: Saunders.

Odum, E. P. 1972. Ecosystem theory in relation to man. In J. A. Wiens, ed., *Ecosystem Structure and Function*, pp. 11–24. Corvallis: Oregon State University Press.

Odum, H. T. 1983. *Systems Ecology: An Introduction.* New York: Wiley.

Odum, W. E., and E. Heald. 1975. The detritus-based food web of an estuarine mangrove community. In L. E. Cronin, ed., *Estuarine Research,* vol. 1, pp. 265–286. New York: Academic Press.

Olesen, B., and K. Sand-Jensen. 1994. Demography of shallow eelgrass (*Zostera marina*) populations—Shoot dynamics and biomass development. *Journal of Ecology* 82:379–390.

Oliver, C. D., and B. C. Larson. 1990. *Forest Stand Dynamics.* New York: McGraw-Hill.

O'Neill, R. V., D. L. DeAngelis, J. B. Waide, and T. F. H. Allen. 1986. *A Hierarchical Concept of Ecosystems.* Princeton: Princeton University Press.

O'Neill, R. V., R. H. Gardner, B. T. Milne, M. G. Turner, and B. Jackson. 1991. Heterogeneity and spatial hierarchies. In J. Kolasa and S. T. A. Pickett, eds., *Ecological Heterogeneity*, pp. 85–96. New York: Springer-Verlag.

Orth, R. J. 1975. Destruction of eelgrass, *Zostera marina*, by the cownose ray, *Rhinoptera bonasus*, in the Chesapeake Bay. *Chesapeake Science* 16:205–208.

Orth, R. J., M. Luckenbach, and K. A. Moore. 1994. Seed dispersal in a marine macrophyte: Implications for colonization and restoration. *Ecology* 75:1927–1939.

Orth, R. J., and K. A. Moore. 1983. Chesapeake Bay: An unprecedented decline in submerged aquatic vegetation. *Science* 222:51–53.

Oviatt, C. 1994. Biological considerations in marine enclosure experiments: Challenges and revelations. *Oceanography* 7:45–51.

Oviatt, C., B. Buckley, and S. Nixon. 1981. Annual phytoplankton metabolism in Narragansett Bay calculated from survey field measurements and microcosm observations. *Estuaries* 4:167–175.

Oviatt, C. A., P. H. Doering, B. L. Nowicki, and A. Zoppini. 1993. Net system production in coastal waters as a function of eutrophication, seasonality, and benthic macrofaunal abundance. *Estuaries* 16:247–253.

Paine, R. T., and S. A. Levin. 1981. Intertidal landscapes: Disturbance and the dynamics of pattern. *Ecological Monographs* 51:145–178.

Parsons, T. R. 1990. The use of mathematical models in conjunction with mesocosm ecosystem research. In C. M. Lalli, ed., *Enclosed Experimental Marine Ecosystems: A Review and Recommendations*, pp. 197–210. New York: Springer-Verlag.

Parsons, T. R., T. A. Kessler, and L. Guanguo. 1986. An ecosystem model analysis of the effect of mine tailings on the euphotic zone of a pelagic ecosystem. *Acta Oceanologica Sinica* 5:425–436.

Parsons, T. R., M. Takahashi, and B. Hargrave. 1984. *Biological Oceanographic Processes*. 3rd ed. New York: Pergamon Press.

Pauly, D., and V. Christensen. 1995. Primary production required to sustain global fisheries. *Nature* 374:255–257.

Peters, R. H. 1983. *The Ecological Implications of Body Size*. Cambridge: Cambridge University Press.

Petersen, J. E., C.-C. Chen, and W. M. Kemp. 1997. Scaling aquatic primary productivity: Experiments under nutrient- and light-limited conditions. *Ecology* 78:2326–2338.

Petersen, J. E., J. C. Cornwell, and W. M. Kemp. 1999. Implicit scaling in the design of experimental aquatic ecosystems. *Oikos* 85:3–18.

Pimm, S. L. 1982. *Food Webs*. London: Chapman and Hall.

Piontkovski, S., and R. Williams. 1995. Multiscale variability of tropical ocean zooplankton biomass. *ICES Journal of Marine Science* 52:643–656.

Platt, T., and K. L. Denman. 1975. Spectral analysis in ecology. *Annual Review of Ecology and Systematics* 6:189–210.

Platt, T., K. H. Mann, and R. E. Ulanowicz. 1981. Thinking in terms of scale: Introduction to dimensional analysis. In K. Platt, H. Mann, and R. E. Ulanowicz, eds., *Mathematical Models in Biological Oceanography*, pp. 112–121. Paris: UNESCO Press.

Popper, K. 1962. *Conjectures and Refutations*. New York: Basic Books.

Powell, T. M., J. E. Cloern, and R. A. Walters. 1986. Phytoplankton spatial distributions in south San Francisco Bay: Mesoscale and small-scale variability. In D. Wolfe, ed., *Estuarine Variability*, pp. 369–386. New York: Academic Press.

Rastetter, E. B., A. W. King, B. J. Cosby, G. M. Hornberger, R. V. O'Neill, and J. E. Hobbie. 1992. Aggregating fine-scale ecological knowledge to model coarser-scale attributes of ecosystems. *Ecological Applications* 2:55–70.

Resetarits, W., and J. J. Fauth. 1998. From cattle tanks to Carolina bays. In W. J. Resetarits Jr. and J. Bernardo, eds., *Experimental Ecology*, pp. 133–151. New York: Oxford University Press.

Rose, G. A, and W. C. Leggett. 1990. The importance of scale to predator-prey spatial correlations: An example of Atlantic fishes. *Ecology* 71:33–43.

Roush, W. 1995. When rigor meets reality. *Science* 269:313–315.

Sand-Jensen, K. 1989. Environmental variables and their effect on photosynthesis of aquatic plant communities. *Aquatic Botany* 34:5–25.

Sanford, L. P. 1997. Turbulent mixing in experimental ecosystem studies. *Marine Ecology Progress Series* 161:265–293.

Sarnelle, O. 1997. *Daphnia* effects on microzooplankton: Comparisons of enclosure and whole-lake responses. *Ecology* 78:913–928.

Scheffer, M., S. Hosper, M.-L. Meijer, B. Moss, and E. Jeppesen. 1993. Alternative equilibria in shallow lakes. *Trends in Ecology and Evolution* 8:275–279.

Schindler, D. W. 1987. Detecting ecosystem responses to anthropogenic stress. *Canadian Journal of Fisheries and Aquatic Sciences* 44 (Suppl. 1):6–25.

Schindler, D. W. 1998. Replication versus realism: The need for ecosystem-scale experiments. *Ecosystems* 1:323–334.

Schneider, D. C. 1994. *Quantitative Ecology: Spatial and Temporal Scaling*. San Diego: Academic Press.

Schneider, D. C., R. Walters, S. Thrush, and P. Dayton. 1997. Scale-up of ecological experiments: Density variations in the mobile bivalve *Macomona liliana*. *Journal of Experimental Marine Biology and Ecology* 216:129–152.

Sheldon, R. W., A. Prakash, and W. H. J. Sutcliffe. 1972. The size distribution of particles in the ocean. *Limnology and Oceanography* 17:323–340.

Shelford, V. E. 1911. Ecological succession: Stream fishes and the method of physiographic analysis. *Biological Bulletin* 21:127–151.

Shepard, S. A., A. J. McComb, D. A. Bulthuis, V. Neverauskas, D. A. Steffensen, and R. West. 1989. Decline of seagrasses. In A. W. D. Larkum, A. J. McComb, and S. A. Shepard, eds., *Biology of Seagrasses: A Treatise on the Biology of Seagrasses with Special Reference to the Australian Region*, pp. 346–393. Amsterdam: Elsevier.

Shirazi, M. A., B. Lighthart, and J. Gillett. 1984. A method for scaling biological response of soil microorganisms. *Ecological Modelling* 23:203–226.

Smith, E. M., and W. M. Kemp. 1995. Seasonal and regional variations in plankton community production and respiration for Chesapeake Bay. *Marine Ecology Progress Series* 116:217–231.

Spencer, M., and P. H. Warren. 1996. The effects of energy input, immigration, and habitat size on food web structure: A microcosm experiment. *Oecologia* 108:764–770.

Steele, J. H. 1985. A comparison of terrestrial and marine ecological systems. *Nature* 313:355–358.

Steele, J. H. 1989. Discussion: Scale and coupling in ecological systems. In J. Roughgarden, R. M. May, and S. A. Levin, eds., *Perspectives in Ecological Theory*, pp. 177–180. Princeton: Princeton University Press.

Steele, J. H., and E. W. Henderson. 1994. Coupling between physical and biological scales. *Proceedings of the Royal Society of London, Series B* 343:5–9.

Strickland, J. D. H. 1967. Between beakers and bays. *New Scientist* 33:276–278.

Sturgis, R. B., and L. Murray. 1997. Scaling of nutrient inputs to submersed plant communities: Temporal and spatial variations. *Marine Ecology Progress Series* 152:89–102.

Suess, E. 1980. Particulate organic carbon flux in the oceans: Surface productivity and oxygen utilization. *Nature* 288:260–263.

Summers, U. 1988. Phytoplankton succession in microcosm experiments under simultaneous grazing pressure and resource limitation. *Limnology and Oceanography* 33:1037–1054.

Swartzman, G. L., F. B. Taub, J. Meador, C. Huang, and A. Kindig. 1990. Modeling the effect of algal biomass on multispecies aquatic microcosms' response to copper toxicity. *Aquatic Toxicology* 17:93–117.

Taub, F. B. 1969. A biological model of a freshwater community: A gnotobiotic ecosystem. *Limnology and Oceanography* 14:136–142.

Threlkeld, S. T. 1988. Planktivory and planktivore biomass effects on zooplankton, phytoplankton, and the trophic cascade. *Limnology and Oceanography* 33:1362–1375.

Threlkeld, S. T. 1994. Benthic-pelagic interactions in shallow water columns: An experimentalist's perspective. *Hydrobiologia* 275/276:293–300.

Thrush, S. F., D. C. Schneider, P. Legendre, R. B. Whitlatch, P. K. Dayton, J. E. Hewitt, A. H. Hines, V. J. Cummings, S. M. Lawrie, J. Grant, R. D. Pridmore, S. J. Turner, and B. H. McArdle. 1997. Scaling-up from experiments to complex ecological systems: Where to next? *Journal of Experimental Marine Biology and Ecology* 216:243–254.

Tilman, D. 1989. Ecological experimentation: Strengths and conceptual problems. In G. E. Likens, ed., *Long-Term Studies in Ecology*, pp. 136–157. New York: Springer-Verlag.

Turner, M. G. 1989. Landscape ecology: The effects of pattern on process. *Annual Review of Ecology and Systematics* 20:171–199.

Turner, M. G., and R. H. Gardner. 1991. Quantitative methods in landscape ecology: An introduction. In M. G. Turner and R. H. Gardner, eds., *Quantitative Methods in Landscape Ecology*, pp. 3–14. New York: Springer-Verlag.

Turner, M. G., and R. V. O'Neill. 1995. Exploring aggregation in space and time. In C. G. Jones and J. H. Lawton, eds., *Linking Species and Ecosystems*, pp. 194–208. New York: Chapman and Hall.

Turpin, D. H., and P. J. Harrison. 1979. Limiting nutrient patchiness and its role in phytoplankton ecology. *Journal of Experimental Marine Biology and Ecology* 39:151–166.

Twilley, R. R., W. M. Kemp, K. W. Staver, J. C. Stevenson, and W. R. Boynton. 1985. Nutrient enrichment of estuarine submersed vascular plant communities. 1. Algal growth and effects on production of plants and associated communities. *Marine Ecology Progress Series* 23:179–191.

Uhlmann, D. 1985. Scaling of microcosms and the dimensional analysis of lakes. *Internationale Revue der Gesamten Hydrobiologie* 70:47–62.

Ulanowicz, R. E., and W. M. Kemp. 1979. Towards canonical trophic aggregations. *American Naturalist* 114:871–883.

Walker, D. I., R. J. Lukatelich, G. Bastyan, and A. J. McComb. 1989. Effect of boat moorings on seagrass beds near Perth, Western Australia. *Aquatic Botany* 36:69–77.

Weber, L. H., S. Z. El-Syed, and I. Hampton. 1986. The variance spectra of phytoplankton, krill, and water temperature in the Antarctic Ocean south of Africa. *Deep-Sea Research* 33:1327–1343.

West, G. B., J. H. Brown, and B. J. Enquist. 1997. A general model for the origin of allometric scaling laws in biology. *Science* 276:122–126.

Wiens, J. A. 1989. Spatial scaling in ecology. *Functional Ecology* 3:385–397.

With, K. A., and T. O. Crist. 1996. Translating across scales: Simulating species distributions as the aggregate response of individuals to heterogeneity. *Ecological Modelling* 93:125–137.

Wofsy, S. C. 1983. A simple model to predict extinction coefficients and phytoplankton biomass in eutrophic waters. *Limnology and Oceanography* 28:1144–1155.

Zieman, J. C., J. W. Fourqurean, and R. L. Iverson. 1989. Distribution, abundance, and productivity of seagrasses and macroalgae in Florida Bay. *Bulletin of Marine Science* 33:293–311.

PART II

SCALING THEORY

CHAPTER 2

Understanding the Problem of Scale
in Experimental Ecology

John A. Wiens

S CALE HAS BEEN CALLED "THE NEW FRONTIER OF ECOLOGY" (Allen and Roberts 1998). Yet Kareiva (1994) referred to space as ecology's final frontier, and Klomp and Lunt (1997) edited an entire book about "frontiers in ecology" that includes, in addition to scale and space, such diverse topics as adaptive management, disturbance ecology, land-water linkages, ecosystem complexity, climate change, ecological economics, and land-use planning. Ecology seems to be full of frontiers. All these topics arguably meet the dictionary definition of a frontier as "a new or undeveloped area of knowledge" or "a new field that offers scope for activity." I suspect that I would not be alone, however, in contending that scale holds a preeminent position among these various frontiers. Issues of scale pervade every area of ecological investigation and compromise every form of ecological application. Awareness of scale problems is not new, of course; several ecologists drew attention to scale issues decades ago (e.g., Greig-Smith 1952; Andrewartha and Birch 1954; Steele 1978; Wiens 1981; Allen and Starr 1982). Widespread recognition of the importance of scale effects and concern about their resolution, however, are relatively recent developments (O'Neill 1989; Steele 1989; Wiens 1989a; Levin 1992; Allen et al. 1993).

The tension between this newfound awareness of scaling issues and the ways ecologists go about gaining knowledge is especially strong for ecological experiments. Experiments have increasingly become the *sine qua non* of ecological research, yet recent books on the design and analysis of ecological experiments (e.g., Hairston 1989; Manly 1992; Scheiner and Gurevitch 1993; Underwood 1997) scarcely mention the confounding effects of scale (see, however, Dutilleul 1998; Resetarits and Bernardo 1998). My goal here is to develop a framework for thinking about the interplay between scale and experiments in ecology. I will do this by addressing four questions:

- What are the essential features of ecological experiments?
- What are the essential features of scaling in ecology?
- How can we approach scaling issues in ecological experiments?
- In a more philosophical vein, how should experiments or considerations of scale enter into the process of doing ecology?

EXPERIMENTS IN ECOLOGY

Ecologists conduct experiments to get at causation, to determine the relationships between pattern and process. As in other sciences, the essence of experimentation is simplification: one controls extraneous sources of variation, leaving only the causal linkages exposed. Our confidence in the power of experiments is based on the premise that nature will in fact give consistent answers to the questions we pose, if only we eliminate the noise of natural variation. That is why we use statistics. Underlying this approach is the belief that the phenomena we observe are ultimately based on simple and general truths. This belief is the foundation of reductionism.

There are two consequences of following this philosophical pathway in experimental science. These have to do with assumptions and scale. Traditionally, experimental ecologists have assumed that the systems they investigate are in equilibrium, are spatially homogeneous, and are closed (Wiens 1984, 1989b; Pickett et al. 1992). In other words, natural variation is fleeting or unimportant. Thus, a single, short-term experiment can be taken to indicate the response of the system to a given treatment; if it were done at another time, it would yield the same result. This may be why ecologists generally do not repeat one another's experiments. Similarly, because spatial variation is assumed to be unimportant, the location of experiments is usually not explicitly considered. This is one of several reasons why replication in space is generally limited (Kareiva and Andersen 1988). And because the system is considered to be closed (or, alternatively, "regional" effects are small in relation to proximate, "local" effects), the surroundings of the experimental sites are rarely measured or included in the experimental design.

Of course, none of these assumptions holds in nature. Real ecological systems are variable in time and space and are influenced by external factors. The magnitude of these effects may vary depending on how the "system" is defined, and the importance of these assumptions varies correspondingly. Different kinds of ecological experiments differ in the degree to which they address reality and therefore embrace (consciously or not) these assumptions. Laboratory or simulation experiments are at one extreme: if properly designed, they may have great internal validity (Manly 1992; Naeem, this volume) but (by design) little direct relation to the natural world. Their dependence on the simplifying assumptions of equilibrium, homogeneity, and closure is absolute. At the other extreme lie field experiments,[1] in which external validity is maximized, but at the cost of sacrificing internal validity (Naeem, this volume). The assumptions are usually violated, but these effects are normally absorbed into the variance or error term of the analysis. Somewhere in between are experimental model systems (Ims et al. 1993; Wiens et al. 1993; Ims 1999) or model ecosystem experiments such as microcosms or mesocosms (Lawton 1995, 1998; Fraser and Keddy 1997; Lawler 1998; Petersen et al. 1999) that relax or adhere to the assumptions to varying degrees. Mesocosm experiments, for example, may be conducted in containers where system closure is complete, or *in situ* in the field where external influences may have free play over the experimental results (Kemp et al., this volume). As Naeem has noted in this volume, no single approach along this spectrum of kinds of experiments is best in fostering ecological understanding (although some experimentalists might disagree!). Our goal as ecologists, however, is to understand how ecological systems function in the real world rather than as theoretical or artificial abstractions, to learn why the patterns of nature are the way they are. Accepting the assumptions of equilibrium, homogeneity, and system closure may facilitate experimental design and logistics, and experiments conducted in this way may help to define what is possible under these conditions. They do not necessarily guarantee, however, that we will have moved any closer to our broader goal.

The other consequence of following the philosophical pathway that leads to reductionism relates to scale, and it has clear repercussions for the design and conduct of experiments. Laboratory and simulation

---

1. I omit consideration of "natural experiments" (Diamond 1986); lacking both controls and repeatability, they do not qualify as true experiments in any formal sense of the term.

experiments are necessarily limited in their spatial extent, and usually also in their temporal duration. These constraints make it easier to accept the assumptions of equilibrium, homogeneity, and closure. Most mesocosm experiments (especially closed ones) are similarly constrained, although the scales may be slightly broader. Field experiments may encompass the broadest range of scales, from less than a square meter to entire lakes, or days to decades (Carpenter et al. 1995; Lodge et al. 1998). Nonetheless, most ecological experiments are conducted at relatively fine scales of space and time (Kareiva and Andersen 1988; Wiens 1989a). To generalize from such experiments one must either ignore scale or assume that it does not matter.

Yet we know that scale *does* matter. Ecological patterns are different when they are considered at different scales. For example, the association of shrubsteppe or forest birds with habitat features (Wiens et al. 1987; Saab 1999), patterns of community structure or diversity (Shmida and Wilson 1985; Fuisz and Moskát 1992; Huston 1994; Rosenzweig 1995), patterns of temporal variation in grassland vegetation (Fuhlendorf and Smeins 1996), the distribution of fungivorous beetles among basidiocarps (Rukke and Midtgaard 1998), or the effects of parasitoids on forest caterpillars (Roland and Taylor 1997) are but a few of many instances in which patterns and relationships change with scale. The recent spate of books on scaling (e.g., Ehleringer and Field 1993; Edwards et al. 1994; Stewart et al. 1996; Quattrochi and Goodchild 1997; van Gardingen et al. 1997; Peterson and Parker 1998) indicates that scale is a topic of considerable interest to ecologists. Constructing diagrams that depict the domains of various phenomena in space and time (e.g., Steele 1978; Wiens 1981; Delcourt et al. 1983; Urban et al. 1987; Marquet et al. 1993; Holling et al. 1996) has become a popular way to express the scaling properties of systems. In order to develop an understanding of how scale relates to experimental ecology and to begin to deal with the problems posed by scale, however, it is necessary to go beyond such heuristic exercises and address what is meant by "scale" and "scaling" in ecology.

SCALE IN ECOLOGY

"Scale" has a good start on contesting "niche" as one of the vaguest yet most often used words in ecology. Like physicists, engineers, astronomers, economists, or cell biologists, ecologists use scale in a variety of ways. For example, O'Neill and King (1998) define scale as the "physical dimensions of observed entities and phenomena," while Schneider (1994) and Gardner (1998) use scale to refer to the resolution or range of measurement of a quantity of interest. Others (e.g., Rahel 1990; Reynolds et al. 1993; Waring 1993; Grime et al. 1997; Hinckley et al. 1998) equate scale with level in an organizational hierarchy, such as leaf to tree stand or cell to landscape. Those interested in spatial pattern may use scale in the cartographic sense, as the ratio of the size of a representation of an area to its actual size (Lam and Quattrochi 1992; Atkinson 1997; Goodchild and Quattrochi 1997). On the other hand, my dictionary (not written by ecologists) defines scale as a graduated or graded series—a gradient on which quantities may be ordered, as in small to large or cold to hot. These definitions have quite different meanings, which translate into different ways of thinking about scale and different perceptions of the problems that scale may pose for ecological investigations.

Scale specifies the measurement domain of a study, but ecologists are often interested in scale in the context of extending the results of a study from its measurement domain to other scales—the process of "scaling." But ecologists also use scaling in different ways. Which of the following, for example, represents ecological scaling: The relationship between brain size or metabolism and body mass? The increase in activity level of an insect with increasing temperature? The increase in species number with area? The relationship between spatial heterogeneity and biodiversity? The relationship between phytoplankton biomass and water depth? The increase in probability of a large disturbance with time? The increase in population persistence with spatial heterogeneity? The increase in feedback complexity with level of organization? The relationship between food-web connectance and species richness? The relationship between leaf photosynthesis and forest-stand production or global carbon budgets?

The answer would seem to be "all of the above." Ecologists deal with relationships, and relationships expressed quantitatively are scalings. This is why ecologists are so fond of regressions, and why some of them (at least) have embraced fractal geometry (Nikora et al. 1999; Solé et al. 1999). But if discussions of scale and scaling in ecology are to have any substance, I believe that we must be more precise and restrictive in our use of these terms. Although not discounting that other aspects of scale in ecology are important, I propose that, in an ecological context, scale should refer to a quantified portion of the spectra of space or time. This portion (what Schneider [1998] calls *scope*) is bounded by *grain*, the resolution of measurements, and *extent*, the upper limit of the range of scales over which measurements are taken (Levins 1968; Wiens 1976, 1990; Allen et al. 1984; O'Neill et al. 1986). "Scaling," then, involves relating measurements made at one scale to those made or predicted at another (i.e., shifting grain and extent to encompass a different portion of the time or space spectrum). The scales at which measurements are taken in space or time are absolute metrics (e.g., 5 ha, 2 h), although references to "fine" or "broad" or "short" or "long" scales are as relative as "small" or "large."[2]

There are at least three reasons for advocating a restrictive use of "scale" and "scaling." First, it makes them operational. Spatial and temporal scale, grain, extent, and changes in scale can all be expressed in quantitative terms on scaled axes. As Allen and Hoekstra (1992), Allen (1998), and O'Neill and King (1998) have cogently argued, considering levels of organization as scales only promotes ambiguity, because levels are arbitrarily defined types, not quantities. Thus, the scale of an organism or a landscape or an ecosystem can vary over orders of magnitude, depending on which organisms one considers or how one delimits landscapes or ecosystems.

The second reason for restricting the scope of scale and scaling as terms is that it makes them specific, reducing vagueness. Thus, although most of the relationships listed above can be quantified, they often reduce to a statement that variable y is some function of the value of variable x. Most ecological studies aim in one way or another to identify such relationships, and the determination of such "scalars" or scaling equations has led to important insights in many areas of ecology. Indeed, these equations are the foundation of predictive ecology. Some of these scalars, such as species-area equations or the allometric relationships between body mass and home range or generation length, relate directly to spatial or temporal scales, but others, such as the relationship between body mass and metabolism or species diversity and ecosystem function, do not. Scalars or scaling equations are useful for expressing the spatial or temporal scale-dependency of ecological patterns or processes (see

---

2. Cartographers use "small" and "large" to refer to map scales, but in an opposite way to how ecologists generally think of these terms. Thus, a small-scale map is one in which the resolution (grain size) is broad. Some have recommended that "fine" and "broad" be used in ecological studies, "small" and "large" when referring to map scales (Turner and Gardner 1991; Silbernagel 1997), although this solution leaves landscape ecology (which deals with both ecology and maps) characteristically confused.

below), as they are for a wide array of other ecological phenomena. But simply generating a quantitative relationship between variables is not the same thing as scale or scaling, as I use them here.

The third reason for being restrictive in the use of these terms is that it enables us to focus directly on how space and time, and changes in the scales of space and time, affect ecological systems. Space or time can be used as the *x*-axis in graphs that depict explicitly the response of an ecological variable to changes in scale. More importantly, using space or time as a dimension in analysis permits one to address directly the traditional assumptions of homogeneity and equilibrium. In a strictly homogeneous or equilibrial system, for example, we can hypothesize that the response variable will not change with changes in scale (figure 2.1a); scale-dependency (figure 2.1b, 2.1c) indicates that the traditional assumptions have been violated (cf. Romme et al. 1998).

Spatial or temporal scaling can take a variety of forms. The absence of a scaling relationship (figure 2.1a) represents the null hypothesis: scale-independence. In this situation, the results of a study or experiment at any scale can be extrapolated to other scales with impunity. A departure from this null expectation indicates scale-dependence. If the form of scale-dependence remains constant over some range of scale, then scaling relationships can be expressed as equations in which the ecological response variable (*y*) is some linear, monotonic function of scale (*x*) (figure 2.1b). The species-area relationship, in which species richness is a

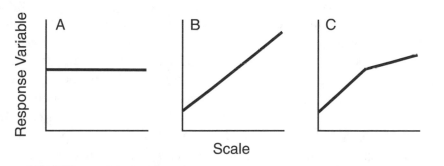

FIGURE 2•1 *Three Scaling Patterns*
Three patterns of scaling of an ecological variable. In A, the magnitude of the response does not change with changing scale; the pattern is scale-independent (note that a varying response could also be independent of scale). B and C depict conditions of scale-dependence, in which the value of the response variable changes monotonically with scale (B) or exhibits a scaling threshold (C).

power function of area, is a well-known example (see Harte et al. 1999). A scaling equation provides the basis for extrapolation from one scale (e.g., area) to another; once the equation is derived, scaling extrapolation is seemingly straightforward.

But scaling relationships are rarely so simple. There are often thresholds in scaling relationships, particular scales at which the processes determining the structure of ecological systems (or the relative importance of different processes) suddenly change (figure 2.1c). These thresholds produce nonlinearities in scaling relationships, causing marked departures from the predictions of simple scaling equations (Krummel et al. 1987; Wiens 1989a; King 1991; Rastetter et al. 1992). The species-area relationship, for example, exhibits such thresholds with changes in scale (Shmida and Wilson 1985; Wiens 1989b:106). Thresholds delimit *domains* of scale (sensu Wiens 1989a), portions of a scale spectrum within which the relation between pattern and process does remain at least proportionately constant (e.g., a power function) and extrapolation is possible. Various procedures are available for detecting scale-dependence or scale thresholds (Gardner 1998; O'Neill and King 1998), but extrapolation across scale thresholds ("transmutation error"; O'Neill 1979; King 1991) remains one of the most vexing problems in ecology.

Soil nutrient dynamics in Australian savanna and desert woodland ecosystems illustrate both within-domain and between-domain (threshold) scaling properties (Ludwig et al. 2000). At a scale of meters, these systems consist of mosaics of patches of perennial tussock grasses (e.g., *Eragrostis eriopoda*) or chenopod shrubs (e.g., *Marieana polypterygia*) interspersed with bare, unvegetated interpatch areas. At somewhat broader scales, groves of woody vegetation (e.g., mulga, *Acacia aneura*) may be interspersed with open intergrove areas. As the scale (i.e., the area) of the vegetated patches increases, soil N is progressively concentrated in these patches, increasing disproportionately the difference in soil N between the patches and the interpatch or intergrove areas. Over a certain range of patch scales, this relationship follows a significant regression—a scaling equation (figure 2.2). A similar equation can be derived at broader scales for the relationship between the patch/ interpatch difference in soil N levels and patch scale in mosaics of monsoon rainforest patches embedded in an interpatch matrix of upland eucalypt savanna in northern Australia (figure 2.2). This relationship, however, is quantitatively different from that derived for the semiarid shrublands and woodlands. Somewhere between the two scale domains a

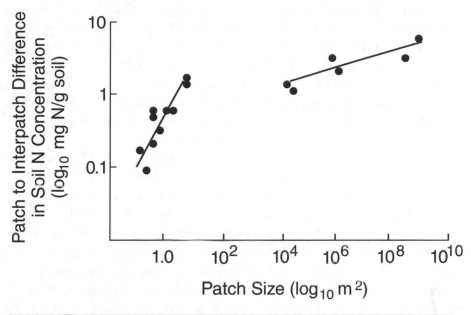

FIGURE 2•2 *Scaling Relationships for Soil Nitrogen*
Scaling relationships between the difference in soil nitrogen levels between vegetation patches and areas between patches (interpatches) and patch scale (area) for local landscape patches of grass or woody vegetation interspersed with interpatch areas of bare ground or grass (left relationship) and regional landscape patches of lowland floodplains or monsoonal rainforest interspersed with interpatch areas of upland eucalypt savanna (right) in the Northern Territory of Australia. Patch area is $\log_{10}$ transformed. (From Ludwig et al. 2000)

threshold has been passed, changing the form of the scaling relationship for soil N concentration.

What mechanisms might underlie such scaling equations? In the semiarid Australian systems, the relationships may be driven by fine-scale hydrology (Tongway and Ludwig 1994; Ludwig et al. 1997). When rain falls in these systems, runoff from the open interpatch areas is high and water (and nutrient) infiltration into the soil is low. When the runoff from such areas meets a vegetated patch, velocity is slowed and infiltration increases. Detritus carried in the runoff is deposited in the patch, where it may contribute to soil N buildup. The increased levels of soil nutrients and greater moisture infiltration within the patches in turn foster plant establishment and growth, increasing patch size. Larger patches are more effective in capturing runoff and detritus and retain litter produced within the patch, further enhancing N concentration. A positive feedback cycle is established, which drives the increased N-concentration differential between patches and interpatches as patch scale increases. The spatial pattern and scaling of the system are, in a sense, self-organizing (Wilson and Agnew 1992; Rietkerk et al. 1997). Similar processes may produce banded vegetation patterns in other arid regions (White 1971; Thiéry et al. 1995) or contribute to the "islands of fertility" that Schlesinger et al. (1990) have documented in semiarid shrublands in New Mexico. The wave-like spatial patterns of *Abies* forests in Japan and North America result from an interaction between windthrow and forest demography, but the patterns are also self-organizing (Sprugel 1976; Sato and Iwasa 1993). In all of these cases there are limits to the sizes of patches, beyond which the self-organizing processes break down—a threshold has been passed.

Such scenarios prompt the speculation that ecological systems might generally evolve so as to "avoid" scale thresholds. Thresholds occur at scales where predictability breaks down, where uncertainty and variability increase (Krummel et al. 1987; Wiens 1989a, 1992). If systems evolve in a way that reduces uncertainty, then they should through time gravitate to operate within scaling domains, where pattern-process relationships are more predictable (cf. MacNally 1999). Physical processes, such as the runoff-runon dynamics of shrub patches or mulga groves, may constrain patterns to particular scales. The biological attributes of the system, such as the growth, reproduction, and mortality dynamics of the plant populations, may then reinforce the pattern and its scaling. This view coincides with thinking about the evolution of self-organizing systems (e.g., Kauffman 1993) and with Holling's arguments on the development of "lumpiness" in the size distributions of species in biological communities (Holling 1992; Holling et al. 1996). Thus, the self-organizing properties of ecological systems may lead to a clustering about certain nodes or scale domains. Because we have so little insight about how ecological systems are scaled, however, this remains an untested (and untestable?) proposition.

DEALING WITH SCALE

Scaling Functions

Clearly, if pattern-process relationships change with scale and if scaling properties of systems vary depending on the kinds of species present in the system, then experiments conducted at different scales, or in different systems, are likely to yield different results. There is no magic solution to the problems posed by scale in ecological experimentation, but an understanding of what determines the scaling of ecological systems in time and space may at least help one design experiments with scale awareness.

Simply stated, the scaling characteristics of a system result from the intersection of the scaling properties and dynamics of the environment with the scale-dependent responses of organisms to that environment. This coupling may be thought of in terms of *scaling functions* (figure 2.3). On the one hand, environmental scaling reflects the fact that ecological processes are played out on a habitat or landscape templet (Southwood 1977) that is not spatially homogeneous. The heterogeneity of landscapes means that ecological interactions change from place to place, and because heterogeneity itself changes with the spatial scale of resolution (Friedel 1994; Dale 1999; Wiens 2000), both scaling equations and scaling thresholds will be affected by the ways in which landscape composition and pattern change with scale. Changing the patchiness of resource

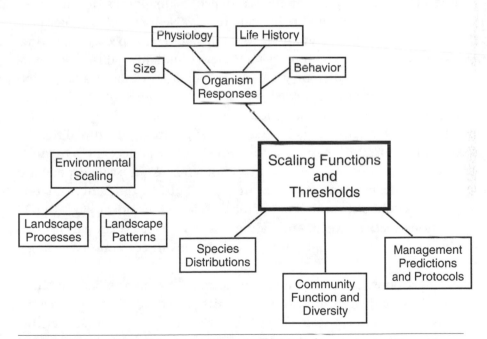

FIGURE 2•3 *Interrelationships Among Factors Affecting Scale*
Features of environmental scaling combine with characteristics of individual organisms and species to determine scaling functions, which in turn affect how species are distributed over a heterogeneous landscape mosaic, how communities are structured, and how management of natural resources should be scaled.

distribution in a landscape, for example, will alter the spatial scale on which organisms using those resources operate (O'Neill et al. 1988; De Roos et al. 1991; Tilman and Kareiva 1997; McIntyre and Wiens 1999).

On the other hand, species respond to the heterogeneity of a landscape over different ranges of scale for different life-history functions (e.g., predator avoidance, mating, social aggregation, foraging). Such process-dependent scaling ranges have been termed "ecological neighborhoods" (Addicott et al. 1987) or "scaling windows" (Riitters et al. 1997); they specify the grain and extent over which habitat or landscape pattern is perceived or "filtered" by a species engaged in a particular activity (Kotliar and Wiens 1990). The scales on which organisms respond to environmental heterogeneity are determined by features of the behavior, physiology, mobility, and life-history of species. As a consequence, the scaling of a species' response to an environment and the scales at which its responses exhibit thresholds may not correspond directly to the scaling of environmental variation revealed by mappings of landscape patterns or analyses of satellite imagery (e.g., O'Neill et al. 1991). Because they differ in ecological and life-history traits, different species occupying the same location may also perceive the environment at different scales of resolution, with different grain-extent windows. Roland and Taylor (1997), for example, provided an elegant demonstration of how differences in body sizes among four species of parasitic flies that attack the forest tent caterpillar (*Malacosoma disstria*) translate into different scales of responses to forest structure, which in turn affect the dynamics of the parasitoid-host systems. Thus, differences in scale-dependent features of life histories may in turn contribute to differences in the demography and distribution of populations (e.g., Van Horne et al. 1997). This among-species diversity in scaling responses accounts in part for the elusiveness of broadly applicable generalizations in community ecology (Lawton 1999).

We make observations or conduct experiments to discern patterns and, ultimately, link them to processes. Both the patterns and the linkages to processes, however, are scale-dependent. It is therefore essential that the window of observation or experiment (its grain and extent) encompass the scales at which the variation of critical environmental features is concordant with those of the ecological phenomena of interest. The distributions of murres (*Uria aalge*) and the fish they prey upon (capelin, *Mallotus villosus*), for example, exhibit different patterns of association at different scales of resolution (i.e., grain). Schneider and Piatt (1986;

Schneider 1994; see also Mehlum et al. 1999) found that there was only a weak match between the distributions of predator and prey at a grain scale of 250 m. As grain was systematically increased to 2.5 km, a pattern emerged in which concentrations of murres occurred immediately adjacent to the largest concentrations of fish. This pattern accords well with the feeding behavior of the birds, which aggregate rapidly over schools of fish, feed, and then begin drifting away from their prey as the fish scatter in response to the predators (Hoffman et al. 1981). This example clearly indicates that if observations or experiments are inappropriately scaled, one is likely to see noise rather than pattern or, worse yet, "patterns" that are artifacts rather than real.

From the beginning of a study, we impose restrictions on the kinds of environments and environmental factors considered and the biotic components and phenomena included. Although the decision about how to bound a study is ultimately arbitrary, driven by our objectives and by logistical constraints, this arbitrariness should not extend to the determination of the scale(s) of the investigation. That is where scaling functions come into play. In order for the notion of scaling functions to provide guidance in the design of experimental studies, however, it must be operational.

Scaling Landscape Patterns

Developing scaling functions requires that we quantify scale-dependencies in landscape patterns. Landscape structure may be characterized by a variety of metrics, such as fractal dimension, contagion, connectivity, the frequency distribution of patch sizes, or the rate of increase in landscape diversity with scale (Wiens et al. 1993; McGarigal and Marks 1995; Riitters et al. 1995; Hargis et al. 1997; Quattrochi and Goodchild 1997; Gustafson 1998; Nikora et al. 1999). Multiscale analyses can be conducted using GIS (Geographic Information System) over a range of pixel resolutions (scales). Geographic window algorithms (Dillworth et al. 1994) or hierarchical Bayesian approaches (Cullinan et al. 1997), which follow hierarchical clustering procedures to aggregate similar patch types, can be used to evaluate scale-dependency in patch context and connectivity, and scale-dependent connectivity measures can also be derived from percolation theory analysis (Keitt et al. 1997). By examining how these various measures of landscape structure change with changes in the scale of resolution (i.e., grain), it may be possible to define scaling equations that hold within specified scale domains and to recognize scales at which thresholds occur. This is the foundation of multiscale analyses of landscape fractal dimension (Krummel et al. 1987) or variance in coverage of land-cover types (O'Neill et al. 1991).

Geostatistical procedures that analyze the structure of spatial variance may also reveal aspects of the scaling of landscape patterns (Deutsch and Journel 1992; Rossi et al. 1992; Meisel and Turner 1998; Fortin 1999). Spatial autocorrelation, for example, may be high among nearby samples, then disappear with increasing scale, then appear again at even broader scales (e.g., Brown et al. 1995; Dale 1999). Such patterns provide a preliminary indication of the scope of scale domains and the scales at which spatial correlation is eroded by other processes (i.e., scaling thresholds). The degree to which variance among sample points is a function of the distance between them (scale) can be measured by semivariance, and a variogram then depicts semivariance as a function of distance (figure 2.4). In such graphs, the *range* indicates the scale beyond which there is no spatial correlation among samples and the *sill* indicates the magnitude of spatially related variation; *nugget* (a term derived from the original applications of these statistics in mining) indicates the sampling error, which includes spatial variation at distances smaller than the smallest sampling distance (i.e., the sampling grain). In practical terms, the range indicates the scale over which there is spatial interdependency for the measured variable (e.g., patch size; see Schlesinger et al. 1996), whereas the nugget indicates the amount of variation whose spatial component (if any) is missed because it falls below the scale of resolution of the sampling grain. Thus, an experiment conducted using a grain and extent that fall within the semivariance range for a given landscape will encompass a qualitatively different form of internal heterogeneity than one with a grain and extent that exceed the range (figure 2.4). A high nugget variance suggests that among-plot variation will be great, which affects the amount of replication required. By indicating something of the scaling of patchiness of systems, variograms can provide some guidance in establishing the grain and extent of experiments. Variograms, however, are not infallible indicators of scale; in particular, they are sensitive to directionality in spatial patterns (anisotropy) and are not amenable to statistical testing (Legendre and Fortin 1989; Legendre 1993; Gardner 1998), so they should be used with care.

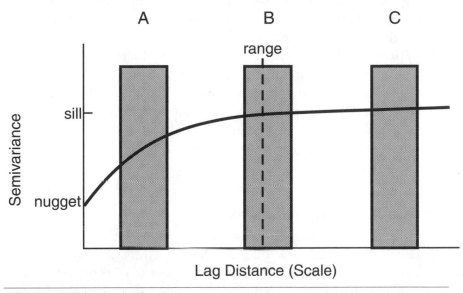

FIGURE 2•4 *Using Semivariance*
A variogram, showing how semivariance changes with increasing distance between sample points (i.e., increasing scale). Nugget, sill, and range are described in the text. (A), (B), and (C) represent the grain-extent domains of three experiments. In (A) the experiment is conducted at a scale in which spatial autocorrelation among sample points is important (much of the variance is spatially related), so the sample points are not really independent of one another and pseudoreplication may be a concern. In (B) the experiment straddles a scaling threshold and includes both spatially correlated and spatially independent samples, potentially resulting in blurred or artifactual patterns. In (C) the scale is broad enough that sample points are statistically independent of one another; the variance is not spatial. In this situation, conventional experimental design and traditional (nonspatial) statistical tests are appropriate. None of these scales of experimentation is necessarily "best," but because the experiments differ in scale relative to the spatial structure of variance, both statistical analysis and interpretation of the results will differ.

Regardless of the approach used to describe landscape structure and assess its scaling properties, it is important that the environmental variables selected for analysis be relevant to the questions asked and the biological attributes of the system. It is all too easy to use remotely sensed imagery in a GIS analysis to derive various landscape metrics (usually having to do with vegetative cover or land use) that can then be subjected to multiscale analysis. This may provide a useful measure of environmental scaling from a human perspective (e.g., Krummel et al. 1987), but it is not necessarily the most relevant for other organisms. For example, Brandon Bestelmeyer and I have been investigating the responses of ant communities to local land use and environmental features at several grassland-desert sites in Colorado and New Mexico. Grazing is the dominant land use, and from a human perspective its effects on the landscapes are dramatic: grazed and ungrazed areas differ markedly in vegetation coverage and structure. To the ants, however, these features are apparently unimportant. Instead, the strongest correlates of diversity and abundance are often features of soil texture, which affect the nest placement of colonies. Although multiscale analyses of vegetation maps may reveal environmental scalings that are relevant to the ants in some areas, they do not do so in others.

SCALING ORGANISM RESPONSES

Ultimately, determining how different kinds of organisms respond to environmental features and to the scale-dependence of those features in space demands an intimate knowledge of their natural history. These details of the biology of individual species provide the basis for developing species-specific scaling functions. In some instances, however, it may be useful to develop scaling functions that can be generalized. In such situations, it may be appropriate to define functional groups that cluster together species that share general features of ecology or life history. Such approaches have been applied, for example, to describe functional groups of plants (Gillison and Carpenter 1997; Lavorel et al. 1997; Smith et al. 1997; Díaz et al. 1999), ants (Bestelmeyer and Wiens 1996; Andersen 1997a, 1997b), and stream invertebrates (Poff 1997). Functional groups may be defined by combining a species inventory for an area of interest with specifications of the ecological and life-history traits of the species in a species x trait matrix and then grouping together (e.g., by clustering algorithms) species that share sets of traits. Keddy (1992; Weiher and Keddy 1995; Keddy and Weiher 1999) has used such trait groups to predict which species may pass through different "environmental filters" to become established in a local community during the process of community assembly. By specifying traits that define the grain-extent (scale) window through which species respond to the environment, such species x trait matrices might be used to define suites of species that share scaling responses based on general features of their ecology and life history. Thus, for example, species that share physiological tolerances, are residents rather than long-distance migrants, are actively searching predators on foliage insects, and have individual home ranges of 1–5 ha might be expected to respond to the scale of environmental or resource patchiness in a given environment in much the same way. Traits such as body size, home-range area, overall vagility, dispersal distance, habitat (patch) selectivity, social organization and spatial aggregation, and foraging behavior have obvious relationships to the grain and extent of environmental response of species, but other features may also be important for particular taxa (see Poff 1997). By integrating the scale windows of organisms with the scale structure of the environment, scaling functions allow us to determine which domains or thresholds of environmental scale or patchiness may be relevant to suites of organisms that are defined functionally rather than taxonomically.

How do such scaling functions relate to ecological experiments? Simply put, they specify the scale context for an experiment. Say that one wishes to conduct an experiment to test the relationship between a specified set of environmental factors and an assemblage of organisms occupying that environment. Scaling functions incorporate both the form of scale-dependency of variation in those environmental factors and the scale windows over which different subsets (functional groups) of the species respond to them. On the other hand, species that have different functional properties may scale the environment in different ways, and interactions among such species can produce complex scale dependence. For example, Englund (1997) found that predator-prey interaction experiments may be sensitive to prey movements at fine spatial scales, but to predator-related mortality at broader scales (see also Thrush 1999). In either case, scaling functions define one or more grain-extent windows that characterize the ecological system being studied. Rather than being carried out at some arbitrarily designated (and usually unrealistically fine) scale(s), experiments should be designed to fall within the range of scale defined by the scaling functions (i.e., the intersection of environmental scaling and organism responses). Ideally, studies should be conducted over the entire range of scale that encompasses the processes of interest in order to derive scaling equations and identify scaling thresholds in responses.

COMMENTS

The Role of Scale and Experiments in Ecology

Scaling functions may help to define an appropriate scale window for ecological experiments, but more often than not this window will fall in the range of scales where experiments are not logistically feasible. What does one do then?

This dilemma can be boiled down to whether one does experiments and ignores scale, or whether one respects scale and abandons experiments. This is an oversimplification, of course, but like many simplifications it brings the central issues into sharp relief. The first issue is whether and how scale should be incorporated into the design of ecological experiments. This issue, in turn, prompts two questions. First, are the results of an experiment conducted at a given scale sensitive to influences from other scales? And second, can the results of such an experiment be applied to the same or similar systems at different scales? The first question is essentially one of system openness, and this must be determined by a careful consideration of the nature (and scale) of the processes that are likely to affect the pattern of interest. The second question deals with extrapolation. If the scaling properties of a system are as depicted in figure 2.1a, then of course scale really doesn't matter and it can safely be ignored. If the situation is like that shown in figure 2.1b, then the results can be extrapolated to other scales, so long as the scaling equation and its domain of applicability are known. If there are thresholds in scaling relationships (figure 2.1c), then extrapolation is more problematic. Of course, if one is not interested in extrapolating the results of an experiment, many (but not all) of these concerns vanish. Rarely, however, are experimentalists uninterested in extending their results to other scales or other systems.

Usually experiments are conducted directly on the system of interest, and the difficulties arise when this system occurs at a relatively broad scale. One solution to this problem is to conduct the experiment on a fine-scale analog of the system of interest. This approach is used liberally in other biological and physical sciences, and it is the premise underlying the use of experimental model systems (EMS) or mesocosm experiments in ecology (Ims et al. 1993; Fraser and Keddy 1997; With 1997; Barrett and Peles 1999; Ims 1999; With et al. 1999). By condensing the phenomena of interest into relatively fine spatial scales, the experiment becomes logistically feasible and perhaps can even be replicated. Such experiments are usually conducted using different species and environments than those of direct interest, leaving one with the problem of extrapolating among *kinds* of systems as well as that of extrapolating among scales. It is not likely, for example, that the EMS experiments we have conducted on beetles in grassland microlandscapes (e.g., Wiens and Milne 1989; Wiens et al. 1997; McIntyre and Wiens 1999) can be directly extended to, say, bison in the same grasslands, or to beetles in desert dune ecosystems. Rather than attempting such direct extrapolations, we view these experiments as indicating in a qualitative manner the ways in which landscape structure can affect ecological systems, and the mechanisms involved. The value of such experiments ultimately depends on the adequacy of the model in relation to the system of interest. Here also a consideration of functional groupings of organisms may be helpful.

Scale enters into ecological experiments in two ways. One is through what might be called "scale awareness." This has been the focus of much of this chapter. To what degree do investigators incorporate an explicit consideration of the scaling properties of environments and the organisms that occupy them into the design of experiments? The second involves a direct assessment of scaling effects—scale is incorporated as an explicit treatment in the experimental design. Both aspects are important, but the need for the second is especially great. If we are to develop an understanding of how scale affects ecological patterns and processes, and of the circumstances in which it is likely to be important or unimportant, experiments on scale should be an important part of our attack on the problem (Petersen et al. 1999). Although not all experiments need to include scale as a treatment, all experiments *should* demonstrate scale awareness.

The second issue arising from the conflict between scale and experimentation has to do with the role of experiments in ecology. Scale problems force us to ask whether experiments are really what we should be doing, or whether we have perhaps become overly attached to a physics-based reductionist approach. We should perhaps take a lesson from evolutionary biologists who are confronted with seemingly intractable problems in dealing with temporal scaling in the form of history. Evolutionary biologists use experiments to get at evolutionary *mechanisms*, the processes by which natural selection produces phenotypic and genotypic change in populations (Endler 1986). Most of the *consequences* of evolution (evolutionary patterns), however, are not directly accessible to experimental investigation. To deal with this problem, evolutionary biologists have devoted considerable effort to formalizing the "comparative method" (Harvey and Pagel 1991) as a means of drawing evolutionary inferences with reasonable confidence.

Ecologists need to reassess the prevailing notion that experiments are the best (or only) way to gain reliable ecological knowledge. Experiments will always be an important way to unravel mechanisms; indeed, this is how we have used our EMS of beetles and microlandscapes. The broad spatial or temporal scales of many ecological patterns and applications, however, invariably constrain experimentation. We need to adopt other approaches to understanding these patterns that, while different from experiments, are no less scientific.

ACKNOWLEDGMENTS

My graduate students and colleagues have helped to shape my thinking about scale in ecological systems, but they bear no responsibility for the result. My research on scaling has been supported by the National Science Foundation (Grants DEB-9207010 and DEB-95-27111) and the Environmental Protection Agency (Grant R 826764-01-0). Tim Allen and Tony King provided numerous comments on an early draft of this chapter, which, if nothing else, forced me to think more carefully about my arguments.

LITERATURE CITED

Addicott, J. F., J. M. Aho, M. R. Antolin, D. K. Padilla, J. S. Richardson, and D. A. Soluk. 1987. Ecological neighborhoods: Scaling environmental patterns. *Oikos* 49:340–346.

Allen, T. F. H. 1998. The landscape "level" is dead: Persuading the family to take it off the respirator. In D. L. Peterson and V. T. Parker, eds., *Ecological Scale: Theory and Applications*, pp. 35–54. New York: Columbia University Press.

Allen, T. F. H., and T. W. Hoekstra. 1992. *Toward a Unified Ecology*. New York: Columbia University Press.

Allen, T. F. H., A. W. King, B. Milne, A. Johnson, and S. Turner, 1993. The problem of scaling in ecology. *Evolutionary Trends in Plants* 7:3–8.

Allen, T. F. H., R. V. O'Neill, and T. W. Hoekstra. 1984. *Interlevel Relations in Ecological Research and Management: Some Working Principles from Hierarchy Theory.* U.S. Department of Agriculture Forest Service General Technical Report RM-110. Fort Collins, Colo.: Rocky Mountain Experiment Station.

Allen, T. F. H., and D. W. Roberts. 1998. Foreword. In D. L. Peterson and V. T. Parker, eds., *Ecological Scale: Theory and Applications*, pp. xi–xiii. New York: Columbia University Press.

Allen, T. F. H., and T. B. Starr. 1982. *Hierarchy: Perspectives for Ecological Complexity.* Chicago: University of Chicago Press.

Andersen, A. N. 1997a. Functional groups and patterns of organization in North American ant communities: A comparison with Australia. *Journal of Biogeography* 24:433–460.

Andersen, A. N. 1997b. Using ants as bioindicators: Multiscale issues in ant community ecology. *Conservation Ecology* [online] 1(1):8.

Andrewartha, H. G., and L. C. Birch. 1954. *The Distribution and Abundance of Animals.* Chicago: University of Chicago Press.

Atkinson, P. M. 1997. Scale and spatial dependence. In P. R. van Gardingen, G. M. Foody, and P. J. Curran, eds., *Scaling-up: From Cell to Landscape*, pp. 35–60. Cambridge: Cambridge University Press.

Barrett, G. W., and J. D. Peles, eds. 1999. *Landscape Ecology of Small Mammals.* New York: Springer-Verlag.

Bestelmeyer, B. T., and J. A. Wiens. 1996. The effects of land use on the structure of ground foraging ant communities in the Argentine Chaco. *Ecological Applications* 6:1225–1240.

Brown, J. H., D. W. Mehlman, and G. C. Stevens. 1995. Spatial variation in abundance. *Ecology* 76:2028–2043.

Carpenter, S. R., S. W. Chisholm, C. J. Krebs, D. W. Schindler, and R. F. Wright. 1995. Ecosystem experiments. *Science* 269:324–327.

Cullinan, V. I., M. A. Simmons, and J. M. Thomas. 1997. A Bayesian test of hierarchy theory: Scaling up variability in plant cover from field to remotely sensed data. *Landscape Ecology* 12:273–285.

Dale, M. R. T. 1999. *Spatial Pattern Analysis in Plant Ecology.* Cambridge: Cambridge University Press.

Delcourt, H. R., P. A. Delcourt, and T. Webb. 1983. Dynamic plant ecology: The spectrum of vegetational change in space and time. *Quaternary Science Review* 1:153–175.

De Roos, A. M., E. McCauley, and W. G. Wilson. 1991. Mobility versus density-limited predator-prey dynamics on different spatial scales. *Proceedings of the Royal Society of London, Series B* 246:117–122.

Deutsch, C. V., and A. G. Journel. 1992. *GSLIB: Geostatistical Software Library and User's Guide.* New York: Oxford University Press.

Diamond, J. 1986. Overview: Laboratory experiments, field experiments, and natural experiments. In J. Diamond and T. J. Case, eds., *Community Ecology*, pp. 3–22. New York: Harper & Row.

Díaz, S., M. Cabido, and F. Casanoves. 1999. Functional implications of trait-environment linkages in plant communities. In E. Weiher and P. Keddy, eds., *Ecological Assembly Rules: Perspectives, Advances, Retreats*, pp. 338–362. Cambridge: Cambridge University Press.

Dillworth, M. E., J. L. Whistler, and J. W. Merchant. 1994. Measuring landscape structure using geographic and geometric windows. *Photogrammetric Engineering and Remote Sensing* 60:1215–1224.

Dutilleul, P. 1998. Incorporating scale in ecological experiments: Study design. In D. L. Peterson and V. T. Parker, eds., *Ecological Scale: Theory and Applications*, pp. 369–386. New York: Columbia University Press.

Edwards, P. J., R. M. May, and N. R. Webb, eds. 1994. *Large-scale Ecology and Conservation Biology*. Oxford: Blackwell Science.

Ehleringer, J. R., and C. B. Field, eds. 1993. *Scaling Physiological Processes: Leaf to Globe*. New York: Academic Press.

Endler, J. A. 1986. *Natural Selection in the Wild*. Princeton: Princeton University Press.

Englund, G. 1997. Importance of spatial scale and prey movements in predator caging experiments. *Ecology* 78:2316–2325.

Fortin, M.-J. 1999. Spatial statistics in landscape ecology. In J. M. Klopatek and R. H. Gardner, eds., *Landscape Ecological Analysis: Issues and Applications*, pp. 253–279. New York: Springer-Verlag.

Fraser, L. H., and P. Keddy. 1997. The role of experimental microcosms in ecological research. *Trends in Ecology and Evolution* 12:478–481.

Friedel, M. H. 1994. How spatial and temporal scale affect the perception of change in rangelands. *Rangelands Journal* 16:16–25.

Fuhlendorf, S. D., and F. E. Smeins. 1996. Spatial scale influence on longterm temporal patterns of a semi-arid grassland. *Landscape Ecology* 11:107–113.

Fuisz, T., and C. Moskát. 1992. The importance of scale in studying beetle communities: Hierarchical sampling or sampling the hierarchy. *Acta Zoologica Hungarica* 38:183–197.

Gardner, R. H. 1998. Pattern, process, and the analysis of spatial scales. In D. L. Peterson and V. T. Parker, eds., *Ecological Scale: Theory and Applications*, pp. 17–34. New York: Columbia University Press.

Gillison, A. N., and G. Carpenter. 1997. A generic plant functional attribute set and grammar for dynamic vegetation description and analysis. *Functional Ecology* 11:775–783.

Goodchild, M. F., and D. A. Quattrochi. 1997. Scale, multiscaling, remote sensing, and GIS. In D. A. Quattrochi and M. F. Goodchild, eds., *Scale in Remote Sensing and GIS*, pp. 1–11. Boca Raton, Fla.: CRC Lewis.

Greig-Smith, P. 1952. The use of random and contiguous quadrats in the study of structure in plant communities. *Annals of Botany* 16:293–316.

Grime, J. P., K. Thompson, and C. W. MacGillivray. 1997. Scaling from plant to community and from plant to regional flora. In P. R. van Gardingen, G. M. Foody, and P. J. Curran, eds., *Scaling-up: From Cell to Landscape*, pp. 105–127. Cambridge: Cambridge University Press.

Gustafson, E. J. 1998. Quantifying landscape spatial pattern: What is the state of the art? *Ecosystems* 1:143–156.

Hairston, N. G., Sr. 1989. *Ecological Experiments: Purpose, Design, and Execution*. Cambridge: Cambridge University Press.

Hargis, C. D., J. A. Bissonette, and J. L. David. 1997. Understanding measures of landscape pattern. In J. A. Bissonette, ed., *Wildlife and Landscape Ecology: Effects of Pattern and Scale*, pp. 231–261. New York: Springer-Verlag.

Harte, J., S. McCarthy, K. Taylor, A. Kinzig, and M. Fischer. 1999. Estimating species-area relationships from plot to landscape scale using species spatial-turnover data. *Oikos* 86:45–54.

Harvey, P. H., and M. D. Pagel. 1991. *The Comparative Method in Evolutionary Biology*. Oxford: Oxford University Press.

Hinckley, T. M., D. G. Sprugel, J. R. Brooks, K. J. Brown, T. A. Martin, D. A. Roberts, W. Schaap, and D. Wang. 1998. Scaling and integration in trees. In D. L. Peterson and V. T. Parker, eds., *Ecological Scale: Theory and Applications*, pp. 309–337. New York: Columbia University Press.

Hoffman, W., D. Heinemann, and J. A. Wiens. 1981. The ecology of seabird feeding flocks in Alaska. *Auk* 98:437–456.

Holling, C. S. 1992. Cross-scale morphology, geometry, and dynamics of ecosystems. *Ecological Monographs* 62:447–502.

Holling, C. S., G. Peterson, P. Marples, J. Sendzimir, K. Redford, L. Gunderson, and D. Lambert. 1996. Self-organization in ecosystems: Lumpy geometries, periodicities, and morphologies. In B. Walker and W. Steffen, eds., *Global Change and Terrestrial Ecosystems*, pp. 346–384. Cambridge: Cambridge University Press.

Huston, M. A. 1994. *Biological Diversity: The Coexistence of Species on Changing Landscapes*. Cambridge: Cambridge University Press.

Ims, R. A. 1999. Experimental landscape ecology. In J. A. Wiens and M. R. Moss, eds., *Issues in Landscape Ecology*, pp. 45–50. Guelph, Ontario: International Association for Landscape Ecology.

Ims, R. A., J. Rolstad, and P. Wegge. 1993. Predicting space use responses to habitat fragmentation: Can voles *Microtus oeconomus* serve as an experimental model system (EMS) for capercaillie grouse in boreal forest? *Biological Conservation* 63:261–268.

Kareiva, P. 1994. Space: The final frontier for ecological theory. *Ecology* 75:1.

Kareiva, P., and M. Andersen. 1988. Spatial aspects of species interactions: The wedding of models and experiments. In A. Hastings, ed., *Community Ecology*, pp. 35–50. New York: Springer-Verlag.

Kauffman, S. A. 1993. *The Origins of Order*. New York: Oxford University Press.

Keddy, P. A. 1992. Assembly and response rules: Two goals for predictive community ecology. *Journal of Vegetation Science* 3:157–164.

Keddy, P., and E. Weiher. 1999. Introduction: The scope and goals of research on assembly rules. In E. Weiher and P. Keddy, eds., *Ecological Assembly Rules: Perspectives, Advances, Retreats*, pp. 1–20. Cambridge: Cambridge University Press.

Keitt, T. H., D. L. Urban, and B. T. Milne. 1997. Detecting critical scales in fragmented landscapes. *Conservation Ecology* [online] 1(1):4.

King, A. W. 1991. Translating models across scales in the landscape. In M. G. Turner and R. H. Gardner, eds., *Quantitative Methods in Landscape Ecology*, pp. 479–517. New York: Springer-Verlag.

Klomp, N., and I. Lunt, eds. 1997. *Frontiers in Ecology: Building the Links*. Oxford: Elsevier Science.

Kotliar, N. B., and J. A. Wiens. 1990. Multiple scales of patchiness and patch structure: A hierarchial framework for the study of heterogeneity. *Oikos* 59:253–260.

Krummel, J. R., R. H. Gardner, G. Sugihara, R. V. O'Neill, and P. R. Coleman. 1987. Landscape patterns in a disturbed environment. *Oikos* 48:321–324.

Lam, N., and D. A. Quattrochi. 1992. On the issues of scale, resolution, and fractal analysis in the mapping sciences. *Professional Geographer* 44:88–98.

Lavorel, S., S. McIntyre, J. Landsberg, and T. D. A. Forbes. 1997. Plant functional classifications: from general groups to specific groups based on response to disturbance. *Trends in Ecology and Evolution* 12:474–478.

Lawler, S. P. 1998. Ecology in a bottle: Using microcosms to test theory. In W. J. Resetarits Jr. and J. Bernardo, eds., *Experimental Ecology: Issues and Perspectives*, pp. 236–253. New York: Oxford University Press.

Lawton, J. H. 1995. Ecological experiments with model systems. *Science* 269:328–331.

Lawton, J. H. 1998. Ecological experiments with model systems: The Ecotron Facility in context. In W. J. Resetarits Jr. and J. Bernardo, eds., *Experimental Ecology: Issues and Perspectives*, pp. 170–182. New York: Oxford University Press.

Lawton, J. H. 1999. Are there general laws in ecology? *Oikos* 84:177–192.

Legendre, P. 1993. Spatial autocorrelation: Trouble or new paradigm? *Ecology* 74:1659–1673.

Legendre, P., and M.-J. Fortin. 1989. Spatial pattern and ecological analysis. *Vegetatio* 80:107–138.

Levin, S. A. 1992. The problem of pattern and scale in ecology. *Ecology* 73:1943–1967.

Levins, R. 1968. *Evolution in Changing Environments*. Princeton: Princeton University Press.

Lodge, D. M., S. C. Blumenshine, and Y. Vadeboncoeur. 1998. Insights and application of large-scale, long-term ecological observations and experiments. In W. J. Resetarits Jr. and J. Bernardo, eds., *Experimental Ecology: Issues and Perspectives*, pp. 202–235. New York: Oxford University Press.

Ludwig, J., D. Tongway, D. Freudenberger, J. Noble, and K. Hodgkinson, eds. 1997. *Landscape Ecology: Function and Management*. Collingwood, Victoria, Australia: CSIRO.

Ludwig, J. A., J. A. Wiens, and D. J. Tongway. 2000. A scaling rule for landscape patches and how it applies to conserving soil resources in savannas. *Ecosystems* 3: 84–97.

MacNally, R. 1999. Dealing with scale in ecology. In J. A. Wiens and M. R. Moss, eds., *Issues in Landscape Ecology*, pp. 10–17. Guelph, Ontario: International Association for Landscape Ecology.

Manly, B. F. J. 1992. *The Design and Analysis of Research Studies*. Cambridge: Cambridge University Press.

Marquet, P. A., M.-J. Fortin, J. Pineda, D. O. Wallin, J. Clark, Y. Wu, S. Bollens, C. M. Jacobi, and R. D. Holt. 1993. Ecological and evolutionary consequences of patchiness: A marine-terrestrial perspective. In S. A. Levin, T. M. Powell, and J. H. Steele, eds., *Patch Dynamics*, pp. 277–304. New York: Springer-Verlag.

McGarigal, K., and B. Marks. 1995. *Fragstats: Spatial Pattern Analysis Program for Quantifying Landscape Structure.* U.S. Department of Agriculture, Forest Service General Technical Report PNW-GTR-351. Portland, Ore.: Pacific Northwest Research Station.

McIntyre, N. E., and J. A. Wiens. 1999. How does habitat patch size affect animal movement? An experiment with darkling beetles. *Ecology* 80:2261–2270.

Mehlum, F., G. L. Hunt Jr., Z. Klusek, and M. B. Decker. 1999. Scale-dependent correlations between the abundance of Brünnich's guillemots and their prey. *Journal of Animal Ecology* 68:60–72.

Meisel, J. E., and M. G. Turner. 1998. Scale detection in real and artificial landscapes using semivariance analysis. *Landscape Ecology* 13:347–362.

Nikora, V. I., C. P. Pearson, and U. Shankar. 1999. Scaling properties in landscape patterns: New Zealand experience. *Landscape Ecology* 14:17–33.

O'Neill, R. V. 1979. Transmutation across hierarchical levels. In G. S. Innis and R. V. O'Neill, eds., *Systems Analysis of Ecosystems*, pp. 59–78. Fairlands, MD: International Cooperative Publishing House.

O'Neill, R. V. 1989. Perspectives in hierarchy and scale. In J. Roughgarden, R. M. May, and S. A. Levin, eds., *Perspectives in Ecological Theory*, pp. 140–156. Princeton: Princeton University Press.

O'Neill, R. V., D. L. DeAngelis, J. B. Waide, and T. F. H. Allen. 1986. *A Hierarchical Concept of Ecosystems.* Princeton: Princeton University Press.

O'Neill, R. V., R. H. Gardner, B. T. Milne, M. G. Turner, and B. Jackson. 1991. Heterogeneity and spatial hierarchies. In J. Kolasa and S. T. A. Pickett, eds., *Ecological Heterogeneity*, pp. 85–96. New York: Springer-Verlag.

O'Neill, R. V., and A. W. King. 1998. Homage to St. Michael; or, why are there so many books on scale? In D. L. Peterson and V. T. Parker, eds., *Ecological Scale: Theory and Applications*, pp. 3–15. New York: Columbia University Press.

O'Neill, R. V., J. R. Krummel, R. H. Gardner, G. Sugihara, B. Jackson, D. L. DeAngelis, B. T. Milne, M. G. Turner, B. Zygmunt, S. W. Christensen, V. H. Dale, and R. L. Graham. 1988. Indices of landscape pattern. *Landscape Ecology* 1:153–162.

Petersen, J. E., J. C. Cornwell, and W. M. Kemp. 1999. Implicit scaling in the design of experimental aquatic ecosystems. *Oikos* 85:3–18.

Peterson, D. L., and V. T. Parker, eds. 1998. *Ecological Scale: Theory and Applications.* New York: Columbia University Press.

Pickett, S. T. A., V. T. Parker, and P. L. Fiedler. 1992. The new paradigm in ecology: Implications for conservation biology above the species level. In P. L. Fiedler and S. K. Jain, eds., *Conservation Biology*, pp. 65–88. New York: Chapman and Hall.

Poff, N. L. 1997. Landscape filters and species traits: Towards mechanistic understanding and prediction in stream ecology. *Journal of the North American Benthological Society* 16:391–409.

Quattrochi, D. A., and M. F. Goodchild, eds. 1997. *Scale in Remote Sensing and GIS*. Boca Raton, Fla.: CRC Lewis.

Rahel, F. J. 1990. The hierarchical nature of community persistence: A problem of scale. *American Naturalist* 136:328–344.

Rastetter, E. B., A. W. King, B. J. Crosby, G. M. Hornberger, R. V. O'Neill, and J. E. Hobbie. 1992. Aggregating fine-scale ecological knowledge to model coarse-scale attributes of ecosystems. *Ecological Applications* 2:55–70.

Resetarits, W. J., Jr. and J. Bernardo, eds. 1998. *Experimental Ecology: Issues and Perspectives*. New York: Oxford University Press.

Reynolds, J. F., D. W. Hilbert, and P. R. Kemp. 1993. Scaling ecophysiology from the plant to the ecosystem: A conceptual framework. In J. R. Ehleringer and C. B. Field, eds., *Scaling Physiological Processes: Leaf to Globe*, pp. 127–140. New York: Academic Press.

Rietkerk, M., F. van den Bosch, and J. van de Koppel. 1997. Site-specific properties and irreversible vegetation changes in semi-arid grazing systems. *Oikos* 80:241–252.

Riitters, K. H., R. V. O'Neill, C. T. Hunsaker, J. D. Wickham, D. H. Yankee, S. P. Timmins, K. B. Jones, and B. L. Jackson. 1995. A factor analysis of landscape pattern and structure metrics. *Landscape Ecology* 10:23–39.

Riitters, K. H., R. V. O'Neill, and K. B. Jules. 1997. Assessing habitat suitability at multiple scales: A landscape-level approach. *Biological Conservation* 81:191–202.

Roland, J., and P. D. Taylor. 1997. Insect parasitoid species respond to forest structure at different spatial scales. *Nature* 386:710–713.

Romme, W. H., E. H. Everham, L. E. Frelich, M. A. Moritz, and R. E. Sparks. 1998. Are large, infrequent disturbances qualitatively different from small, frequent disturbances? *Ecosystems* 1:524–534.

Rosenzweig, M. L. 1995. *Species Diversity in Space and Time*. Cambridge: Cambridge University Press.

Rossi, R. E., D. J. Mulla, A. G. Journel, and E. H. Franz. 1992. Geostatistical tools for modeling and interpreting ecological spatial dependence. *Ecological Monographs* 62:277–314.

Rukke, B. A., and F. Midtgaard. 1998. The importance of scale and spatial variables for the fungivorous beetle *Bolitophagus reticulatus* (Coleoptera, Tenebrionidae) in a fragmented forest landscape. *Ecography* 21:561–572.

Saab, V. 1999. Importance of spatial scale to habitat use by breeding birds in riparian forests: A hierarchical analysis. *Ecological Applications* 9:135–151.

Sato, K., and Y. Iwasa. 1993. Modeling of wave regeneration in subalpine *Abies* forests: Population dynamics with spatial structure. *Ecology* 74:1538–1550.

Scheiner, S. M., and J. Gurevitch, eds. 1993. *Design and Analysis of Ecological Experiments*. London: Chapman and Hall.

Schlesinger, W. H., J. A. Raikes, A. E. Hartley, and A. F. Cross. 1996. On the spatial pattern of soil nutrients in desert ecosystems. *Ecology* 77:364–374.

Schlesinger, W. H., J. F. Reynolds, G. L. Cunningham, L. F. Huenneke, W. M. Jarrell, R. A. Virginia, and W. G. Whitford. 1990. Biological feedbacks in global desertification. *Science* 247:1043–1048.

Schneider, D. C. 1994. *Quantitative Ecology: Spatial and Temporal Scaling.* New York: Academic Press.

Schneider, D. C. 1998. Applied scaling theory. In D. L. Peterson and V. T. Parker, eds., *Ecological Scale: Theory and Applications*, pp. 253–269. New York: Columbia University Press.

Schneider, D. C., and J. F. Piatt. 1986. Scale-dependent correlation of seabirds with schooling fish in a coastal ecosystem. *Marine Ecology Progress Series* 32:237–246.

Shmida, A., and M. V. Wilson. 1985. Biological determinants of species diversity. *Journal of Biogeography* 12:1–20

Silbernagel, J. 1997. Scale perception—From cartography to ecology. *Bulletin of the Ecological Society of America* 166–169.

Smith, T. M., H. H. Shugart, and F. I. Woodward, eds. 1997. *Plant Functional Types: Their Relevance to Ecosystem Properties and Global Change.* Cambridge: Cambridge University Press.

Solé, R. V., S. C. Manrubia, M. Benton, S. Kauffman, and P. Bak. 1999. Criticality and scaling in evolutionary ecology. *Trends in Ecology and Evolution* 14:156–160.

Southwood, T. R. E. 1977. Habitat, the templet for ecological strategies. *Journal of Animal Ecology* 46:337–365.

Sprugel, D. G. 1976. Dynamic structure of wave-generated *Abies balsamea* forests in the North Eastern United States. *Journal of Ecology* 64:889–911.

Steele, J. H. 1978. Some comments on plankton patches. In J. H. Steele, ed., *Spatial Pattern in Plankton Communities*, pp. 11–20. New York: Plenum.

Steele, J. H. 1989. Scale and coupling in ecological systems. In J. Roughgarden, R. M. May, and S. A. Levin, eds., *Perspectives in Ecological Theory*, pp. 177–180. Princeton: Princeton University Press.

Stewart, J. B., E. T. Engman, R. A. Feddes, and Y. Kerr, eds. 1996. *Scaling-up in Hydrology Using Remote Sensing.* Chichester, U.K.: Wiley.

Thiéry, J. M., J.-M. D'Herbès, and C. Valentin. 1995. A model simulating the genesis of banded vegetation patterns in Niger. *Journal of Ecology* 83:497–507.

Thrush, S. F. 1999. Complex role of predators in structuring soft-sediment macrobenthic communities: Implications of changes in spatial scale for experimental studies. *Australian Journal of Ecology* 24:344–354.

Tilman, D., and P. Kareiva, eds. 1997. *Spatial Ecology: The Role of Space in Population Dynamics and Interspecific Interactions.* Princeton: Princeton University Press.

Tongway, D. J., and J. A. Ludwig. 1994. Small-scale resource heterogeneity in semi-arid landscapes. *Pacific Conservation Biology* 1:201–208.

Turner, M. G., and R. H. Gardner. 1991. Quantitative methods in landscape ecology: An introduction. In M. G. Turner and R. H. Gardner, eds., *Quantitative Methods in Landscape Ecology*, pp. 3–14. New York: Springer-Verlag.

Underwood, A. J. 1997. *Experiments in Ecology.* Cambridge: Cambridge University Press.

Urban, D. L., R. V. O'Neill, and H. H. Shugart Jr. 1987. Landscape ecology. *BioScience* 37:119–127.

van Gardingen, P. R., G. M. Foody, and P. J. Curran, eds. 1997. *Scaling-up: From Cell to Landscape.* Cambridge: Cambridge University Press.

Van Horne, B., G. S. Olson, R. L. Schooley, J. G. Corn, and K. P. Burnham. 1997. The effects of drought and prolonged winter on Townsend's ground squirrels in shrubsteppe habitats. *Ecological Monographs* 67:295–315.

Waring, R. H. 1993. How ecophysiologists can help scale from leaves to landscapes. In J. R. Ehleringer and C. B. Field, eds., *Scaling Physiological Processes: Leaf to Globe*, pp. 159–166. New York: Academic Press.

Weiher, E., and P. A. Keddy. 1995. The assembly of experimental wetland plant communities. *Oikos* 73:323–335.

White, L. P. 1971. Vegetation stripes on sheet wash surfaces. *Journal of Ecology* 59:615–622.

Wiens, J. A. 1976. Population responses to patchy environments. *Annual Review of Ecology and Systematics* 7:81–120.

Wiens, J. A. 1981. Scale problems in avian censusing. *Studies in Avian Biology* 6:513–521.

Wiens, J. A. 1984. On understanding a non-equilibrium world: Myth and reality in community patterns and processes. In D. R. Strong Jr., D. Simberloff, L. G. Abele, and A. B. Thistle, eds., *Ecological Communities: Conceptual Issues and the Evidence*, pp. 439–457. Princeton: Princeton University Press.

Wiens, J. A. 1989a. Spatial scaling in ecology. *Functional Ecology* 3:385–397.

Wiens, J. A. 1989b. *The Ecology of Bird Communities. I. Foundations and Patterns*. Cambridge: Cambridge University Press.

Wiens, J. A. 1990. On the use of "grain" and "grain size" in ecology. *Functional Ecology* 4:720.

Wiens, J. A. 1992. Ecological flows across landscape boundaries: A conceptual overview. In A. J. Hansen and F. di Castri, eds., *Landscape Boundaries: Consequences for Biotic Diversity and Ecological Flows*, pp. 217–235. New York: Springer-Verlag.

Wiens, J. A. 2000. Ecological heterogeneity: An ontogeny of concepts and approaches. In M. J. Hutchings, E. A. John, and A. J. A. Stewart, eds., *The Ecological Consequences of Habitat Heterogeneity*, pp. 9-31. Oxford: Blackwell Science.

Wiens, J. A., and B. T. Milne. 1989. Scaling of "landscapes" in landscape ecology, or, landscape ecology from a beetle's perspective. *Landscape Ecology* 3:87–96.

Wiens, J. A., J. T. Rotenberry, and B. Van Horne. 1987. Habitat occupancy patterns of North American shrubsteppe birds: The effects of spatial scale. *Oikos* 48:132–147.

Wiens, J. A., R. L. Schooley, and R. D. Weeks Jr. 1997. Patchy landscapes and animal movements: Do beetles percolate? *Oikos* 78:257–264.

Wiens, J. A., N. C. Stenseth, B. Van Horne, and R. A. Ims. 1993. Ecological mechanisms and landscape ecology. *Oikos* 66:369–380.

Wilson, J. B., and A. D. Q. Agnew. 1992. Positive-feedback switches in plant communities. *Advances in Ecological Research* 23:263–336.

With, K. A. 1997. Microlandscape studies in landscape ecology: Experimental rigor or experimental rigor mortis? *US-IALE Newsletter* 13(1):13–16.

With, K. A., S. J. Cadaret, and C. Davis. 1999. Movement responses to patch structure in experimental fractal landscapes. *Ecology* 80:1340–1353.

CHAPTER 3

The Nature of the Scale Issue in Experimentation

Timothy F. H. Allen

S O LONG AS THE MODELS WE USE AND TEST INVOKE A SIMPLE system, then the myth of objectivity is as good a myth as any. I use it every day as I assert that my office is real, and it will be there when I get back. But the myth of objectivity fails for complex systems, the ones that matter most in the modern world. Failure comes because models that invoke a complex system require scientists and their conversants each individually to make a large number of arbitrary decisions. Since experiments are one of the more focused devices of ecology, they invite the naive to assert objectivity in experimentation. This chapter offers an antidote to the myth of objectivity in experiments when multiple scales of observation and analysis invoke a complex ecological system.

For simple systems, everyone agrees as to what is structure versus behavior, discrete versus continuous, qualitative versus quantitative, and meaningful versus blindly dynamical. There is no such unanimity for complex systems. Dealing with a complex system requires each discussant to be explicit as to a personal assignment of significance and meaning of the things that are recognized as important. Nature does not tell us how to make those decisions, but experimentation is a powerful device for making those decisions explicit. By taking responsibility for those judgments, we can avoid rude tautology and muddled semantic arguments. By showing what experiments do and what they do not, this chapter offers a set of intellectual devices for shouldering responsibility so as to yield high quality decisions. Scientists have several modes of operation: observation, modeling, calibration, inference, among others. Experimentation is one such a mode where the ecologist takes a particular posture, and in this chapter I lay out the subtleties that are distinctive for experimentation in the realm of multiple scales and complexity.

Throughout this chapter the notion of assumptions is crucial. There are assumptions embodied in levels of analysis as well as in the more explicit scaling of experimental protocol. The assumptions are stylistically different for experimentation in modeling as opposed to experiments using physical model systems. I therefore must dissect modeling from physical experiments, highlighting congruence among as well as the differences between them. Experiments are by design highly focused, and so are not suitable devices for description. Early in the chapter I move on to offer caveats about premature experimentation before adequate description. Proper description of the complex situation makes obvious the critical questions around which the experiment should be scaled. In all of this, assumptions are the vehicle for the discussion.

Scaling is central to models, because all models are scaling devices. Experiments become focused by being explicit to the point of compulsion as to the scales in use. Precision scaling yields repeatable results. But results can appear so unequivocal that one is tempted to reify the level of analysis as the level of true analysis. It is apt to remember scientists learn most when models and experiments fail, so the full power of experimentation is in identifying which scales and levels of analysis do not work. The bottom line is that experiments facilitate changing the scale so that new levels of analysis can be identified easily, quickly, and confidently. Accordingly, a large part of the second half of this chapter gives a specific example of a series of experiments on emergent properties on a vegetational system. The experiments did not inform us as to the full working of vegetation, but initial failure to predict the outcome did allow us to see that the level of analysis we had been using was inappropriate. We therefore chose a more apt level of analysis, and achieved prediction at a higher level. Precision of scaling in experimentation makes it a particularly powerful device in choosing which passage we should take next through the labyrinth of decisions presented by each complex system.

Before beginning our discussion, let us first acknowledge the modern state of experimentation in ecology. Experiments have become a larger part of ecology over the last quarter century. The ascendancy of experimentation has been a mixed blessing. On the positive side, there has been increased rigor and precision. On the negative side, narrow studies consume much effort and resources, resulting in little that is significant or generalizable. A director of a major funding agency commented recently how refreshing it was to have an experiment inspired and designed at the outset to probe a specific ecological question. His implied misgivings about the mainstream of experimental proposals he sees are well placed. Usually an experiment is contrived so as to use as a vehicle some ecological site that is already at hand, with the ecological question as a passenger and not the driver. Much of the increased use of experimentation in ecology has turned on validation of simple models that often are not well tied to the material system in question. For example, it was only after a decade of flaccid experimentation on competition that Simberloff was able to put a stop to it, by pointing out that much theory was tautological or untestable, and no competition had been unequivocally demonstrated (Connor and Simberloff 1979; Simberloff and Boecklen 1981; Simberloff and Connor 1981; Simberloff 1983). As Simberloff (1982) said of competition, "the theory has caused a generation of ecologists to waste a monumental amount of time," mostly doing experiments. So experimentation has brought a certain precision to some parts of ecology, but not all of the increased effort on experimentation has been well spent. What is there to do about it?

ASSUMPTIONS AND PREDICTIONS

Experimentation plays a special role in science, but model building must precede experimentation. Even so, models are not in themselves the point of science, for they are only formalizations of what really matters, that is, the assumptions. Experimentation tests the assumptions that underlie the model.

The central issue in all science is prediction. There is a naive realist philosophy prevalent in biology and ecology asserting that the agenda is to determine the ultimate truth of a situation (Ahl and Allen 1996). That is a relatively difficult position to defend, even for philosophers, let alone for scientists unschooled in the arts of contemplation. Further, naive realism has the disadvantage of setting up a reference system, reality beyond observation, which is undefined, and would not be recognizable if it was captured in observation. By contrast, reliance on prediction attaches the enterprise to something that is most explicitly defined. By turning away from an undefined reality as the central issue and making prediction the goal, science can be made manageable. To aim at the outset for ultimate truth ties the scientist to an undefined reference, and opens the door to muddled thinking. If one wishes to assert, after the fact of the investigation, a belief that predictive capacity takes one closer to the truth, it does no harm. Even so, there is no excuse for the tautology that what I see through the lens of my model is real, and that reality justifies the model.

Predictions may be the central issue for science, but they can always be cast in terms of assumptions. Many of the classic assumptions that underlie whole sectors of physics are demonstrably false, as in the perfect gas particle and the frictionless pendulum. It seems reasonable to admit that the presumed material system is functionally infinite in its intricacy. Given that admission, all assumptions fail to capture the entire system, and so all are, in that sense, false. Assumptions appear as an essential part of human understanding that cannot be avoided, no matter how much we may wish to escape them. Even the simple act of naming something makes assumptions about the bounds of the entity in question, and asserts arbitrary limits regarding the class to which it belongs. With the verity of "correct" assumptions cast as beside the point, what then is the point of assumptions? It is that one needs to distinguish between the assumptions that lead to prediction, as opposed to those that one cannot afford to make because they cause predictions of the model to fail. One is not interested in the verity of assumptions, but rather in which untruths we can afford to use in models (Allen 1997).

ANALOG AND DIGITAL EXPERIMENTATION

Rosen (1991) distinguishes between two aspects of models. The first is the formal model, which is a set of scale-relative statements. A set of scale-relative statements is independent of any particular spatiotemporal scale in the material system. As a result, both large and small systems can be encoded into and out of the same formal model. One such formal model is the set of equations for aerodynamics that link together model airplanes with full scale prototypes (the horizontal connections across figure 3.1). This linking of two material systems leads to the second aspect of modeling. With successful encoding of two material systems into and out of a formal model, the two material systems become models of each other, as in a model airplane (the vertical connections on the right side of figure 3.1). The reciprocity of the big system as a model of what might be called a "model airplane" becomes clear from the point of view of toy manufacturers: the "real" DC-10 becomes the model for the toy.

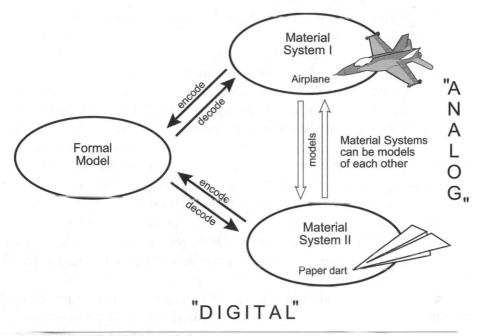

FIGURE 3•1 *A Concrete Image of Abstract Concepts*

A formal model is a set of scale-relative statements that can be applied to particular material systems. Any two material systems that can be encoded into and out of a given formal model are identified as being different only with regard to scale. The two material systems become models of each other. The relationships of the material systems to the formal model is an encoding into symbols, and so is digital in nature. The manner in which the material systems are models of each other is through material equivalence, capturing some aspects of an analog relationship.

The formal model operates through symbols, such as the equations for drag in aerodynamics. This symbolic coding is akin to the digital representation in digital computers or digitally coded music on a compact disc. The manner in which the two material systems work as models of each other is not digital, in that the physicality of the two systems is what links them together. There is no formal symbolic representation involved in this physical equivalence between material systems, much as a vinyl long-playing record does not employ symbols to link the recording session to the sound coming out of the stereo. As distinct from digital recordings on compact discs, analog recordings use patterns etched physically into the grooves of the LP or patterns of polarity on magnetic tape. The physical wiggles in the analog recording medium carry the signal. When two material systems are taken as models of each other, that modeling relationship is akin to that captured in analog recordings. In this sense, I call the link in figure 3.1 through the formal model a *digital representation*, whereas the physical equivalence of the two material systems I call *analog*.

The formal model is purely symbolic and conceptual, working through, and limited by, a series of definitions. By contrast, the two material systems as models of each other involve one material system in all its intricacy as a representation of another. Once the formal model has established the equivalence between the two material systems, the relationship between them takes on a life of its own that goes beyond the formal model's definitions. The consistency of the relationship between the two material systems is dependent on physical constancies like the surface properties of water or the effect of gravity. Many of those relationships are symbolically recorded in the formal model, but when the one material system is used directly as a model for the other, the symbolic representation is eclipsed by the material equivalence. The material equivalence is infinitely richer than the symbolic equivalence.

Clearly, experimentation in ecology uses these two different modes of modeling that I call digital and analog. Each mode of experimentation leads to distinctive styles of experimentation. Computer simulations test the formal model by experimentation with its parameters. Sometimes the changes in the model are in response to anomalous behavior in the formal model relative to the material system being modeled. This amounts to a calibration of the model as anomalies are incorporated. Each run with new parameters is an experimental test of the encoding and decoding processes to and from the formal model. A model run on a digital computer can in principle be run on a large analog computer, but most ecological models these days are performed on digital machines. This is a clue that experimental simulation is performing a formal operation that studies the linguistic coding of the formal model. Strong inference (Platt 1964) does not come easily from this sort of testing, but powerful and unlikely simulations are achieved, with the benefits and insights they bring.

On taking advantage of the analog equivalence between formally linked material systems, a very different sort of experimentation emerges. It uses a model material system as the experimental device to test aspects of the material system of interest. The experiment can be performed on the model system, when experimentation with the system of interest is difficult, impractical, or impossible. Here the assembly-rule work of Jim Drake, using plankton in flasks, is a surrogate for larger ecological systems, where control of the assembly process is out of reach (see Drake [1990], for discussion of general principles through nonexperimental correlation studies of lakes, but Drake [1991] for in-flask experiments). When whole lakes are manipulated by Steve Carpenter and his colleagues (e.g., Carpenter et al. 1995), the experimental lake is a physical model of all similar lakes. Note that in whole lake experiments there is no obvious way to translate the results to community assembly. Therefore, modeling with small, easily manipulated systems such as Drake's (1991) flasks is most valuable. The rescaling in the application of results from the model to material systems of interest is often taken as a given. The experiments that use a convenient material model often take for granted the formal model that links the experimental system to the commonplace systems of interest. The unspecified formal model is substituted by an informal counterpart that is only intuited. Allen et al. (1993) point out that physicists generally use a quantitative formal for mapping to and from the material system, but ecologists usually start with some sort of qualitative narrative that is later translated into formal terms from which measurements may be predicted. The qualitative narrative could be called an informal model (E. J. Rykiel, personal communication, 1994). The assumptions about the informal model in this mode of experimentation are usually justifiable, but are often not explicitly justified in a translation to a formal model.

EXPERIMENTS AND DESCRIPTION

If simulations and physical experiments are two different approaches to experimentation, there is a third approach that moves beyond the analog and digital distinction. In this last method, the experimental system itself appears to be the system of interest. In the eighteenth and nineteenth centuries, a country vicar might have displayed interest in only the immediate environs, the local vegetation, the animals and their habits, and the people of the countryside, with little interest in generalizing from discoveries in the local setting. The larger significance was only spelled out as parables for virtuous human behavior embodied in nature. In the 1877 edition to Gilbert White's *The Natural History and Antiquities of Selborne*, Frank Buckland (1877) wrote in the preface, "Gilbert White's writings are colored throughout with that right tone of feeling which recognizes the work of the great Creator in everything, both large and small." The modern experiments in the tradition of country vicars are focused on a particular region and system, the understanding of which is taken as an end in itself. This is the sort of experimentation that so frustrated the funding agency director, cited earlier. There is reference to larger issues, but the results are not generally encoded to make the link to other systems or the general condition to which the results might apply. Sometimes there is a model to which lip service is paid, but as often as not it is left implicit and informal.

In this last approach, the experimenters appear often to be barely cognizant of the relationships between material and formal models. These appear to be the experimenters to which Simberloff (1982) refers when he says, "These investigations are strongly in the hypothesis-testing tradition, but rather than testing general theory they test specific predictions about specific systems, and they seem to arise as much out of intense curiosity about these systems as a desire to find general patterns or laws of nature." This is the area of greatest increase in experimentation in ecology, and its lack of any sophisticated theory of modeling is the Achilles heel of the experimental movement across ecology at large. In the most profitable examples of this type, the system is well known and described, perhaps some regional system of streams, an island system, or some well established nature reserve. In less satisfactory examples of this mode of experimentation, the system is often not well described. While reviewing grant proposals, I have observed that experiments appear too often as a ritual display of rigor in situations that are so poorly described that experimentation is premature. Experimentation is precisely not the mode of science that describes systems, because experiments narrow the discourse so as to leave as little room as possible for equivocation. In a system that is insufficiently described, the scientist does not know enough to be able to identify the critical issues on which an experiment might turn.

The most powerful use of experimentation is in very austere settings. Experiments are useful in calibrating the encoding to the formal model. They are also helpful in testing whether or not the decoding to the material system works in an interesting way. A further area where experiments are useful is in probing the limits of modeling of two material systems, so as to identify explicitly the form of the formal model that links them. Here the scientist looks at the scaling operations that the formal model embodies. The reason for the compulsion that must surround experimental design is the need to scale and bound both the material system at hand and the one for which it is surrogate. Experimentation is not effective at describing systems, where it is asked to scale the system and identify its type. In insufficiently described or undefined situations, the burden of scaling the comparisons overwhelms, or more often trivializes, the experimental protocol. The rise of careerism in science invites addressing only the large issues. Large issues often involve many levels of analysis, each with its own set of scales. Going for the big ecological idea and the cute journal title after the colon has led to experimentation in situations where there is too much ambiguity of scale for the strategy of experimentation in the first place. It is to this premature experimentation that Simberloff (1982) refers when he characterizes interspecific competition theory as generating "predictions that are either practically untestable, by virtue of unmeasurable parameters or unrealized assumptions, or trivially true."

ASSUMPTIONS AND WHAT IS REASONABLE

In experimentation, assumptions and results may be either reasonable or otherwise. This leads to a two-way table of reasonable and unreasonable assumptions as opposed to reasonable and unreasonable results (figure 3.2) (Allen 1997). Much ecological experimentation makes reasonable assumptions and gets reasonable results. Although this does present confirmatory information, it does not lead easily to strong inference (Platt 1964). Pielou (1974) baldly stated that one has no business erecting null hypotheses that cannot reasonably be expected to stand. Her objection is often disregarded in the reasonable-assumptions/reasonable-results quadrant (figure 3.2), where the null hypothesis is a straw man. The experiment there is often expected to give reasonable results, and it does so, merely confirming conventional wisdom. An example of value here might be experimentation inside simulations making iterative adjustments to reasonable assumptions until the simulation corresponds to field observation.

Achievements in the merely confirmatory quadrant (figure 3.2) are modest, but the immature scientist might mistake making reasonable assumptions aimed at getting reasonable results for the whole scientific endeavor. Such a view imagines science as confirmation that certain models are true. My colleague in the Department of Physics in Madison, Wisconsin, Robert March (personal communication, 1998), was present in 1954 in the lab in Berkeley when Fermi told a junior scientist that he did not think the experiment to find the antiproton would work. Theory indicated that the antiproton must exist. The young scientist asked whether that meant that Fermi anticipated negation of the antiproton, a failure to demonstrate it. "No," replied Fermi, going on to say that he thought the experiment would show, albeit for the first time, that the antiproton did exist, just as everyone would expect. For Fermi, an experiment working meant disconformation, leading to unexpected results. Clearly, the young scientist was used to making reasonable assumptions and getting reasonable results when an experiment "worked."

The quadrant (figure 3.2) of reasonable assumptions yielding unreasonable results that are taken seriously is normal science (Kuhn 1970) functioning well. Normal science refutation demonstrates that the preferred model cannot work in the manner that a reasonable extension of it would indicate. If the assumptions in the preferred model were true (of course, none ever are) then there would be a reasonable outcome, which fails to come to pass. A common use of normal science refutation occurs when the scientist cranks the handle on some conventional model that is being used as a tool to make decisions and fix distinctions. A case in point might be the use of molecular systematics to sort out ambiguity in phylogenetic relationships derived from alpha taxonomy based on morphology. Less focused, but still useful enough, are refutations of assumptions of which the scientist is unaware. The assumptions are taken for granted, but are only recognized in hindsight, as when molecular systematics establishes an unexpected relationship by the happenstance. For instance, some group of plants might be discovered to have unsuspected southern hemisphere affinities. The refuted assumption is a northern hemisphere affinity that was unquestioned and unstated. However, such soft refutation leaves much to be desired, for there is a big difference between happening to refute a model, rather than aiming to refute it. Aiming to refute is normal science at its best, and experimentation is a good mode to achieve just that.

Sometimes the agenda of the scientist is to work in the confirmation quadrant (figure 3.2) making reasonable assumptions to get reasonable results, but a refutation arises by happenstance. The aggressive scientist takes advantage of such gifts, but that is all too rare. The less aggressive scientist facing happenstance refutation all too often puts a small patch on the preferred model. The prevailing view survives, elaborated by a correction term or weasel word. Only when there is finally a shift in paradigm will the correction terms be shoveled aside. As the refutation that underlies the corrections is finally taken seriously, the old cluttered

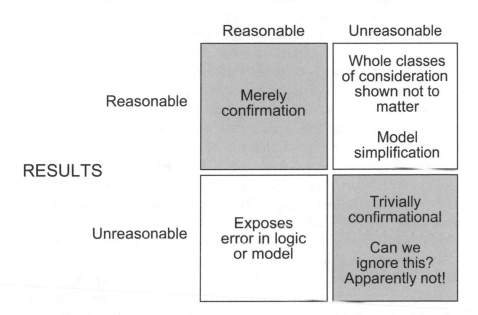

ASSUMPTIONS

	Reasonable	Unreasonable
Reasonable	Merely confirmation	Whole classes of consideration shown not to matter Model simplification
Unreasonable	Exposes error in logic or model	Trivially confirmational Can we ignore this? Apparently not!

(RESULTS on left axis)

FIGURE 3•2 *Assumptions Can Affect Results*
The point of science is not so much to find out what is real, but rather to investigate ourselves as we observe and try to understand. No assumption is strictly speaking true, so truth is beside the point. The focus needs to be on which assumptions can we afford to make, and still be able to predict. Assumptions may be reasonable or not, and may yield results that make the assumption appear true, for the moment, or immediately false. The two-by-two table indicates the mode of scientific inquiry that comes with combinations of reasonable assumptions and reasonable results, and their respective converses.

paradigm is rejected. Suppressing happenstance rejection of the preferred model is a price that must be paid when normal science holds on its course, preserving the paradigmatic model.

The opposite quadrant (figure 3.2) to reasonableness all around is unreasonable assumptions yielding unreasonable results. This quadrant embodies experiments whose results might yield, "Duh, what did you expect?" It amounts to stupid-in/stupid-out. In fact one must go through this quadrant often to reach one of the other two particularly valuable results. Sometimes unreasonable assumptions do give reasonable results. This sort of outcome is most desirable, in that it indicates whole rafts of assumptions that can be cut adrift. For instance, the FORET stand simulator (Shugart and West 1977) makes gross assumptions about species behavior and models in terms that are not just false, but are on the face of it false, any way you might look at it. For example, FORET ignores all horizontal spatial placement of trees in a forest. The model essentially puts all the trees on top of each other on the very same spot in the middle of the plot. And yet the model comes up with very reasonable results. The message is that the overriding factors in a forest are associated with vertical placement, to the significant exclusion of position on the ground.

The best of population biology works with patently false assumptions, aiming to achieve the reasonable result from unreasonable assumptions. In 1993 Monica Turner gave a colloquium in Madison, Wisconsin, on her model for ungulates in Yellowstone Park. My colleague in the Department of Zoology, Tony Ives, commented the next day that he would have had those elk moving at the speed of light. Only after an unreasonable result would he then begin to slow them down to realistic rates of movement. He never implemented the plan, but it had a certain logic. If reasonable results could come from elk modeled as riding on photons, all the reasonable, field-calibrated assumptions about ungulate movement and food preference could be dropped, and the model could be made much simpler. One reason to seek simpler models is that they are computationally more tractable, and so can be run countless times across large universes of discourse. This is what is happening when FORET is used as a subroutine in matching vegetation to models for global climate change.

EXPERIMENTATION TO ACHIEVE NEW LEVELS OF ANALYSIS

Social scientists appear generally to be more philosophically sophisticated than ecologists, and for social scientists, level of analysis is almost everything. Significantly, experiments are for identifying the appropriate level of analysis, given the question at hand and its scope. The issue of level of analysis is generally ignored in the naive realist approaches that are mainstream in ecology. If the agenda of the realist is to find out the truth of the ultimate reality of a situation as closely as possible, then there is only one truth. For such an ecologist, level of analysis recedes as an issue. The shades and shifts in meaning that occur with a change of level of analysis become moot when the one truth rides roughshod over all meaning. When models fail in an anomaly, the cause of apparent incompatibility is often a matter of level of analysis. Thus the search in experimentation is less a matter of ultimate truth, and is more a matter of discovering powerful levels of analysis. If one wishes to believe in hindsight that more powerful levels of analysis are closer to the truth then that does no harm, because the issue of choosing between levels of analysis is by that time past. However, ignoring the issue of level of analysis is a fatal flaw, because the level is then chosen by the happenstance of the first expectations of the experimenter, with little consideration of alternative levels.

In a recent set of experiments performed in my own laboratory (Havlicek 1999), the discovery was not facts about the material system, but the appropriate level of analysis. The point of visiting these experiments is less the verity of the findings and more an example of the thought process underlying experimentation when level of analysis is explicit. There is an emerging paradigm that life on earth is an elaboration of structure that emerged in response to the temperature gradient from the warm planet surface to cold outer space (Schneider and Kay 1994). Compared with other biomes, tropical forests have the lowest seasonal long wave radiation coming from them. That signal indicates that the tops of those ecosystems are cold. The signal comes from the top of very high clouds, but the forest is responsible for the clouds being there. The reason that tropical forest regions are the coolest when viewed from space is that those forests are elaborate and efficient energy dissipaters. In the Pacific Northwest of the United States, airplane fly-over of the Andrews Long Term Ecological Research site indicates that the vegetation surface is coldest in order of old growth, then even-aged young plantations, burned areas, and roads (Schneider and Kay 1994). Also, unpublished measurements made by Stephen Murphy of the University of Waterloo show soybean vegetation is cooler in field conditions of increasing nitrogen. Akbari et al. (1999) report that old fields are cooler than hay fields, which are cooler than lawns. All these results on such different systems are consonant with the model of life as a heat dissipating system. With such disparate systems as evidence for the usefulness of the paradigm, level of analysis needs to be made explicit.

To probe the emerging paradigm, we wished to put an ecosystem in a can to see if we could explain the general phenomenon of vegetational elaboration and ecosystem cooling. So we put soybeans in wind tunnels with full accounting of water uptake, biomass, wind speed, humidity, leaf area index, and surface temperature measured by infrared radiation. The preliminary results informed us that we were thinking at the wrong level of analysis.

At the outset we thought that elaboration of vegetational structure was the key, because bigger, more structurally elaborate vegetation appeared cooler across a variety of types from rain forest biomes to soybean fields. This led us to a prediction that heterogeneous vegetation would be cooler than homogeneous vegetation in the wind tunnel. We made heterogeneous vegetation out of plants of different sizes and physiological performance, and made the homogeneous vegetation out of plants of only one size and consistent physiology. The variety among plants was achieved by growing some plants in slow wind and some in fast wind. It is well known that growth in high wind produces stockier, smaller plants with a capacity to transpire at a high rate. Indeed, our fast-air plants grew as expected and were predictably different from the leafy, lanky plants grown in slow wind. The homogeneous vegetation grown in fast wind was cooler in fast air, but became warmer in slow air (figure 3.3). The explanation is that less wind lowers the capacity of plants to cool through latent heat of vaporization. By contrast, the homogeneous vegetation grown in slow wind was cooler in slow air, but warmer in fast air (figure 3.3). The explanation here is that at higher wind speeds, the plants acclimatized to slow wind are not plumbed to take advantage of high-wind, evaporative cooling, and so warm up through input from the sensible heat of fast moving, warm air.

To keep all else as equal as possible we grew all plants over the same time period. We then mixed the two types of plant in an even-aged sward to make a more structurally elaborate vegetation. Our expectation was that the more structurally elaborate vegetation would be cooler. However, the mixed vegetation was warmer than all the other contrived vegetations in almost all conditions (figure 3.4, upper panel). Our first prediction, that more elaborate even-aged vegetation would be cooler, was an unmitigated failure. When we mixed our two types of plant of the same age, the tall plants were slow-wind reared, whereas the short plants were high-wind reared. As air moved over and around the vegetation that was the aggregate of two types of even-aged vegetation, each type of plant was in the wrong environment for its history. Slow-wind reared plants were high in the canopy and so were up in the wind, whereas plants reared in fast wind were down in the slow air of the vegetation interior.

The failure to achieve cooling in the mixed, even-aged vegetation led us to abandon even-aged vegetation. We predicted that we could achieve the expected cool vegetation if we could switch the overstory for the understory and vice versa. We grew the fast-wind acclimatized plants earlier, so that they had time to grow tall and function as a proper overstory in the mixture (figure 3.4, lower panel). When tall plants grown in fast wind are mixed with short plants grown in slow wind, the resulting vegetation is more

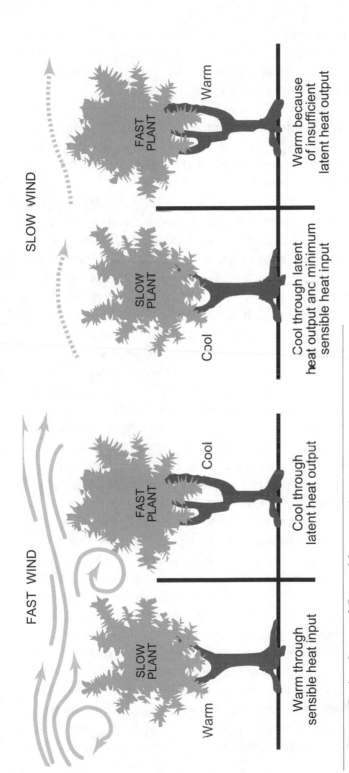

FIGURE 3•3 *Previous States and Current Measurements*

Plants were grown in either fast or slow wind. The labels of "fast plant" and "slow plant" refer to the wind speed in which the plants were reared and to which they were acclimatized. The temperatures of experimental vegetation were measured using the infrared radiation of the upper radiative surface of the vegetation. Vegetations measured in the environment in which the plants were grown were cooler than in the environment in which the other respective vegetation was grown. Fast-wind grown plants were cool in fast wind but warm in slow wind. Slow-wind grown plants were cool in slow wind and warm in fast wind.

natural, in that the overstory was grown in fast winds that overstory plants normally encounter. Also, the short plants in the understory were plants grown in slow wind, as plants in natural understory would be. The uneven-aged mixed vegetation was indeed cooler, as we predicted.

The failure of more structurally elaborate even-aged vegetation to demonstrate emergent cooling (figure 3.4, upper panel) led us to realize that the appropriate units of analysis are not individual plants collected together to make vegetation. The cooler temperature of the vegetation in the lower panel of figure 3.4 was a new prediction made in the light of the failure of our first prediction that merely mixing any sorts of different plants would produce cool vegetation. The value of the results are not so much in exposing the truth of vegetational cooling, but in forcing us to think about the level of analysis when aggregating organisms to make a higher ecological unit. It appears that the units to be aggregated to make complex vegetation must include the history of the organism as part of the elaboration. If we view wind as a stress, then the high-wind plants are forced to deploy resources away from investment in growth capacity, and toward a capacity to deal with the stress. Following this logic, vegetation made of trees is cooler because the units of aggregation have not only diverted more resources into water transport tissue in a plastic response to wind in the treetops, but also they are genetically programmed through an evolutionary history to divert carbon into transpiration capability. From these experiments it emerges that elaborate vegetation has two separate components: complicatedness and true complexity. First there is mere elaboration of structure, with no elaboration of organization. Second there is an elaboration of organization. Elaborate structure involves, in the absence of more elaborate organization, only a large number of symmetric relationships, and manifests only complicatedness. On the other hand, elaborate organization involves a large number of asymmetric relationships, where upper-level components constrain and protect lower-level components in a display of true complexity (Allen et al. 1999). The increased cooling of the uneven-aged, more natural vegetation comes from part of the system being able to withstand and block external stress, high wind. In this cooler-mixed vegetation, the plants that are physiologically of an overstory type are indeed properly situated in the overstory. As systems self-organize over time, they ameliorate the effect of stressors by developing resistance and resilient relationships. The experimental uneven-aged, cool vegetation mimics a natural self-organized system. Complex aggregates, as opposed to those that are merely complicated, cannot be understood without knowledge of their

Even-aged "Unnatural" mix

WRONG HISTORY–WARM

Merely Complicated

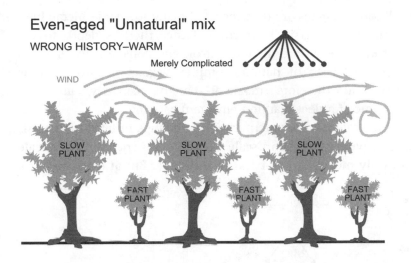

Uneven-aged "Natural" mix

RIGHT HISTORY–COOL

Properly Complex

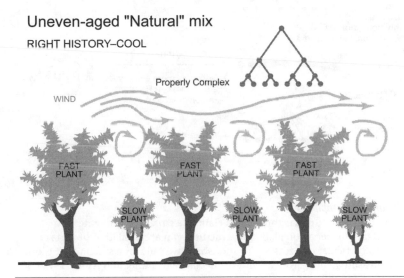

FIGURE 3•4 *Complex or Complicated*

The labels of "fast plant" and "slow plant" refer to the wind speed in which the plants were reared and to which they were acclimatized before testing. The two heights of plants in the top "Even-aged" panel come from the plants grown in slow wind growing faster and being taller, all plants being the same age. The two heights of plants in the bottom "Uneven-aged" panel come from the plants grown in fast wind being older and so taller. In the even-aged mixture of the top panel, the vegetation is unnatural because the history of the under- and overstory is wrong. There the overstory plants were grown in slow wind characteristic of understory conditions. In the uneven-aged mixture of the bottom panel, the vegetation is natural because the history of the under- and overstory is as in nature, where the first plants to grow feel the full force of the wind. In the natural mixture, overstory plants were raised in high wind, characteristic of overstory conditions. Measurement results show that the unnatural vegetation is warmer than are pure stands of its constituent types of plant.

history. Complicated systems are described as having low wide hierarchies. Expressing complex systems in hierarchical terms, there is depth to the system, so that upper levels are the context of, and constrain lower levels (figure 3.5). The behavior of natural complex vegetation should show minimal stress and therefore maximal cooling capacity, and our experimental mimic of it does manifest emergent cooling.

With regard to scale in experimentation, the change in level of analysis is commensurate with a change in scale. Sometimes a change in the scale of the system only means more or less of the same. But at other times, a

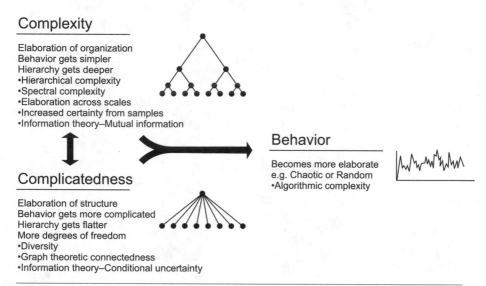

Complexity

Elaboration of organization
Behavior gets simpler
Hierarchy gets deeper
•Hierarchical complexity
•Spectral complexity
•Elaboration across scales
•Increased certainty from samples
•Information theory–Mutual information

Complicatedness

Elaboration of structure
Behavior gets more complicated
Hierarchy gets flatter
More degrees of freedom
•Diversity
•Graph theoretic connectedness
•Information theory–Conditional uncertainty

Behavior

Becomes more elaborate
e.g. Chaotic or Random
•Algorithmic complexity

FIGURE 3•5 *Explaining Complexity*
The general notion of complexity has three parts: elaborate structure, elaborate organization, and elaborate behavior. Elaborate structure is an accumulation of parts and connections with no particular or necessary relationships; we call this a system that is only complicated. The measure here is diversity and connectance (Gardner and Ashby 1970). Elaborate organization involves asymmetric relationships between parts, the upper level parts being the context of the lower level parts. We call this an organizationally complex system. The measure here invokes scalar issues and hierarchical order manifested as number and distinctiveness of levels. Behavioral elaboration is the outcome of elaborate structure and elaborate organization above. Since structurally complicated systems have more degrees of freedom, they exhibit elaborate, complicated behavior that is difficult to predict. Since organizationally complex systems impose constraints and context on low level parts, these truly complex systems suppress degrees of freedom of parts, and so the system behavior is simple and predictable. Most systems are some mixture of structurally complicated and organizationally complex, so the predictability and the degree of elaboration of behavior capture yet another version of complicatedness and complexity. The measure here is algorithmic complexity.

change in scale leads to a change in quality. Hegel stated that with a large enough change, the change is one of quality rather than quantity. This is associated with the phenomenon of emergence. Certainly, larger changes tend to be of quality as opposed to quantity, but qualitative change can occur under small changes across critical boundaries (Rosen 1989). In our experimental soybean vegetation, the entities in question change from being merely structural plants to organisms whose structure is a manifestation of an included history. By changing the grain to include the history of the lower-level entities that are being aggregated, we can see them as working components in a larger whole, as opposed to each plant being just an item in a pile.

The distinction between continuous scaling changes and changes in logical type is critical (Rosen 1989). If science is less about truth and more about prediction and powerful perception, then the type of thing being observed is critical. A useful change in type amounts to a change in level of analysis. The point is not that the lower level of analysis is essentially wrong, it is only less powerful and applies to fewer interesting situations. For example, phlogiston was "[u]sed in the 18th Century, to name a supposed substance emitted during combustion and calcification of metals: the 'food of fire' or 'inflammable principle'" (Bynum et al. 1981). It is mistaken to view phlogiston as a wrong element, even though it does not fit with the contemporary model that characterizes fire as rapid oxidation. Phlogiston is not wrong, it is just negative oxygen. As a conceptual device, negative oxygen is valid, but it is not a generally useful concept. Clearly, seeing the role of oxygen in positive terms with regard to fire is a more powerful model, a superior level of analysis that fits easily into the wide, modern intellectual framework that is chemistry. "Where a modern chemist sees a gain or loss of oxygen, early chemists saw an inverse loss of gain of phlogiston" (Bynum et al. 1981). Eventually, enough experiments on fire were performed for the emergence of the modern view with its new level of analysis and its own logical types. The power of experimentation is in the rescaling of the lower-level entities, so as to shift the level of analysis to one which is more powerful and generalizable. Thus, bottom line, experimentation is all about scale.

COMMENTS

So, there are two take-home messages with regard to scale and experimentation. First, at a modest level, in the cause of normal science, scaling in experimentation is a matter of calibration. Without calibration, science would grind to a halt. As the size of the material system changes, it will require new values for scaling variables if the system is to be generalized to the class to which it has been assigned. One needs to recalibrate the physiology of an organism if it is to be seen in terms that are equivalent to some other organism of very different size. This aspect of experimental scaling is crucial for science to be generalizable enough to function.

Second, scale in experimentation is a tool that leads the observer to change the model and its assumptions. All this is to achieve a more powerful level of analysis. Changes in scale are continuous. The strategy laid out by Rosen (1989) is for one to change the scale continuously until it forces a new, more effective type into the analysis. The critical insights come from failure of the model to behave in a continuous fashion under continuous changes in scale. That failure forces a change of assumptions that redefines the character of the critical entities in the analysis.

The scale changes force an instability into the model. In the example that Rosen (1989) uses to lay out his ideas, there is only one instability. He analyzes an elaboration of the gas laws, the van der Waals equation, as it recognizes phase changes in the system. The simple gas laws assert a relationship between pressure, P, volume, V, and temperature, T, in the equation $PV = rT$, where r is a constant for any given species of gas, say oxygen versus ammonia. The simple equation indicates one can increase the pressure on a gas indefinitely, and all that will happen is that the volume will decrease while the temperature goes up. Of course, gasses under sufficient pressure liquefy, and that is missing from the simple equation, but not from the van der Waals equation. The critical change in behavior occurs when a slightly greater change from an initial PVT condition leads to a new set of system constraints coming as if from nowhere. A particular gas under different environmental conditions will have different volumes. The character of a liquid is not only different from a gas; in Rosen's (1989) terms it is dissimilar, and requires a new set of principles in order to deal with it.

In biological systems, unlike Rosen's (1989) single instability, one finds many instabilities, and that is what leads to the facility of hierarchical models in biology. At the edge of a species one finds an instability, a gap where continuous rescaling of the system to accommodate new examples breaks down. In convenient vertebrate species, but much less often in plants, the gap at the edge of species corresponds to a critical threshold in breeding capacity, sometimes from a total incapacity to breed to a full capacity to breed. There is a similar discontinuity at the edge of most genera; it is not based on current breeding, but on historical failures to breed long ago. The exceptions are beside the point, for the rule is discontinuity between taxa, and different types of discontinuity at the edges of higher and lower taxonomic levels.

In ecology, there are similar patterns of discontinuity, which tempt a classification of community types, or guilds. Whether or not one finds such ecological classifications useful, there are discontinuities in ecology on which turn our central models. The distinction between food chains and food webs is a case in point. Look close enough (at a scale fine-grained enough) at a food chain and one often finds that an upper-level consumer is significantly feeding at more than one level. That breaks down the integer properties of trophic levels, and one is forced into food web analysis. The naive realist would assert that the food web model is closer to reality than the food chain, but by now the reader of this chapter might be questioning the utility of that point of view. In fact one does not stop using old paradigms because they are further from reality, one instead uses them for what they have always been good at predicting. For instance, we do not build bridges with quantum mechanics, but use the Newtonian model, which works just fine. So if one is concerned with energy flow and energy loss, food chains work well. Alternatively, the differently grained food web model addresses questions at which it is adept (energy flow generally not being one of them). So experimentation looking for anomalies is a way of finding scalar instabilities, which generate new types of analysis (Allen et al. 1993). Scale is thus the device that experimentation uses to force an anomaly to the fore. But the point of it all is the recognition and enlightenment that comes when the old types and levels of analysis fail. The continuous, quantitative aspects of science are associated with scale differences, but quantification and scaling are only devices, tools in the hand of the experimenter.

The hallmark of experimentation is that the control is explicit and equivalent to the test situation. Whenever a scientist makes a statement, there is always an implied "as opposed to what?" The experimental control lays out the "as opposed to what" factor in stark terms. A new definition changes the "what" and therefore the "as opposed to" component. More significant than a change in definition is a change in logical type, for that leads us to new classes of distinction and definition (Ahl and Allen 1996). Scientific advances occur when a new distinction is noticed, such as the need to distinguish ecosystem from community (Tansley 1935) or diversity from complexity (Van Voris et al. 1980). When scale changes force us into new definitions, with logical types that take us to a more powerful level of analysis, science moves forward in an important way. Experimentation uses scale powerfully, and it is the general issue of scaling that allows the focused protocol of experimentation its ultimate generality and utility in ecology at large. Thus scale is crucial to keep clear what is equivalent, so that we can recognize what is importantly and fundamentally dissimilar.

ACKNOWLEDGMENTS

I wish to thank John Norman and Tanya Havlicek for their permission to publish our experimental results on soybeans in the sweeping terms presented here before the full technical publication of Tanya Havlicek's masters thesis on her windtunnel work. On a draft of this chapter, Bob Ulanowicz and Vic Kennedy made extensive, detailed comments, which were most helpful in guiding revisions.

LITERATURE CITED

Ahl, V., and T. F. H. Allen. 1996. *Hierarchy Theory: A Vision, Vocabulary, and Epistemology*. New York: Columbia University Press.

Akbari, M., S. Murphy, J. J. Kay, and C. Swanton. 1999. Energy-based indicators of ecosystem health. In D. Quattrochi and J. Luvall, eds., *Thermal Remote Sensing in Land Surface Processes*. Ann Arbor, Mich.: Ann Arbor Press.

Allen, T. F. H. 1997. Community ecology. In S. I. Dodson, T. F. H. Allen, S. R. Carpenter, A. R. Ives, R. L. Jeanne, J. F. Kitchell, N. E. Langston, and M. G. Turner. *Ecology*, pp. 315–383. Oxford: Oxford University Press.

Allen, T. F. H., A. W. King, B. T. Milne, A. Johnson, and S. Turner. 1993. The problem of scaling in biology. *Evolutionary Trends in Plants* 7:3–8.

Allen, T. F. H., J. Tainter, and T. W. Hoekstra. 1999. Supply-side sustainability. *Systems Research and Behavioral Science* 16:403–427.

Buckland, F. 1877. Preface in Gilbert White, *The Natural History and Antiquities of Selborne*, pp. vii–xii. London: Macmillan.

Bynum, W. F., E. J. Browne, and R. Porter, eds.1981. *Dictionary of the History of Science*. Princeton: Princeton University Press.

Carpenter, S. R., D. L. Chistensen, J. J. Cole, K. Cottingham, X. He, J. Hodgeson, J. F. Kitchell, S. Knight, M. L. Pace, D. Post, D. Schindler, and N. Voichick. 1995. Biological control of eutrophication in lakes. *Environmental Science and Technology* 29:784–786.

Connor, E., and D. Simberloff. 1979. The assembly of species communities: Chance or competition? *Ecology* 60:1132–1140.

Drake J. 1990. The mechanics of community assembly and succession. *Journal of Theoretical Biology* 147:213–233.

Drake J. 1991. Community-assembly mechanics and the structure of an experimental species ensemble. *American Naturalist* 137:1–26.

Gardner, M. R., and W. R. Ashby. 1970. Connectance of large dynamic (cybernetic) systems: Critical values for stability. *Nature* 228:784.

Havlicek, T. 1999. Ecosystem responses to stress and complexity as evidenced from experiments on a small-scale vegetative system. Master's thesis, University of Wisconsin.

Kuhn, T. S. 1970. *The Structure of Scientific Revolutions*. Chicago: University of Chicago Press.

Platt, John R. 1964. Strong inference. *Science* 146:347–353.

Pielou, E. C. 1974. *Population and Community Ecology: Principles and Methods*. New York: Gordon & Breach.

Rosen, R. 1989. Similitude, similarity, and scaling. *Landscape Ecology* 3:207–216.

Rosen, R. 1991. *Life Itself: A Comprehensive Inquiry into the Nature, Origin, and Fabrication of Life*. New York: Columbia University Press.

Schneider, E. D., and J. J. Kay. 1994. Life as a manifestation of the second law of thermodynamics. *Mathematical Computational Modeling* 19:25–48.

Shugart, H. H., and D. West. 1977. Development of an Appalachian deciduous forest succession model and its application to the assessment of the impact of the chestnut blight. *Journal of Environmental Management* 5:161–179.

Simberloff, D. 1982. The status of competition theory in ecology. *Annales Zoologici Fennici* 19:241–253.

Simberloff, D. 1983. Competition theory, hypothesis-testing, and other community ecology buzzwords. *American Naturalist* 122:626–635.

Simberloff, D., and W. Boecklen. 1981. Santa Rosalia reconsidered: Size ratios and competition. *Evolution* 35:1206–1228.

Simberloff, D., and E. F. Connor. 1981. Missing species combinations. *American Naturalist* 118:215–239.

Tansley, A. G. 1935. The use and abuse of vegetational concepts and terms. *Ecology* 16:284–307.

Van Voris, P., R. V. O'Neill, W. R. Emanuel, and H. H. Shugart. 1980. Functional complexity and functional stability. *Ecology* 61:1352–1360.

CHAPTER 4

Spatial Allometry
Theory and Application to Experimental and Natural Aquatic Ecosystems

David C. Schneider

THE INCREASING USE OF SCALE REPRESENTS AN ENDURING CHANGE IN the way that ecology is pursued; it may well emerge as a unifying concept (Steele 1991; Allen and Hoekstra 1992; Levin 1992). The burgeoning interest in the topic is readily quantified. The phrase "spatial scale" appears for the first time in the journal *Ecology* in 1972, then appears in one or two articles per year until 1983. The phrase then appears in 7 articles in 1984, 15 in 1989, and 25 in 1994. Recognition of the need for multiscale analysis grew rapidly in the 1980s as it became clear that

- spatial and temporal patterns depend on the scale of analysis (Platt and Denman 1975; Allen and Starr 1982);

- spatial and temporal variability are linked at multiple scales (Valentine 1973; Haury et al. 1978; Shugart 1978);

- biological interactions with the environment occur at multiple scales (Harris 1980; O'Neill et al. 1986);

- the relation of populations to resources depends on spatial scale in groups as different as terrestrial plants (Greig-Smith 1952), marine plankton (Smith 1978; Mackas and Boyd 1979), and mobile predators (Schneider and Piatt 1986);

- population processes do not occur at scales convenient for investigation (Dayton and Tegner 1984); and

- environmental problems arise through propagation of effects across scales (Ricklefs 1990; May 1991). In the 1990s, quantification of multiscale patterns accelerated (Turner and Gardner 1991; Schneider 1994; O'Neill and King 1998).

The standard way of including spatial scale in ecological research is to define subsystems within a system (Allen and Starr 1982; O'Neill et al. 1986). Scaling to larger areas occurs by summation, with correction factors introduced as needed (Rastetter et al. 1992; Wiens et al. 1993). Once enough has been learned about a system, it often becomes possible to use power laws, which consider the scaling of rates to non-Euclidean lengths, areas, and volumes. This spatially allometric approach, based on the principle of incomplete similitude (Barenblatt 1996), produces predictive scalings that can be tested against new data, then used to scale from an experiment or survey to questions of larger scope. A general account of allometric rescaling can be found elsewhere (Schneider 1998). This chapter demonstrates spatially allometric rescaling for both experimental and natural aquatic systems.

Multiscale spatial analysis can be traced back to agricultural trials eight decades ago (Mercer and Hall 1911), but for population and community ecology, multiscale analysis is more recent. The need for explicit treatment of spatial scale in ecological research design was recognized two decades ago for surveys (Wiens 1976; Smith 1978) and a decade ago for manipulative experiments (Dayton and Tegner 1984; Ricklefs 1987). Experiments (typically at a small scale) cannot be extrapolated directly to larger scale questions, but widespread recognition of this problem has been relatively recent (Carpenter et al. 1995). The problem of scale-up goes unmentioned in book-length treatments of ecological experiments (Hairston 1989; Underwood 1997, but see Resetarits and Bernado 1998). The problem of scaling up from experiments to conclusions relevant to patterns and processes at a larger scale was addressed by Thrush and 21 co-researchers (1997).

Several solutions have been proposed for the problem of scaling from controlled experiments (necessarily at a limited scale) to target questions (usually at regional or global scales). One is to undertake deliberately large-scale experiments (Likens 1985; Schindler 1987). This precludes multiple experimental units, and so standard hypothesis testing cannot be applied to such experiments (Eberhardt 1976; Hurlbert 1984). Analysis of statistical uncertainty, however, can still be undertaken to constrain interpretation (Carpenter 1990), to control for some sources of chance variation (Jassby and Powell 1990; Dutilleul 1993), and to place a probability level on an outcome (Reckhow 1990) or on an estimate of risk (Sutor 1996), rather than declaring a decision against a fixed error rate.

For experiments where adequate replication is possible, statistical inference remains the best-known and only widely practiced solution to the problem of scaling up from limited data. Spatial scale-up via statistical inference requires an adequate number of randomly assigned experimental units, to infer from a sample to the larger scale population. When too few units are available, or cannot be drawn randomly, Hurlbert (1984) recommends against statistical analysis. Recognizing that a small number of experimental units will reduce the power of inference from experimental units alone, Eberhardt and Thomas (1991) recommend that manipulative experiments be combined with larger-scale surveys (e.g., Schneider 1978). Legendre et al. (1997) combined surveys with experimental results by estimating a *density surface* at a large scale, then embedding a manipulative experiment within this surface, thus obtaining more information than random placement of experimental units (Thrush et al. 1997).

Intuitively, one expects to scale to larger areas by summation and averaging, but this approach is subject to aggregation error (O'Neill and Rust 1979; O'Neill and King 1998), resulting in biased estimates of larger scale rates and densities. The dynamics of smaller scale components are not necessarily the same as the sum of the components (e.g., Welsh et al. 1988). To address the problem, Rastetter et al. (1992) recommend that summation over small spatial units be used as a first approximation; if this proves inaccurate, then sums weighted by the inverse of the variance are recommended to reduce the influence of highly variable subunits. These weighted sums are then calibrated against larger scale data (e.g., satellite imagery) to obtain an empirical correction factor.

Iterative cycling between experimental results and a larger scale model has been recommended repeatedly. Platt (1986) and Wessman (1992) recommended a cycle of observation, model testing, and model revision to address the problem of connecting locally measured photosynthetic rates to larger scale patterns of chlorophyll concentration depicted in satellite imagery. Wiens et al. (1993) recommended direct scale-up from experiment, as a first approximation, with detail added as needed to reduce aggregation error. These authors prefer the use of judgment, rather than inferential statistics, to compare large-scale measurements with expectations computed from small-scale investigations. Rastetter et al. (1992) recommended a similar approach, but rely on statistical weighting and correction factors. Cycling between larger scale models and smaller scale mechanistic studies occurs either by the "scale-down paradigm" or by the "scale-up paradigm" (Root and Schneider 1995). Rastetter et al. (1992) and Wiens et al. (1993) used the *scale-up paradigm*, in extrapolating from small-scale mechanistic studies to the larger system. The *scale-down paradigm* employs a model of the entire system to calculate local values of larger scale processes, such as climatically induced variation in light levels and total photosynthesis. There are however, few examples of either paradigm (Root and Schneider 1995).

The classical solution to the problem of change in scale has been to define two systems (such as a model boat and full-scale prototype), define groups of similar measurement units, then work out the scaling relation from model to prototype relative to these groups. The most commonly used groups, called "dimensions," are Mass (M), Time (T), Euclidean lengths (L), areas (L^2), and volumes (L^3). If the model and prototype are completely similar (Barenblatt 1996) then the scale-up is readily computed. For example we expect a tenfold increase in the length of the prototype relative to the model to result in a 10^3 fold increase in volume.

$$\frac{\text{Volume}_{\text{prototype}}}{\text{Volume}_{\text{model}}} = \left(\frac{\text{Length}_{\text{prototype}}}{\text{Length}_{\text{model}}}\right)^3 \qquad \textbf{(EQ 4•1)}$$

We further expect a 10^3-fold increase in mass, provided model and prototype are constructed of the same material and have the same density. The scaling relation based on the principle of complete similarity of mass with volume and length is Mass $\cong$ Volume $\cong$ Length^3.

Dimensional analysis based on complete similarity (Barenblatt 1996) has a long history in physics (Bridgman 1922). It is routinely used to scale observation to theory in physical oceanography (Gill 1982); it is used occasionally in biological oceanography (Mann and Lazier 1991), limnology (Uhlmann 1985; Schneider and Haedrich 1989), and the analysis of enclosed aquatic systems (Uhlmann 1985). It has been notably successful in engineering (Taylor 1974), where it is used to ensure that models have the same dynamical behavior as full-scale prototypes. The classic apparatus of dimensional analysis has been more successful with engineered than with natural objects, because the dynamics of a natural system are incompletely known (Kline 1965) and because dynamics are spatially heterogeneous or "lurch" rather than ticking along in clocklike fashion.

Patchy or episodic dynamics result in *aggregation error* (O'Neill and Rust 1979), where large-scale structures or dynamics cannot be computed directly from local scale measurements. In ecology, this problem has been addressed by hierarchy theory (Allen and Starr 1982). Using Herbert Simon's concept of *nearly decomposable levels*, the hierarchical method describes structure or computes dynamics at two (or more) levels, then compares the results. O'Neill and King (1998, table 1.1) list 20 such studies, mostly terrestrial. This approach has proved useful as a first step in understanding the effects of change in scale. What has proved difficult is defining the space and time scales that separate "nearly decomposable levels" if more than one variable is considered.

Episodic dynamics and non-Euclidean structures (such as landscapes surfaces) result in failure of classical dimensional methods based on complete similarity (Barenblatt 1996). In geophysics, this problem has been addressed by using a power law to compute from one scale to another according to the principle of incomplete similarity (Barenblatt 1996). A familiar example of incomplete similarity is the fractal relation of coastline length cL to Euclidean length L. If we take large steps along a coastline, then count them as if on a straight line, we obtain a length of $cL(L_{large})$. If we take small steps, then count again as if on a line, we obtain a length $cL(L_{small})$ that exceeds $cL(L_{large})$. The relation of the two measures of coastline length is

$$\frac{cL(L_{large})}{cL(L_{small})} = \left(\frac{L_{large}}{L_{small}}\right)^{\beta}$$ (EQ 4•2a)

In briefer form:

$$cL(L) = \alpha \cdot L^{\beta}$$ (EQ 4•2b)

In yet briefer form:

$$cL(L) \cong L^{\beta}$$ (EQ 4•2c)

The exponent β is not an integer, as it would be for completely similar Euclidean objects. Mandelbrot coined the term "fractal" to describe non-Euclidean objects where β is not an integer.

Incomplete similarity relates some property, quantified as a cumulative frequency distribution $S(x)$ relative to the measure x according to a power law:

$$S(x) = \alpha x^{\beta}$$ (EQ 4•3)

This relation says that when x is rescaled (say by a factor of 2), the cumulative frequency distribution $S(x)$ remains proportional to x^β. If x is some geometric measure (length, area, etc.) then the relation is described as being spatially allometric. If the full frequency distribution is unknown, we can still use summary information from the distribution. For example, we could use the total number of species $S(x_{max})$ in lakes of area x_{max}; these totals are expected to scale as $x_{max}{}^\beta$. The total number of cichlid fish species in six African lakes, as reported by Ricklefs and Schluter (1993:359) scales as $A^{1.152}$, as estimated by reduced major axis regression (Sokal and Rohlf 1995) of log species number on log lake area. Consequently, a lake with twice the area of another lake is expected to have $2^{1.152}$ = 2.22 times as many species, not twice as many.

Species-area curves are not the only example of incomplete similarity in biology. Allometric scaling of form and function to body size (Schmidt-Nielsen 1984) relies on the principle of incomplete similarity. Allometric scalings have a long history in biology, extending back to Thompson's landmark book of 1917 on form and function (Thompson 1961), though without the mathematical apparatus (Barenblatt 1996) that extends classical dimensional analysis to incompletely similar systems.

Incomplete similarity is widely applicable in ecology, but rarely used. This is due in part to the lack of a generic recipe for applying the principle, with examples. Here I use a series of examples to demonstrate the application of incomplete similarity to the biology of both natural systems and enclosed systems, or *cosms*. This can be called spatial allometry, by analogy with body size allometry, which also relies on incomplete similarity.

The next two sections outline the quantitative basis for spatial allometry, beginning with the concepts of scaled quantities and scope, then moving to the concepts of similarity and scaling functions. The subsequent four sections present spatial allometry as applied to a series of published examples: primary production in lakes, fish production harvest from lakes, biomass accumulation in mesocosms, and primary production in mesocosms. Spatial allometry is then extended from the analysis of a single rate to the analysis of concomitant rates. This extension is illustrated with a published analysis of adult juvenile interactions in benthic habitats.

SCALED QUANTITIES AND SCOPE

A well-defined quantity is not just a set of numbers. It has a name, a statement of the method and conditions for measurement, a set of number generated according to the statement, units on one of several types of measurement scale, and a symbol. An example is the area of six African lakes reported by Ricklefs and Schluter (1993).

Procedural Statement	Name	Symbol	Numbers	· Units
Ricklefs and Schluter (1993:359)	Lake area	$A =$	$\begin{bmatrix} 69484 \\ 32893 \\ 28490 \\ 6410 \\ 5346 \\ 2150 \end{bmatrix}$	$\cdot\ km^2$

These areas are for Lakes Victoria, Tanganyika, Malawi, Turkana, Albert, and Edward respectively.

Measurements occur on one of four types of measurement scale: nominal (yes/no), ordinal (ranks), interval (arbitrary zero such as time of day), and ratio (true zero such as hours elapsed since noon). Only ratio scale quantities can be rescaled according to the principles of complete or incomplete similarity.

A single definition of *scale* is neither desirable nor possible. It is useful to distinguish two components of scale, namely, grain and extent (Wiens 1989). Equivalent terms are inner and outer scale, and resolution and range (Schneider 1994). Scope is defined for research designs, for models, and for rates as the ratio of the range to the resolution (Schneider 1994). This definition is consistent with the physiological literature, where scope refers to the ratio of the maximum to minimum rate. An example is the scope of lake area for the set of six African lakes: 69484 km^2 / 2150 km^2 = 32.3 (Lake Victoria / Lake Edward). This is the area of the largest lake in units of the smallest lake, rather than in km^2. Another example is the scope of lake area in the data set assembled by Ryder (1965): 82400 km^2 / 3 km^2 = 27460 (Lake Superior / Lake Heming). Scopes are computed from artificial isolates (e.g., quadrat samples) as well as from true isolates (lakes, mesocosms). For example, five measurements of lake color, each with area (10 m)^2 will together have a scope of 500 m^2 / 100 m^2 = 5 for the sample, and a scope of 69484 km^2 / 100 m^2 = 6.9·10^8 for the population of all possible samples from the surface of Lake Victoria. If the five measurements were contiguous, then the scope of the sample remains 5, but the scope becomes 69,484 km^2 / 500 m^2 = 1.4·10^8 for the population of all possible blocks of five measurements from the surface of Lake Victoria.

Scope computations are useful in comparing studies with respect to scale. An example is the greater scope of the lake areas analyzed by Ryder (1965) than by Ricklefs and Schluter (1993). Scope computations are useful in comparing surveys, experiments, and model computations. Figure 4.1a shows the spatial and temporal scope of the mesocosm experiments reported by Chen et al. (1997). Tanks ranged from 0.1 m^2 to 10 m^2 in horizontal area; most measurements were made three times a week for up to 16 weeks, with two runs (summer and fall) in a year. The length of the line on the horizontal logarithmic scale (figure 4.1a) represents the ratio of largest to smallest tank; the length of the line on the vertical logarithmic scale represents the ratio of the experimental duration to minimum period between measurements. For comparative purposes the scale of a relevant question has been added: What causes anoxic episodes in Chesapeake Bay? (Officer et al. 1984). The spatial scale is 6 x 10^{10} m^2 for the entire Bay (M. Kemp, Horn Point Laboratory, Cambridge, Maryland, personal communication), with a time scale taken to be biweekly to a year.

Figure 4.1b shows the scope diagram for the marsh mesocosm experiments at the same laboratory (J. Dewar, Horn Point Laboratory, personal communication). Series I mesocosms (sandy fringe systems) consist of 2 ft by 10 ft plots (1.8 m^2) that will run for three years. Series II mesocosms (interior organic marsh systems) consists of 3 ft by 20 ft plots (6 m^2) that have run for two years, possibly continuing to three years (shown as arrow in figure 4.1b). The marsh ecosystem in Chesapeake Bay has an area of 7.985·10^4 ha, reported as 197,236 acres by the U.S. Environmental Protection Agency (Chesapeake Bay Program 1997). The time scales of interest for the entire bay are one to five years (J. Dewar, personal communication).

Figure 4.1c shows the scope diagram for the mesocosm experiments at the Marine Ecosystems Research Laboratory (MERL) on Narragansett Bay, at the University of Rhode Island. Fourteen tanks, each 2.5 m^2 in area, were sampled weekly for water chemistry during experiments that ran up to two and a half years (S. Nixon, University of Rhode Island,

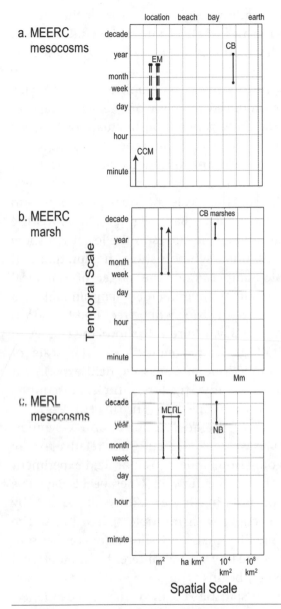

FIGURE 4•1 *Mesocosms for Three Ecosystems*

Scope diagrams comparing mescosms to system scale questions for Chesapeake Bay (CB), Chesapeake Bay marshes, and Narragansett Bay (NB). (a) Water column mesocosms experiments (EM) compared with system scale questions (CB), to planet earth, and to coffee cup mesocosms (CCM) distributed to workshop participants. (b) Marsh mesocosm experiments compared with system scale questions (Chesapeake Bay marshes); upward arrow indicates experiment still in progress. (c) Mesocosm experiments (MERL) compared with system scale questions (NB). Spatial scale shown as areas (m^2, etc.) and as diameters (m, km, and Mm = 10^3 km). MEERC = Multiscale Experimental Ecosystem Research Center; MERL = Marine Ecosystems Research Laboratory.

Narragansett, personal communication). Figure 4.1c shows the area of one tank (left side of box) and the area of all 14 tanks (right side of box). For comparison, Narragansett Bay (250 km^2) is shown with relevant time scales, taken to be one to 10 years.

Scope diagrams have a long history in oceanography, extending back to Steele's (1978) frequently reproduced diagram of the scope of ship surveys relative to phytoplankton and zooplankton fluctuations in space and time. These diagrams show the change in scope in going from experiments to field surveys or relevant questions at the scale of ecosystems. Figure 4.2a shows the change in scope in going from a single measurement in a 1 m^2 plot to the scale of questions posed to environmental biologists (e.g., Chesapeake Bay). Figure 4.2b shows the change in scale within a typical survey. The circles in the lower left show the area and duration of samples; the x symbols show the duration and area of the set of all possible samples (i.e., the target of statistical inference for a survey). Even for a survey covering many km^2 and taking a week to execute, a substantial change in scope will be required to address questions at the scale of Chesapeake Bay. Figure 4.2c shows the change in scope in going from the MERL tanks to questions at the scale of Narragansett Bay. It also shows the scope of a massive field experiment that was carried out by dozens of investigators, and that required thousands of hours to collect and process the data (Thrush et al. 1997). As with the survey diagram, dots show the area sampled, and x symbols show the area from which samples were taken. Figure 4.2d relative to 4.2c shows the change in scope in going from a mesocosm or field experiment to a spatially explicit model of material flow in Chesapeake Bay. The spatial resolution (left side of box in figure 4.2d) will be set by computational constraints in a model with multiple spatial units. The spatial and temporal scopes of satellite images are shown for comparison.

Changes in scope are accomplished by several different computational tactics. Coarse-graining (figure 4.3) begins with a series of units, usually contiguous, computes a summary statistic (mean, variance, covariance, etc.) for some property at the finest resolution, then aggregates adjacent units, allowing recomputation of the property of interest across a smaller number of larger units. This tactic has been used for spatial units in terrestrial ecology (Greig-Smith 1952), biological oceanography (Platt and Denman 1975), and limnology (figure 4.3). Coarse-graining was used to quantify the multiscale relation of plants to habitat variables such as soil nutrients (Greig-Smith 1952; Kershaw 1957), the relation of zooplankton

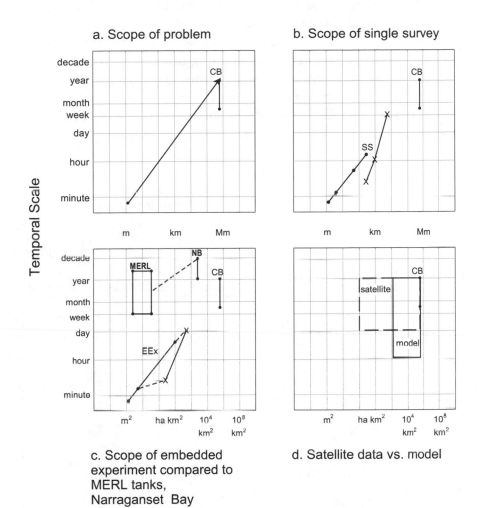

a. Scope of problem

b. Scope of single survey

c. Scope of embedded
experiment compared to
MERL tanks,
Narraganset Bay

d. Satellite data vs. model

FIGURE 4•2 *Empirical Comparisons*
Scope diagrams comparing system scale questions in Chesapeake Bay (CB) and Narragansett
Bay (NB) with surveys, experiments, and spatially explicit models. (a) Comparison with
single observation. (b) Comparison with single survey; dots show area and time measured
directly; x shows target of inference, such as plot sampled. (c) Comparison with mesocosm
experiments (MERL tanks) and with embedded experiment (EEx); dots show area and time
measured directly; x shows target of inference as plot sizes and durations. (d) Comparison
with a year-long archive of satellite data and with a spatially explicit model. Spatial scale
shown as areas (m^2, etc.) and as diameters (m, km, and Mm = 10^3 km).

patchiness to phytoplankton patchiness (Mackas and Boyd 1979; Weber
et al. 1986), the relation of mobile predators to schooling prey (Schneider

and Piatt 1986; Rose and Leggett 1991), and the relation of benthic megafauna to seafloor habitat (Schneider et al. 1987).

Another computational tactic is lagging, where a summary statistic is computed for some property of neighboring units, then recomputed at successively larger distances between units. This tactic is used in oceanography and limnology, for either contiguous or isolated samples (figure 4.3). For contiguous units, coarse-graining and lagging express the same information, though in differing formats and notation.

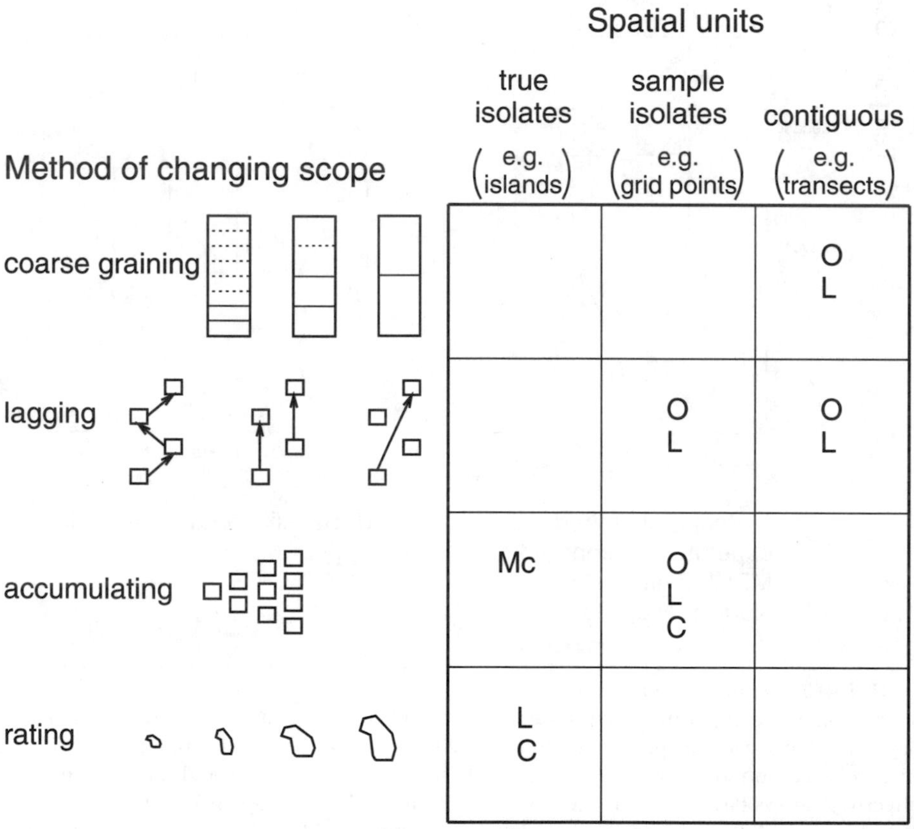

Spatial units

Method of changing scope	true isolates (e.g. islands)	sample isolates (e.g. grid points)	contiguous (e.g. transects)
coarse graining			O L
lagging		O L	O L
accumulating	Mc	O L C	
rating	L C		

FIGURE 4•3 *Options for Multiscale Analysis*
Tactics for multiscale analysis by changing spatial scope, applied to isolated systems, sample isolates, and contiguous units. All 12 combinations are possible. Published examples are common for combinations marked O = oceanography, L = limnology, C = isolated mesocosm experiments, Mc = multicosms (linked mesocosms).

Another computational tactic is accumulating units to increase scope (figure 4.3). If units are equal in size, accumulation results in diagonal increases in scope as in figures 4.2b and 4.2c. If units are sample isolates, accumulation increases scope and reduces uncertainty in studies carried out in oceans, lakes, and enclosed systems (figure 4.3). Accumulation of contiguous units will also increase scope, but computations are not usually made on this basis (figure 4.3). Accumulation of isolated systems will increase scope, but here again computations are not usually made on this basis. An example of increasing scope by accumulating isolated systems would be the construction of multicosms by connecting enclosed systems.

For true isolates (lakes, islands, cosms) the usual computation tactic for changing scope is to assemble units of different size, then rate a variable with respect to size (diameter, area, or volume). In the example of the data set assembled by Ryder (1965), fish catches range from 1 tonne yr^{-1} = 10^6 g yr^{-1} = 1 Mg yr^{-1} (Lake Heming) to 17300 Mg yr^{-1} (Lake Erie). The rating of fish catch against lake size is readily visualized as a plot of catch relative to lake area. Another example is rating (scaling) of cichlid species number to lake area, again visualized as a plot. Rating is commonly used in limnology and for enclosed systems (figure 4.3).

SIMILARITY, SCALING THEORY, AND SCALING FUNCTIONS

The principle of incomplete similarity allows us to use simple units (clock time, Euclidean lines, planes, and volumes) to measure any quantity, no matter how crooked in shape, heterogeneous, patchy, or episodic in dynamics. We cannot assume that the measurements we obtain will follow Euclidean rules: a coastline measured as 1 km in units of 1 km will not have a length of 1000 m if measured in units of 1 m. We can, however, assume that the measurements we obtain will depend in some regular way on the units that are used. This regularity, called incomplete similarity, is expressed as an exponent that relates the scope of one quantity to another. To measure a coastline we need two rulers, one a multiple of the other. The scope of a 1 km ruler, relative to a 1 m ruler, is L_{big}/L_{little} = 1000. The scope of the number of steps counted off by each ruler is N_{big}/N_{little}. Within this scope the number of steps increases as the step size decreases. The relation of step number to ruler length is expressed by a power law:

$$\frac{N_{\text{big}}}{N_{\text{little}}} = \left(\frac{L_{\text{big}}}{L_{\text{little}}}\right)^{-\beta}$$

(EQ 4•4)

If the scaling is complete ($\beta = 1$), then $N_{\text{big}}/N_{\text{little}} = 1{:}1000$. The number of steps will increase from 1 to 1000, a familiar trade-off between ruler length and number of steps that applies only to Euclidean objects. For fractal objects the scaling is incomplete ($\beta = 1.3$ for a typical coastline), and $N_{\text{big}}/N_{\text{little}} = 1{:}1000^{1.3} = 1{:}7943$. That is, the number of steps will increase from 1 to nearly 8,000, if the ruler length is shortened from 1 km to 1 m.

Scaling theory says power laws will also apply to the dynamics of systems with incompletely similar geometry. An example familiar to most biologists will be the incomplete similarity of respiration ($\dot{E}$, in kJ day^{-1}) to body mass (M, in kg). A power law relation is expected because respiratory exchange occurs across the fractal surface of the lung, which is incompletely similar to body volume and body mass (Pennycuick 1992; Schneider 1994). For the *mouse to elephant curve*, the scope of respiration is $\dot{E}_{\text{elephant}}/\dot{E}_{\text{mouse}}$, a number that measures elephant respiration in mouse units. The scope of the corresponding masses, $\dot{M}_{\text{elephant}}/\dot{M}_{\text{mouse}}$, measures elephant mass in mouse units. Within this scope, metabolic rate scales as mass$^{0.723}$ (Calder 1984).

$$\frac{\dot{E}_{\text{elephant}}}{\dot{E}_{\text{mouse}}} = \left(\frac{\dot{M}_{\text{elephant}}}{\dot{M}_{\text{mouse}}}\right)^{\beta}, \ \hat{\beta} = 0.723$$

(EQ 4•5)

The hat over the symbol represents the estimated value of the parameter under the hat. This scaling is incomplete or allometric ($\beta \neq 1$) and hence a 100-fold increase in mass will increase metabolic rate by $100^{0.723} = 28$ rather than by a factor of 100. Such a calculation applies only within the scope of energy exchange (for elephants in mouse units) and scope of mass (again for elephants in mouse units). Extension beyond the scope of the measure (mass) assumes no change in scaling relation, that is, that no scaling thresholds exist (Wiens, this volume). This scaling of respiration to body mass has been verified repeatedly; theoretical explanations based on skeletal strength in terrestrial organisms (MacMahon 1973) have now been supplanted by scaling theory for exchange across fractal surfaces (Pennycuick 1992; Schneider 1994). West et al. (1997) provide a theoretical explanation for why respiratory surfaces should be fractal.

Another example of scaling by incomplete similarity is annual fish catch ($\dot{M}$, in kg yr^{-1}) relative to lake area (A, in ha) for the data set assembled by Ryder (1965). The scaling, as estimated by least squares regression (Schneider and Haedrich 1989), was

$$\frac{\dot{M}_{Erie}}{\dot{M}_{Heming}} = \frac{A_{Erie}}{A_{Heming}}, \ \hat{\beta} = 0.84 \qquad \textbf{(EQ 4•6)}$$

The spatial scaling is allometric; a hundred-fold increase in the surface area of a lake can be expected to increase fish catch by a factor of $100^{0.84}$ = 48, rather than by 100. This scaling is empirical, and so cannot be applied outside the set (i.e., Canadian shield lakes) used to estimate it. It also cannot be applied outside the scope of the lake areas used to develop the relation.

The generic expression for the spatial scaling of a rate $\dot{Q}(L)$ within a spatial scope of L/L_0 is

$$\frac{\dot{Q}(L)}{\dot{Q}(L_0)} = \left(\frac{L}{L_0}\right)^{\beta} \qquad \textbf{(EQ 4•7a)}$$

This can be rewritten in shorter form as a spatial scaling function:

$$Rate(L) = k \cdot L^{\beta} \qquad \textbf{(EQ 4•7b)}$$

where k stands for $Rate(L_0) \cdot L_0^{-\beta}$. For the Ryder (1965) data set on fish catch, the scaling function as estimated by regression (Schneider and Haedrich 1989) was $\dot{M} = 20.9 \ A^{0.84}$.

The exponent of a scaling function based on complete similarity is Euclidean ($\beta = 1 - D$), where $D = 1, 2,$ or 3 for lines, areas, or volumes respectively. The scaling based on incomplete similarity is fractal (Mandelbrot 1977), referring to points scattered along a line ($0 < D < 1$), to a convoluted line within a plane ($1 < D < 2$), or to a convoluted surface embedded within a volume ($2 < D < 3$).

The generic expression for scaling any rate $\dot{Q} = dQ/dt$ to another quantity ($Y = $ mass, length, entities, time, etc.) within a defined scope Y/Y_0 according to the principle of incomplete similarity is

$$\frac{\dot{Q}(Y)}{\dot{Q}(Y_0)} = \left(\frac{Y}{Y_0}\right)^{\beta} \qquad \textbf{(EQ 4•8a)}$$

This expression can be shortened to the following scaling function:

$$\dot{Q}(Y) = k \cdot Y^\beta \qquad \text{(EQ 4•8b)}$$

where k stands for $\dot{Q}(Y_0) \cdot Y_0^{-\beta}$. The scope is no longer explicit, making it easy to overlook that the scaling function will not apply beyond defined limits (Wiens, this volume).

The next four sections demonstrate the application of incomplete similarity and scaling functions for isolated systems—lakes and mesocosms. The goal is to test ideas expressed as scaling functions. The exponents are estimated by regression, but our interest is in the reliability of the estimate, rather than tests of whether the exponent differs from zero. Hence criteria other than significance tests are used to evaluate these estimates. First, do the data support a power law relation? This can be judged from a plot of residuals versus fitted values. Second, is the estimate heavily influenced by a single value? This can also be judged from the plot of residuals versus fitted values. Third, does the estimate of the scaling exponent depend on the regression technique? This can be judged by comparison of the reduced major axis (RMA) estimate with the more commonly encountered least squares estimate. The RMA estimate assumes both variables are measured with error and is generally recommended over least squares regression for allometric relations (Sokal and Rohlf 1995).

APPLICATION: PRIMARY PRODUCTION IN LAKES

Fee (1979) examined the relation of primary production to lake size in the Experimental Lakes Area in northern Ontario. Of 19 lakes, 11 were untreated controls, and one was untreated in the first of 3 years. Fee (1979) reported several morphometric variables: lake surface area (A_o, in ha), maximum depth (z_{max}, in m), mean depth ($\bar{z}$, in m), epilimnion depth (z_e, in m), and the ratio of epilimnion area to epilimnion volume (A_e/V_e, in m^{-1}). From this data Fee (1979) developed and tested a scaling relation between production (P, in g–C yr^{-1} m^{-3}) and the surface/volume ratio A_e/V_e. Uhlmann (1985) developed scaling functions for Fee's concept of production in relation to lake morphometry, assuming complete similarity. This assumption is not warranted for lakes, which are fractal objects with crooked shorelines and irregular bottom shape. It was thus of interest to re-examine this work within the larger theory of incomplete similarity, including fractal objects.

Fee (1979) presented data on production per unit area (P/A_o, in g–C yr^{-1} m^{-2}) and production per unit volume (P/V_e, in g–C yr^{-1} m^{-3}), from which I calculated epilimnion volume (V_e, in m^3), epilimnion area (A_e, in m^2), lake volume (V, in m^2), and total production (P_{tot}, in g–C yr^{-1}):

$$V_e = (P/A_o)/(P/V_e)\cdot A_o \cdot 10^4 \qquad \text{(EQ 4•9)}$$

$$A_e = V_e \cdot (A_e/V_e) \qquad \text{(EQ 4•10)}$$

$$V = A_o \cdot \bar{z} \cdot 10^4 \qquad \text{(EQ 4•11)}$$

$$P_{tot} = V_e \cdot (P/V_e) \qquad \text{(EQ 4•12)}$$

Scaling relations for completely similar lakes, as worked out by Fee (1979) and Uhlmann (1985), are

$$A_o/V \cong L^2 \cdot L^{-3} = L^{-1} \qquad \text{(EQ 4•13a)}$$

$$A_e/V_e \cong L^{1.5} \cdot L^{-2.5} = L^{-1} \qquad \text{(EQ 4•14a)}$$

$$P_{tot} \cong A_e^{\;1} \qquad \text{(EQ 4•15a)}$$

$$P_{tot}/V_e \cong (A_e/V_e)^1 \qquad \text{(EQ 4•16a)}$$

The first scaling relation assumes that the ratio of surface area to lake volume is a constant proportional to the inverse of average depth $\bar{z}$ in Euclidean length units L^1. This relation will hold if volume and area are measured with the same spatial resolution. This scaling relation assumes completely similar or "Euclidean" lakes, rather than fractal lakes with convoluted geometries, where the average value of depth depends on the horizontal spatial resolution used in making the measurement. The ratio of volume to area will not scale with depth if depth is measured at some other resolution. The ratio of volume to area will not be expected to scale completely with D_{max}, because this measure is at the coarse scale of the entire lake. Nor will the volume/area ratio scale with depth d computed from the cross-sectional area: $A^{1/2} = d = (\bar{z} \cdot r)^{1/2}$, where r is lake radius. The cross-sectional area will depend on the horizontal spatial resolution (amount of detail) in measuring the depth contour.

The second scaling relation assumes that the ratio of epilimnion area to epilimnion volume scales as average epilimnion depth, even if epilimnion area and volume do not scale directly with lake surface area and volume. Uhlmann (1985) assumed that the ratio of epilimnion volume to total volume scales as depth$^{1/2}$, based on vertical mixing.

The third relation expresses Fee's (1979) idea that total production by a lake scales directly with the area of the benthos shallow enough to be illuminated. This scaling relation assumes a constant vertical mass flux from the illuminated benthos (F_e, in units of mass per unit time per unit area). The fourth relation re-expresses the third, in units of primary production per unit volume. As with the first two scaling relations, complete similarity is assumed. However, incomplete similarity may well hold because benthic surface area is a convoluted surface, and hence the measured value will depend on the units used.

The four scaling relations are first written as scaling functions based on complete similarity:

$$V = \alpha \cdot A_o^{1.5} \qquad\qquad k = L^3(L^{-2})^{1.5} \quad \alpha = k \qquad\qquad \textbf{(EQ 4•13b)}$$

$$V_e = \alpha \cdot A_e^{1} \qquad\qquad k = L^3(L^{-2})^{1.5} \quad \alpha = k \qquad\qquad \textbf{(EQ 4•14b)}$$

$$P_{tot} = \alpha \cdot F_e \cdot A_e^{1} \qquad\qquad k = L^0(L^{-2}L^2)^{-1} \quad \alpha = kM^1T^{-1} \qquad\qquad \textbf{(EQ 4•15b)}$$

$$P_{tot}/V_e = \alpha \cdot F_e \cdot (A_e/V_e)^{1} \qquad k = L^{-3}(L^{-2}L^{-1})^{-1} \quad \alpha = kM^1T^{-1} \qquad \textbf{(EQ 4•16b)}$$

The geometric scaling factor k shows the exponents for Euclidean lengths (L^1), areas (L^2), and volumes (L^3). The coefficient α shows the dimensional exponents for mass (M) and time (T). The scaling functions are then rewritten to include factors permitting incomplete similarity.

$$V = \alpha \cdot A_o^{\beta} \cdot A_o^{1.5} \qquad\qquad k = L^3 \cdot (L^2)^{-1.5-\beta} \quad \alpha = k \qquad\qquad \textbf{(EQ 4•13c)}$$

$$V_e = \alpha \cdot A_e^{\beta} \cdot A_e^{1.5} \qquad\qquad k = L^3 \cdot (L^2)^{-1.5-\beta} \quad \alpha = k \qquad\qquad \textbf{(EQ 4•14c)}$$

$$P_{tot} = \alpha \cdot F_e \cdot A_e^{\beta} \cdot A_e^{1} \qquad\qquad k = L^0 \cdot (L^2)^{-1-\beta} \quad \alpha = k M^1 T^{-1} \quad \textbf{(EQ 4•15c)}$$

$$P_{tot}/V_e = \alpha F_e \cdot (A_e/V_e)^{\beta} \cdot (A_e/V_e)^{1} \quad k = L^{-3} \cdot (L^2L^{-3})^{-1-\beta} \alpha = k M^1 T^{-1} \textbf{(EQ 4•16c)}$$

For each function the spatially allometric scaling factor k has been worked out. The coefficient α again shows the dimensional exponents, which this time allow incomplete similarity. The exponent β will be zero if complete similarity applies; it will deviate from zero if incomplete similarity applies. The coefficient α and the combined exponent (for example $A^\beta \cdot A^1 = A^{\beta+1}$) are estimated from the data, first by least squares regression. In some cases the regression coefficient will be k, the geometric scaling factor; in others the regression coefficient α includes the geometric factor k along with other dimensions. To see if a straight line relation is tenable, the residuals from this regression are plotted against the fitted values. The combined exponent is also estimated by reduced major axis (RMA) regression, to check whether the estimate depends on method used. If the estimates differ, the RMA estimate is used to avoid the assumption that the regression variable is measured without error. For RMA regression, the following pair of relations will hold:

$$Y \cong X^\beta \text{ and } Y^{1/\beta} = X$$

The same relation will not hold for exponents estimated by least squares regression, to the degree that the correlation coefficient is less than unity.

The three scaling exponents $A^{\beta+1}$, $A_e^{\beta+1}$, and $(A_e/V_e)^{\beta+1}$ were initially estimated by least squares regression, that is, a straight line relation on a log-log plot of the regression variables. A plot of residual versus fitted values showed that the least squares regression line was heavily influenced by a single value (lake 228, with an area far larger than the other 11 lakes). The regression line was then estimated by reduced major axis regression, with residuals again plotted against fitted values. This plot showed that the exponent was poorly estimated by RMA regression because of the influence of lake 228. When this lake was removed (11 lakes, 24 cases), the straight line regression on a log-log scale was appropriate for all four functions, as judged from the plot of residual versus fitted values. The estimate based on reduced major axis (RMA) regression was below that for least squares regression, substantially so in the case of the function relating epilimnion volume V_e to area A_e. The estimates based on RMA regression were adopted to avoid the inconsistencies that arise in assuming that the explanatory (independent) variable was measured without error. The scaling functions based on RMA regression were

$$V = \hat\alpha \cdot A_o^{1.678} \qquad \alpha = e^{8.964} m^3 ha^{-1.678} \qquad \hat\beta = 0.18 \qquad \textbf{(EQ 4•13d)}$$

$$V_e = \alpha \cdot A_e^{1.449} \qquad \alpha = e^{-2.8561} m^3 (m^2)^{-1.449} \qquad \hat{\beta} = -0.051 \quad \text{(EQ 4•14d)}$$

$$P_{tot} = \alpha \cdot A_e^{0.976} \qquad \alpha = e^{4.04317} g\text{–}C \cdot yr^{-1} (m^2)^{-0.976} \; \hat{\beta} = -0.024 \quad \text{(EQ 4•15d)}$$

$$P_{tot}/V_e = \alpha \cdot (A_e/V_e)^{0.826} \alpha = e^{3.3998} g\text{–}C \cdot yr^{-1} m^{-3} m^{-0.826} \; \hat{\beta} = -0.17 \quad \text{(EQ 4•16d)}$$

The analysis confirmed Fee's (1979) concept that production scales with epilimnion area. The scaling is completely similar ($\beta = 0$) despite the non-Euclidean scaling of lake volume to area ($\hat{\beta} = 0.22$). Production per unit volume does not appear to scale completely with the area/volume ratio for the epilimnion. Production per unit volume P_{tot}/V_e was not a linear function A_e/V_e, as judged from a plot of the residuals versus fitted values for the linear regression of P_{tot}/V_e on A_e/V_e. Close examination of figure 3 in Fee (1979) bears out the judgment that production per unit volume is not related in a linear fashion to the ratio A_e/V_e. The incomplete scaling ($\hat{\beta} = -0.17$, see above) needs to be tested against another data set, preferably with a wider range of lake sizes than in this analysis.

APPLICATION: FISH CATCH FROM LAKES

Ryder (1965) examined the relation between fish catch and lake size, in light of the well-known inverse relation between production and lake depth. Schneider and Haedrich (1989) developed a series of scaling functions based on Ryder's (1965) data, found one function that was completely similar, then tested this function against fish catch data from Jones and Hoyer (1982). The scaling relations for completely similar lakes were

$$V \cong \bar{z} \cdot A_o \cong (L^2)^{1/2} (L^2)^1 = (L^2)^{3/2} = L^3 \qquad \text{(EQ 4•17a)}$$

$$\dot{M} \cong (TDS/\bar{z}) \cong L^{-1} \qquad \text{(EQ 4•18a)}$$

$$\dot{M} \cong F_{radial} \cdot Circ \cong L^0 \qquad \text{(EQ 4•19a)}$$

$$\dot{M} \cong F_{nutr} \cdot A_o \cong L^0 \qquad \text{(EQ 4•20a)}$$

$$\dot{M} \cong F_{benth} \cdot V^{2/3} \cong L^0 \qquad \text{(EQ 4•21a)}$$

The first scaling relation states that lake volume scales as the product of the area of the top surface (A_o, in m^2) and average depth $\bar{z}$ (in m). The second, due to Ryder (1965), is that fish catch ($\dot{M}$, in kg yr^{-1}) scales as the ratio of total dissolved solids (TDS, in ppm) and averaged depth ($\bar{z}$, in m). The third scaling is that trip times and boat capacity limit fish catch; hence catch scales with radial flux of fish (F_{radial}, in kg km^{-1} day^{-1}) toward the shore (*Circ*, in km). The fourth is that catch scales as upward nutrient flux (F_{nutr}, in g-N m^{-2} yr^{-1}) into the euphotic zone with area A_o. The fifth is that catch scales as nutrient flux from the benthos (F_{benth}, in g-N m^{-2} yr^{-1}) into the lake volume.

The scaling relations are next written as scaling functions that assume complete similarity, using the conventions of standard dimensional analysis:

$$V = \alpha \cdot A_o^{1.5} \qquad\qquad \alpha \cong L^3 (L^2)^{-1.5} = L^0 \qquad\qquad \textbf{(EQ 4•17b)}$$

$$\dot{M} = \alpha \cdot (TDS/\bar{z})^1 \qquad\qquad \alpha \cong L^0 (L^0 L^{-1})^{-1} = L^1 \qquad\qquad \textbf{(EQ 4•18b)}$$

$$\dot{M} = \alpha \cdot F_{radial} \cdot A^{1/2} \qquad\qquad \alpha \cong L^0 (L^{-1} L^1)^{-1} = L^0 \qquad\qquad \textbf{(EQ 4•19b)}$$

$$\dot{M} = \alpha \cdot F_{nutr} \cdot A_o \qquad\qquad \alpha \cong L^0 (L^{-2} L^2)^{-1} = L^0 \qquad\qquad \textbf{(EQ 4•20b)}$$

$$\dot{M} = \alpha \cdot F_{benth} \cdot V^{2/3} \qquad\qquad \alpha = L^0 (L^{-2} L^2)^{-1} - L^0 \qquad\qquad \textbf{(EQ 4•21b)}$$

These scaling relations can be written more generally as scaling functions that include spatial factors k permitting either complete ($\beta = 0$) or incomplete similarity ($\beta \neq 0$):

$$V = \alpha \cdot A_o^\beta \cdot A_o^{1.5} \qquad k = L^3 (L^2)^{-1.5-\beta} \quad \alpha = k \qquad \textbf{(EQ 4•17c)}$$

$$\dot{M} = \alpha \cdot (TDS/\bar{z})^\beta \cdot (TDS/\bar{z})^1 \quad k = L^0 (L^{-1})^{-1-\beta} \quad \alpha = k \cdot M^1 \cdot T^{-1} \quad \textbf{(EQ 4•18c)}$$

$$\dot{M} = \alpha \cdot (A^{1/2})^\beta \cdot A_o^{1/2} \qquad k = L^0 (A^{1/2})^{-1-\beta} \quad \alpha = k \cdot F_{radial} \cdot c_t \cdot c_l \ \textbf{(EQ 4•19c)}$$

$$\dot{M} = \alpha \cdot A^\beta \cdot A_o^1 \qquad k = L^0 A^{-1-\beta} \qquad \alpha = k \cdot F_{nutr} \cdot c_m \qquad \textbf{(EQ 4•20c)}$$

$$\dot{M} = \alpha \cdot (V^{2/3})^\beta \cdot V^{2/3} \qquad k = L^0 (V^{2/3})^{-1-\beta} \quad \alpha = k \cdot F_{benth} \cdot c_m \quad \textbf{(EQ 4•21c)}$$

The coefficient α includes conversion factors to take into account the use of variables that differ in units of time ($c_t = 365$ day/yr), length ($c_l = $

km/1000 m), and mass (c_m = 1000 g/kg). It includes the flux terms F_{radial}, F_{nutr}, and F_{benth} because these are considered invariant with respect to lake geometry. Rewriting the scaling functions in this fashion makes it evident that Ryder's (1965) scaling contains a hidden production term ($\alpha = kM^1 T^{-1}$) not present in the other three scalings.

Linear regressions (on log-log scales) were acceptable for all the scaling functions, based on examination of the plot of residuals versus fitted values from least squares and RMA regression. The RMA estimates of the scaling functions were as follows:

$$V = \alpha \cdot A^{1.343} \qquad \alpha = e^{-4.4435} \, m^3 \, (m^2)^{-1.343} \qquad \hat{\beta} = -0.16 \quad \textbf{(EQ 4•17d)}$$

$$\dot{M} = \alpha \cdot (TDS/\bar{z})^{0.522} \; \alpha = e^{0.1866} \, kg \, yr^{-1} \, (m^{-1})^{-0.522} \quad \hat{\beta} = -0.48 \quad \textbf{(EQ 4•18d)}$$

$$\dot{M} = \alpha \cdot A^{0.863} \qquad \alpha = e^{1.729} \, kg \, yr^{-1} \, m^{-1} \, (m^2)^{-0.863} \, \hat{\beta} = 0.73 \quad \textbf{(EQ 4•19d)}$$

$$\dot{M} = \alpha \cdot A^{0.863} \qquad \alpha = e^{1.729} \, kg \, yr^{-1} \, m^{-2} \, (m^2)^{-0.863} \, \hat{\beta} = -0.14 \quad \textbf{(EQ 4•20d)}$$

$$\dot{M} = \alpha \cdot V^{0.642} \qquad \alpha = e^{4.583} \, kg \, yr^{-1} \, m^{-2} \, (m^3)^{-0.642} \, \hat{\beta} = 0 \quad \textbf{(EQ 4•21d)}$$

The RMA estimates of β differ little from the least squares estimates, which were 1.32, 0.4451, 0.84, 0.84, and 0.60 respectively for the exponents (Schneider and Haedrich 1989). In the third scaling relation the allometric exponent β was computed from the regression estimate as follows: $(1/2)(\hat{\beta}+1) = 0.863$, hence $\ddot{\beta} = 0.73$. In the scaling of catch to lake volume, the allometric exponent β was computed as $(2/3)(\hat{\beta} +1) = 0.642$, from which $\hat{\beta} = -0.037$, or $\beta = 0$ to the nearest tenth.

The first scaling indicates that the lakes used by Ryder (1965) differ in geometry from those used by Fee (1979). Ryder's (1965) lakes are flatter objects whose volume does not increase in Euclidean proportion to area. A doubling in area increases volume by $2^{1.343} = 2.54$, smaller than the increase of $2^{1.5} = 2.83$ expected in Euclidean lakes. Fee's (1979) lakes are rounder objects where a doubling in area increase volume by $2^{1.678} = 3.2$ rather than the Euclidean factor of 2.83. This difference in geometry may be due to the small size range of lakes used by Fee (1979) compared with Ryder (1965).

The scaling according to flux out of the benthos into a volume of water was completely similar with lake volume, despite the incomplete similarity of lake volume with area of the top surface of the water. This analysis is less than completely convincing because an observed scaling relation $\dot{M} \cong A^{0.84}$ was used as a guide in developing the scaling with volume. The scaling function was tested against a second data set, of annual fish catches from lakes and reservoirs in Missouri and Iowa (Jones and Hoyer 1982). The least squares estimate of the scaling exponent for this data set, $V^{0.77}$, did not differ significantly from the theoretical scaling of $V^{2/3}$ (t = 0.77, df = 21, p = 0.22). The theoretical scaling held up when tested against new data, not used in developing the theoretical scaling. This scaling, because it was tested against new data, could reasonably be used to predict changes in fish catch due to changes in reservoir volume at any mid-latitude location, not just Canadian shield lakes or in the vicinity of Missouri and Iowa.

APPLICATION: BIOMASS ACCUMULATION IN MESOCOSMS

Experimental mesocosms are widely used to investigate causal mechanisms under controlled conditions. One important limitation on their use is biomass accumulation on walls (Eppley et al. 1978). This is expected to scale with wall area, and hence increase in severity as wall area increases relative to water volume in tanks that are small, or columnar in shape. In the absence of container shape effects, the expected scalings are that the wall biomass total (B_{wall}, in units of mg chl a tank^{-1}) will scale with wall area,

$$B_{wall} \cong A_{wall} \qquad \text{(EQ 4•22)}$$

that biomass per unit area (B_{wall}/A_{wall} = mg chl a m^{-2}) will rise to a maximum value (B_{max}, in units of mass per unit area) that is independent of wall area,

$$B_{wall}/A_{wall} \cong B_{max} \cdot A_{wall}^{0} \qquad \text{(EQ 4•23)}$$

or that biomass per unit volume (B_{wall}/V) will rise to a maximum fixed by exchange rates (B_{exch}, in units of mass per unit time and unit area) between tank walls and the water. Consequently, biomass per unit volume will scale inversely with the surface to volume ratio, which in turn is proportional to the inverse of the tank radius,

$$B_{wall}/V_{tank} \cong B_{exch} \cdot (A_{wall}/V_{tank}) \cong B_{exch} \cdot r^{-1} \qquad \text{(EQ 4•24)}$$

where $r = V_{tank}/A_{wall}$ (r, in m^1). Chen et al. (1997) tested this in tanks of different size and shape. The scaling relations for completely similar tanks are

$$V_{tank} \cong A_{base} \cdot \bar{z} \cong (L^2)^1 (L^2)^{1/2} = (L^2)^{3/2} \qquad \text{(EQ 4•25a)}$$

$$V_{tank} \cong A_{wall} \cdot r/2 \cong (L^2)^1 (L^2)^{1/2} = (L^2)^{3/2} \qquad \text{(EQ 4•26a)}$$

$$B_{wall}/A_{wall} \cong B_{max} \cdot A_{wall}^0 \cong L^0 \qquad \text{(EQ 4•27a)}$$

$$B_{wall}/V_{tank} \cong B_{exch} \cdot (A_{wall}/V_{tank})^1 \cong L^0 \qquad \text{(EQ 4•28a)}$$

The corresponding scaling functions, including factors allowing incomplete similarity, are

$$V_{tank} = \alpha \cdot A_{base}^{\beta} \cdot A_{base}^{1.5} \qquad k = L^3 \cdot (L^2)^{-1.5} \cdot (L^2)^{-\beta} \quad \alpha = k \qquad \text{(EQ 4•25b)}$$

$$V_{tank} = \alpha \cdot A_{wall}^{\beta} \cdot A_{wall}^{1.5} \qquad k = L^3 \cdot (L^2)^{-1.5} \cdot (L^2)^{-\beta} \quad \alpha = k \qquad \text{(EQ 4•26b)}$$

$$B_{wall}/A_{wall} = \alpha \cdot B_{max} \cdot A_{wall}^{\beta} \qquad k = L^{-2} \cdot (L^{-2})^{-1} (L^2)^{-\beta} \quad \alpha = k \, M^1 \quad \text{(EQ 4•27b)}$$

$$B_{wall}/V_{tank} = \alpha \cdot B_{exch} \cdot (A_{wall}/V_{tank})^{1+\beta} \quad k = L^{-3} \cdot (L^2)^{-1} (L^{-1})^{-1-\beta} \; \alpha = k \, M^1 \; \text{(EQ 4•28b)}$$

The scaling parameters were estimated from data reported by Chen et al. (1997, table 1; figure 5a). Least squares regression was used because the tank geometry was fixed by the experiment, rather than being measured with error, as was the case with lake geometry. Straight line regression on a log-log scale was acceptable, judging from plots of residual versus fitted values. The parameter estimates for the scaling functions were

$$V = \alpha \cdot A_{base}^{1.176} \qquad \alpha = e^{-0.0009} m^3 (m^2)^{-1.176} \quad \hat{\beta} = -0.324 \text{(EQ 4•25c)}$$

$$V = \alpha \cdot A_{wall}^{1.681} \qquad \alpha = e^{-2.1272} m^3 (m^2)^{-1.681} \quad \hat{\beta} = 0.181 \quad \text{(EQ 4•26c)}$$

$$B_{wall}/A_{wall} = \alpha \cdot A_{wall}^{0.3378} \qquad \alpha = e^{-2.24} m^{-2} (m^2)^{-0.3378} mg \, Chl\underline{a} \; \hat{\beta} = 0.34 \quad \text{(EQ 4•27c)}$$

$$B_{wall}/V_{tank} = \alpha \cdot (A_{wall}/V_{tank})^{0.5618} \alpha = e^{3.20} m^{-3} (m^{-1})^{-0.5618} mg \, Chl\underline{a} \; \hat{\beta} = -0.44 \quad \text{(EQ 4•28c)}$$

The first two scaling exponents ($\hat{\beta} = -0.32$, $\hat{\beta} = 0.18$) quantify the incompletely similar geometry of the mesocosms, due to the differing shapes of the five types of tanks, for which the ratio of radius to height ranged from 0.17 (series A tanks) to 1.78 (series E tanks). These tanks were flatter than Ryder's (1965) lakes; they were far flatter than Fee's (1979) experimental lakes. For comparison, a doubling in surface area (same as base area) increases volume by $2^{1.176} = 2.3$, compared to factors of 2.5 (Ryder's lakes), 2.8 (completely similar tanks or lakes), and 3.2 (Fee's lakes).

The third scaling exponent ($\hat{\beta} = 0.34$) quantifies the incomplete scaling of algal biomass with wall area in tanks that vary in height and shape. Doubling wall area does not double the wall biomass. Rather, a doubling in wall area results in a $2^{1.34} = 2.53$ times the algal biomass in that entire tank, or $2^{0.34} = 1.27$ times greater algal biomass per unit area. Wall biomass increases with increasing wall area, but not in the simple one-to-one fashion expected from complete similarity. The fourth scaling exponent ($\hat{\beta} = -0.44$) quantifies the incomplete similarity between biomass per unit volume of water, and the surface to volume ratio of tanks that differ in height and shape. Biomass per unit volume was inversely related to the surface-volume ratio of the tank, rather than directly proportional as expected from complete similarity.

The scaling functions relating wall biomass to wall area and to surface-volume ratio, like the scaling of fish catch with lake surface area, are empirical rather than having been developed from theory. Three theoretical scaling were then examined:

$$B_{wall} \cong tFlux_{NPS} = Flux_{NPS}{\cdot}A_{base} \qquad \text{(EQ 4•29a)}$$

$$B_{wall} \cong tFlux_{light} = zPAR{\cdot}A_{top} \qquad \text{(EQ 4•30a)}$$

$$B_{wall} \cong tFlux_{NPS} = Flux_{NPS}{\cdot}V_{tank} \qquad \text{(EQ 4•31a)}$$

The first scaling was that the wall biomass total (B_{wall}, in mg) depends on vertical flux of nutrients per unit area ($Flux_{NPS}$, in mg m^{-2} s^{-1}) out of the sediment, and hence depends on benthic area (A_{base}, in m^2). The next scaling was that biomass accumulation depends on total light delivery, which in turn depends on surface area at the top of the tank (A_{top}, m^2) and on tank depth (z, in m^1). The vertically integrated light delivery ($zPAR$, in μE m^{-1} s^{-1}) will depend on photosynthetically active radiation (PAR) flux at the surface (PAR_0, in μE m^{-2} s^{-1}) and on the attenuation coefficient (kd, in % m^{-1}):

$$zPAR = \int_0^z Par_0\, e^{-kd{\cdot}z} = Par_0{\cdot}kd^{-1}{\cdot}(1 - e^{-kd{\cdot}z}) \qquad \text{(EQ 4•32)}$$

The third scaling was that the total wall biomass depends on the flux between tank surfaces and water volume (V_{tank}, in m^3).

The three scaling relations were then written as scaling functions based on complete similarity:

$$B_{wall} = \alpha \cdot A_{base} \qquad \text{(EQ 4•29b)}$$

$$B_{wall} = \alpha \cdot PAR_0 \cdot kd^{-1} \cdot (1 - e^{-kd \cdot z}) \cdot A_{top} \qquad \text{(EQ 4•30b)}$$

$$B_{wall} = \alpha \cdot (V_{tank})^{2/3} \qquad \text{(EQ 4•31b)}$$

In the second scaling function, incident light PAR_0 varied by a factor of $(182/107) = 1.7$ among tanks, light attenuation kd varied by a factor of $(1.35/0.57) = 24$, and percent of available light absorbed $(1 - e^{-kd \cdot z})$ varied by a factor of $(0.74/0.38) = 1.9$, based on computations from data in table 1 of Petersen et al. (1997). Incorporating these factors into the scaling parameter α was not warranted. Instead, these three quantities were used to compute the wall biomass scaled to incident light, to attenuation of light, and to percent of total light absorbed:

$$B_{wall}/tL = B_{wall} \cdot (PAR_0 \cdot kd^{-1} \cdot (1 - e^{-kd \cdot z}))^{-1} = \alpha \cdot A_{top} \qquad \text{(EQ 4•33)}$$

The three scaling functions were then rewritten to include factors that permit incomplete similarity:

$$B_{wall} = \alpha \cdot A_{base}^{1} \cdot A_{base}^{\beta} \qquad k = L^0 \cdot (L^2)^{-1-\beta} \qquad \alpha = kM^1 \qquad \text{(EQ 4•29c)}$$

$$B_{wall}/tL = \alpha \cdot A_{top}^{1} \cdot A_{top}^{\beta} \qquad k = L^{-1} \cdot (L^2)^{-1-\beta} \qquad \alpha = kE^{-1}M^1T^{-1} \ \text{(EQ 4•30c)}$$

$$B_{wall} = \alpha \cdot V_{tank}^{2/3} \cdot (V_{tank}^{2/3})^{\beta} \quad k = ((L^3)^{2/3})^{-1-\beta} \quad \alpha = kM^1 \qquad \text{(EQ 4•31c)}$$

The units for tL^{-1} are $(m^{-2}(m^{-1})^{-1}L^0)^{-1} = m^1$ and hence the geometric factor k for the second function includes L^{-1}. The coefficient α includes dimensions for mass (M), light (E), time (T), and geometry (L^1), (L^2), and (L^3).

These scaling functions were evaluated by least squares regression; the linearity assumption (on a log-log plot) was judged acceptable from the plot of residuals versus fitted values. The parameter estimates were

$$B_{wall} = \alpha \cdot A_{base}^{0.8593} \quad \alpha = e^{3.3924} \ mg \ (m^2)^{-0.8593} \qquad \hat{\beta} = 0.14 \ \text{(EQ 4•29d)}$$

$$B_{wall}/tL = \alpha \cdot A_{top}^{-0.26} \quad \alpha = e^{-1.4762} \ mg \ \mu mol \ s^{-1}m^{-1}(m^2)^{0.26} \hat{\beta} = -1.26 \text{(EQ 4•30d)}$$

$$B_{wall} = \alpha \cdot V_{tank}^{0.7747} \quad \alpha = e^{3.9334} \, mg \, (m^3)^{-0.7747} \quad \hat{\beta} = 0.16 \text{ (EQ 4•31d)}$$

The scaling exponent β for the first and second scaling functions were computed by setting the regression estimate equal to $\beta +1$. The scaling exponent β for the third scaling functions was estimated by setting the regression estimate equal to $(2/3)(1+\beta)$, then solving for β. The residual versus fit plots were judged acceptable for the first and third functions, and questionable for the second function due to the influence of tank type E, which had the largest radius. The estimate of the parameters for the second scaling function was thus not considered reliable. The departure from complete similarity was about the same for the first and third scalings. The conclusion from this departure is that some other factor needs to be considered in developing scaling relations for wall biomass in experimental mesocosms. The failure of wall area to keep pace with tank volume in large tanks appears to be important because the degree of incomplete similarity ($\hat{\beta} = 0.14$, Eq 29d) relative to tank base area and to tank volume ($\hat{\beta} = 0.16$, Eq 31d) resembles that for wall area relative to volume ($\hat{\beta} = 0.18$, Eq 26c). Shading effects, which increase with production, may be important in limiting biomass buildup in larger and deeper tanks.

APPLICATION: PRIMARY PRODUCTION IN MESOCOSMS

It was then of interest to examine a scaling relation for primary production. The total primary production (tGP, in units of g $O_2 h^{-1} tank^{-1}$) was calculated from the data in figure 5b in Chen et al. (1997) as $tGP = GPP \cdot A_{wall} = (g \, O_2 m^{-2} h^{-1})(m^2/tank)$. This includes production in the water and on the wall. Several scaling relations could have been examined. The one chosen for analysis was for production in relation to the volume of the system, with total mass flux $tFlux$ taken as the limiting constant:

$$tGP \cong tFlux = Flux \cdot V_{tank}^{2/3} \tag{EQ 4•34a}$$

The scaling function, with allowance for incomplete similarity, is

$$tGP = \alpha \cdot V_{tank}^{2/3} \cdot (V_{tank}^{2/3})^{\beta} \quad k = L^0 \cdot (L^3)^{2/3})^{-1-\beta} \quad \alpha = kM^1 T^{-1} \tag{EQ 4•34b}$$

This was tested for both summer and fall data, using 10 observations taken from figure 5b in Chen et al. (1997). Analysis of covariance was used to test for uniform scaling across the two seasons:

$$ln(tGP) = \beta_0 + \beta_V \cdot ln(V_{tank}) + \beta_{ssn} \cdot ssn + \beta_{V*ssn} \cdot ln(V_{tank}) \cdot ssn + \varepsilon \quad \text{(EQ 4•35a)}$$

$$SS: 26.3315 = \quad + 25.1502 \quad + 0.0714 \quad + 0.0153 \quad + 1.0945 \text{ (EQ 4•35b)}$$

Season is a categorical variable ($ssn = 0$ = summer, $ssn = 1$ = fall). The sums of squares (partitioned sequentially in the order listed) are shown beneath each term in the model. The interaction variance is small relative to the total SS or the error SS, and hence the summer and fall scaling exponents can be considered as homogeneous. The scaling of total primary production (wall + water column) with tank volume, as estimated by least squares regression (both seasons combined) is

$$tGP = \alpha \ V_{tank}^{0.764} \quad \alpha = e^{-2.325} g \ O_2 \ hr^{-1} (m^3)^{-0.764} \quad \hat{\beta} = 0.15 \quad \text{(EQ 4•34c)}$$

The observed scaling exponent of $V^{0.764}$ was close to the expected scaling of $V^{2/3}$, with the degree of divergence ($\hat{\beta} = 0.15$) from complete similarity being nearly the same as that for the scaling of wall area relative to tank volume ($\hat{\beta} = 0.18$).

SCALING THEORY: SPATIAL ALLOMETRY FOR ANTAGONISTIC RATES

Incomplete similarity and power laws are to be expected in systems with incompletely similar (fractal) geometry. But how is it that incomplete similarity and power laws arise in Euclidean tanks? And how is it that incompletely similar structures such as lungs, lakes, or landscapes arise in the first place? According to scaling theory (Wilson 1971), power law behavior appears when exponential rates, acting antagonistically, drive a system toward a critical point where one rate episodically reverses another. The best publicized example is the Bak sandpile (Bak et al. 1988) in which grains drop from a point, piling up until potential energy gained at a regular rate is released spasmodically as the kinetic energy of avalanches that are usually small, sometimes large, and occasionally huge. Scaling theory applies to an astonishing variety of animate and inanimate systems—from heartbeat intervals and lung inflation to gaps in rainforests, earthquakes, crystal growth, watershed structure. These are not applications by analogy. These are demonstrations of power law behavior for systems that meet the conditions where scaling theory applies: exponential rates acting in an opposite but not simultaneously equal fashion.

The space (or time) scale at which two antagonistic rates are of the same order of magnitude is called the *critical scale*. Classical dimensional analysis derives this as a single fixed number (e.g., Skellam 1951; Kierstead and Slobodkin 1953) but recent work (Schneider et al. 1997) shows that the ratio of two antagonistic rates, expressed as a ratio, changes with scale. In order to identify critical scales and which of several concomitant processes prevail at any given space and time scale, the following procedure was developed (Horne and Schneider 1994; Schneider et al. 1997):

1. State the quantity of interest;

2. Derive a conservation equation to identify component rates;

3. Obtain a scaling function for each rate, then combine these to identify which rate prevails at any scale of interest;

4. If scaling functions are not available, renormalize the terms in the equation to each other, forming all possible ratios. Choose pertinent ratios, compute them at multiple space and time scales using available data, and plot them with contour lines separating space and time scales where numerator processes prevail over denominator processes, or vice versa.

These plots are then used to identify the space and time scales at which one rate prevails over another, as well as the space and time scales at which critical phenomena, incomplete scaling, and power law behavior are expected.

APPLICATION: ADULT-JUVENILE INTERACTIONS IN BENTHIC COMMUNITIES

For benthic populations inhabiting sand and mud, Thorson (1966) proposed-density dependent interaction due to inhibition of juvenile recruitment by adults. Experimental studies began a decade later, when several mechanisms were proposed (Young and Rhoads 1971; Woodin 1976). A meta-analysis of 54 experiments (Olafsson et al. 1994) showed that inhibition in bivalves occurred as often as tolerance, and that inhibition was infrequent unless densities exceeded ambient levels in the field. These experiments, carried out in the field and in laboratory tanks, were necessarily at small scales covering limited areas. The meta-analysis did not evaluate the role of spatial scale. This was investigated during an extended (six-week) workshop organized by the New Zealand National Institute of Water and Atmospheric Research (Thrush et al. 1997). Analysis of component rates, as described above, was carried out (Schneider et al. 1997) before undertaking a multiscale field experiment on the tellinid bivalve *Macomona liliana*.

One quantity of interest was rate of change in the density of adult *M. liliana* ($\dot{N} \equiv dN/dt$, in units of organisms m^{-2} day^{-1}). In writing the conservation equation for component rates, it was convenient to work with per capita rates $N^{-1}\dot{N}$:

$$N^{-1}\dot{N} = r + F \qquad \text{(EQ 4•36)}$$

where F is the lateral flux into or out of finite areas, while r is the net rate of change due to births and deaths:

$$r = b - d \qquad \text{(EQ 4•37)}$$

Both the demographic rate r and the kinematic rate F have units of % m^{-2} day^{-1}. The kinematic rate is partitionable into a passive component due to drift with the fluid (F_{fl}, in units of % m^{-2} day^{-1}) and active component due to locomotion (F_{loc}, in units of % m^{-2} day^{-1}). Scaling functions were not available for these rates, so pertinent ratios were formed and plotted. One such ratio was d/F_{loc} the ratio of mortality relative to lateral flux due to fossorial locomotion while foraging at low tide. Figure 4.4 shows the space and time scales at which mortality prevails ($d/F_{fl} > 1$) and the scales at which lateral movement prevails ($d/F_{fl} < 1$). The figure compares hand drawn contours (Schneider et al. 1997) to contours obtained by T. Bult using Monte Carlo techniques (cf. Schneider et al. 1999). Both the rough and the intensive computation show that experiments at typically small scales will pick up kinematic effects. An experiment at the scale of at least 100 m would be needed to detect a change in mortality, given the observed rates of movement in this species.

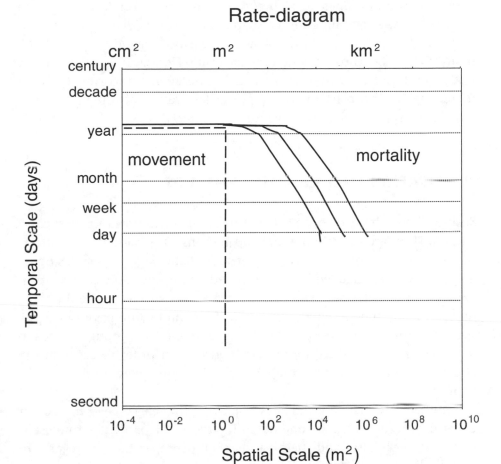

FIGURE 4•4 *Biological Effects*
Scales where mortality prevails over lateral movement ($d/F_{loc} > 1$) and where lateral movement prevails over mortality ($d/F_{loc} < 1$) for adult stages of the bivalve *Macomona liliana*. Dashed line from Schneider et al. (1997). Solid lines computed with Monte Carlo methods by T. Bult (cf. Schneider et al. 1999).

Figure 4.4 also depicts the spatial scale at which the exponential mortality and flux rates are the same order of magnitude. The left most solid contour is that for $d/F_{loc} = 10$, the middle contour is for $d/F_{loc} = 1$, and the right most contour is for $d/F_{loc} = 1/10$. At these scales, under the influence of antagonistic exponential rates of about the same magnitude, one expects the dynamics to give rise to complex structure, including spatial autocorrelations that decay according to a power law (Wilson 1971), rather than more quickly at exponential rates. For adult *M. liliana*, autocorrelation is expected to decay exponentially at spatial scales far from the contour lines in figure 4.4. Autocorrelation is expected to become a power law function of spatial scale in the vicinity of contour lines. Spatial autocorrelations of adult density (Legendre et al. 1997, figures 2a and 2e) show exponential decay at scales (lags) up to 180 m but not beyond. At lags greater than 180 m the decay in autocorrelation with increasing lag is consistent with a power law, reflecting the presence of larger scale spatial structure (Legendre et al. 1997).

APPLICATION TO MESOCOSM ANALYSIS

Scaling theory has not been applied to antagonistic rates in mesocosm studies. There are several potential applications. One is identifying effects that need to be controlled in an experiment. Rate diagrams such as figure 4.4 can be compared with the proposed space and time scales of an experiment to determine whether two rates are expected to be the same order of magnitude in the experiment. If a contour line passes near the space and time scales occupied by the experiment, then thought needs to be invested in designing appropriate controls, in order to separate the effects of the two rates that will be of roughly equal magnitude near the contour line.

Another potential application is displaying the relevance of mesocosm experiments to field experiments and surveys. Of particular interest is whether mesocosm and field studies are comparable in regard to pertinent ratios. Mesocosm and field studies are comparable if they lie on the same side of a contour line. Studies are not comparable if separated by a contour line, marking a reversal in the relative importance of two rates.

A third potential application of concomitant rate analysis is designing mesocosm studies of larger organisms, such as benthic macrofauna and predatory fish. Kinematics will become important for benthic organisms capable of lateral movement, and for any organism that can swim at speeds exceeding the local flow due to mixing in a tank. Swimming speed relative to local flow will be a pertinent ratio in any mesocosm study including nonsessile organisms.

A fourth potential application is identifying and verifying theoretical scalings. Scaling functions that are inconsistent with experimental results can be discarded. However, more than one scaling function may be consistent with an experimental result. When this happens, the ratio of pertinent rates at the scale of the experiment may enable a scaling function to be discarded. Alternatively, plots of pertinent ratios can be used to suggest a new experiment that is capable of rejecting one scaling in favor of another.

Incomplete similarity is merely a descriptive technique unless we can specify the conditions leading to a power law relation between the rate Q and some other quantity Y. Scaling theory states two conditions leading to power law relations. Incomplete similarity results if the quantity Y is self similar (fractal) rather than Euclidean (Schneider 1994). Incomplete similarity also results if the rate Q is the outcome of exponential production and survival rates:

$$ln(Q_{produced}/Q_o) = r \cdot t \qquad \text{(EQ 4•38)}$$

$$ln(Q_{surviving}/Q_o) = -z \cdot t \qquad \text{(EQ 4•39)}$$

hence

$$Q_{produced}/Q_o = (Q_{surviving}/Q_o)^{-r/z} \qquad \text{(EQ 4•40)}$$

This power law behavior is expected at time and space scales where production rate r and loss rate z are of about the same magnitude. Scaling theory provides the apparatus for developing testable hypotheses concerning the appearance of power law behavior in natural systems as well as experimental systems such as mesocosms.

COMMENTS

Scope diagrams (figure 4.1) make it evident that spatial scaling is the major challenge in connecting mesocosm studies to target questions. The degree of temporal scaling is small, with scopes of the order of 10^1, compared to the degree of spatial scale-up, where scopes are of the order of 10^4. How might scaling across this gap be accomplished? Increasing mesocosm size is expensive, and unlikely to narrow the gap by much. Other types of research are available, including field experiments, surveys, and models (figure 4.3). Surveys can bridge the gap with their large spatial and temporal scopes, but suffer from uncontrolled sources of variation and inability to separate concomitant effects. Field experiments can only partially bridge the gap. They cover larger areas than mesocosm studies, but still typically fall short of the spatial scale of target questions (figures 4.1, 4.2). Model studies easily incorporate the scale of target questions, but because of limitations on computer capacity, cannot be extended to spatial resolutions that overlap mesocosm studies. A grid-scale on the order of 10 m or less is hardly feasible, even with current computing power. Temporal resolution is readily extended down to days or even hours, making it easier for model studies to link with field experiments than with mesocosm studies (figure 4.2). Graphic display of the problem suggests mesocosm studies are best linked to field experiments, rather than bridging directly to target questions. The overlap in time scales and the ability to control variables suggest that mesocosm studies be verified against field experiments, which can then be linked to target questions via surveys (Legendre et al. 1997; Thrush et al. 1997) and models (Root and Schneider 1995; McArdle et al. 1997).

Mesocosm studies could potentially be linked to target questions via scaling functions. However, such functions would need to bridge four orders of magnitude, likely crossing scales where concomitant rates reverse in their importance. Any scaling function developed from a mesocosm study would need to be tested against a survey, the only source of data at large scales.

The scale-down paradigm could also be used to link larger scale target questions to mesocosm studies. Scaling functions developed from global or regional data could be tested across a range of mesocosm sizes. Two different spatial scalings have been developed for fish catch, one at a global scale (catch per unit area C/A, in kg ha^{-1} yr^{-1} $\cong$ Area^0, though with large scatter, Nixon 1988), and another at a regional scales for lakes (C, in units of Mg yr^{-1} $\cong$ Volume$^{2/3}$) in lakes with fractal scalings of surface to volume (Schneider and Haedrich 1989). It would be interesting to develop these and other predictions for mesocosms, then test them.

Mesocosm studies can be used to verify rates observed at larger scales in natural systems. Nixon (1997) scaled primary production (*PrP*, in units of g-C m^{-2} yr^{-1}) to dissolved inorganic nitrogen (*DIN*, in units of mol m^{-2} yr^{-1}) in 9 mesocosm experiments and 10 natural systems. The resulting scaling was

$$PrP = 10^{2.332} \, DIN^{0.442} \qquad \text{(EQ 4•41)}$$

The spatial scope of this scaling function is of the order of 10^6 km^2 / 13 m^2 = 10^{12}, the ratio of the area of the shelf ecosystems in this study (e.g.,

Sargasso Sea) to the area of the tank. This relation contains a fractional mass exponent ($mol^{0.442}$) and it is empirical, hence it would need to be applied with caution. It would be interesting to re-analyze this relation as a spatially allometric function to obtain a more generally applicable scaling relation.

Mesocosm studies might also be linked to target questions at larger scales by distorting ratios of pertinent rates so that ratios in the mesocosm match those at larger scales. This is the engineering approach (Taylor 1974) that has been used successfully to scale from laboratory models to reliable prototypes of Euclidean structures (bridges, highways, ships, etc.) where dynamical equilibrium is a necessity. In natural systems scaling theory (equations 8b, 38, 39, and 40) should prove more useful than the restrictive apparatus of dimensional analysis, which cannot handle non-Euclidean shapes, non-equilibrium dynamics, and dimensions such as contact rates, which are required for biological systems.

The principle of similitude is the common origin of body size allometry, scaling functions, and dimensional analysis. There are several impediments to adoption of the principle in ecology. The restrictive apparatus of dimensional analysis is only one. Another is the habit, surprisingly common in ecology (Schneider 1994), of stripping away units, reducing scaled quantities to unscaled numbers. There can be little hope of adequate quantitative work, and no hope of successfully incorporating scale, as long as this habit remains widespread. A contributing factor is the nearly universal practice of omitting units and dimensions from statistical analyses, making them inappropriate for multiscale analysis.

One final impediment is perhaps best illustrated by a brief history of the theoretical scaling function for fish catch. The impetus for this study came from a practical question, that of how to calculate the expected increase in fish yield by enlarging a lake to a reservoir. The initial approach was to search the Ryder (1965) data set for the best empirical function, with maximum explained variance and fewest explanatory variables. No clear result emerged from repeated analysis, executed over several months. An impending deadline forced a radical revision in method. Once I adopted the predictive method, a clear result emerged after two days of work. It was hard work developing scaling functions from concepts, but required less time than developing empirical functions. One can always slog through to the "best" empirical function.

But it continues to be a pleasing shock, accompanied by a sense of discovery, when a theoretical scaling function survives a test against data.

ACKNOWLEDGMENTS

Funding for research on scaling theory was provided through a continuing grant (since 1986) from the Natural Sciences and Engineering Research Council (Ottawa). The manuscript was improved substantially by comments from J. Petersen, J. Wiens, C.-C. Chen, M. Lewis, V. Kennedy, and an anonymous reviewer. I thank the organizers of the workshop for arranging travel to Maryland to participate.

LITERATURE CITED

Allen, T. F. H., and T. W. Hoekstra. 1992. *Toward a Unified Ecology*. New York: Columbia University Press.

Allen, T. F. H., and T. B. Starr. 1982. *Hierarchy*. Chicago: University of Chicago Press.

Bak, P., C. Tang, and K. Weisenfeld. 1988. Self organized criticality. *Physical Reviews* A 38:364–374.

Barenblatt, G. I. 1996. *Scaling, Self-similarity, and Intermediate Asymptotics*. Cambridge: Cambridge University Press.

Bridgman, P. W. 1922. *Dimensional Analysis*. New Haven: Yale University Press.

Calder, W. A. 1984. *Size, Function, and Life History*. Cambridge, Mass.: Harvard University Press.

Carpenter, S. R. 1990. Large-scale perturbations: Opportunities for innovation. *Ecology* 71:2038–2043.

Carpenter, S. R., S. W. Chisholm, C. J. Krebs, D. W. Schindler, and R. F. Wright. 1995. Ecosystem experiments. *Science* 269:324–327.

Chen, C.-C., J. E. Petersen, and W. M. Kemp. 1997. Spatial and temporal scaling of periphyton growth on walls of estuarine mesocosms. *Marine Ecology Progress Series* 155:1–15.

Chesapeake Bay Program. 1997. *Chesapeake Bay Wetlands: The Vital Link Between the Watershed and the Bay*. U.S. Environmental Protection Agency 903-R-97-002. Annapolis, Md.: Chesapeake Bay Program.

Dayton, P. D., and M. J. Tegner. 1984. The importance of scale in community ecology: A kelp forest example with terrestrial analogs. In P. W. Price, C. M. Slobodchikoff, and W. S. Gaud, eds., *A New Ecology: Novel Approaches to Interactive Systems*, pp. 457–481. New York: Wiley.

Dutilleul, P. 1993. Spatial heterogeneity and the design of ecological field experiments. *Ecology* 74:1646–1658.

Eberhardt, L. L. 1976. Quantitative ecology and impact assessment. *Journal of Environmental Management* 42:1–31.

Eberhardt, L. L., and J. M. Thomas. 1991. Designing environmental field studies. *Ecological Monographs* 61:53–73.

Eppley, R. W., P. Koeller, and G. T. Wallace. 1978. Stirring influences the phytoplankton species composition within enclosed columns of coastal seawater. *Journal of Experimental Marine Biology and Ecology* 32:219–239.

Fee, E. J. 1979. A relation between lake morphometry and primary productivity and its use in interpreting whole-lake eutrophication experiments. *Limnology and Oceanography* 24:401–416.

Gill, A. E. 1982. *Atmosphere-Ocean Dynamics*. New York: Academic Press.

Greig-Smith, P. 1952. The use of random and contiguous quadrats in the study of the structure of plant communities. *Annals of Botany* 16:293–316.

Hairston, N. G. 1989. *Ecological Experiments: Purpose, Design, and Execution*. Cambridge: Cambridge University Press.

Harris, G. P. 1980. Temporal and spatial scales in phytoplankton ecology. Mechanisms, methods, models, and management. *Canadian Journal of Fisheries and Aquatic Sciences* 37:877–900.

Haury, L. R., J. S. McGowan, and P. Wiebe. 1978. Patterns and processes in the time-space scales of plankton distributions. In J. Steele, ed., *Spatial Pattern in Plankton Communities*, pp. 277–327. New York: Plenum.

Horne, J. K., and D. C. Schneider. 1994. Analysis of scale-dependent processes with dimensionless ratios. *Oikos* 70:201–211.

Hurlbert, S. H. 1984. Pseudoreplication and the design of ecological field experiments. *Ecological Monographs* 54:187–211.

Jassby, A. D., and T. M. Powell. 1990. Detecting changes in ecological time series. *Ecology* 71:2044–2052.

Jones, J. R., and M. V. Hoyer. 1982. Sportfish harvest predicted by summer chlorophyll-*a* concentration in midwestern lakes and reservoirs. *Transactions of the American Fisheries Society* 111:176–179.

Kershaw, K. A. 1957. The use of cover and frequency in the detection of pattern in plant communities. *Ecology* 38:291–299.

Kierstead, H., and L. B. Slobodkin. 1953. The size of water masses containing plankton blooms. *Journal of Marine Research* 12:141–147.

Kline, S. J. 1965. *Similitude and Approximation Theory*. New York: McGraw-Hill.

Legendre, P., S. F. Thrush, V. J. Cummings, P. K. Dayton, J. Grant, J. E. Hewitt, A. H. Hines, B. H. McArdle, R. D. Pridmore, D. C. Schneider, S. J. Turner, R. B. Whitlatch, and M. R. Wilkinson. 1997. Spatial structure of bivalves in a sandflat: Scale and generating processes. *Journal of Experimental Marine Biology and Ecology* 216:99–128.

Levin, S. A. 1992. The problem of pattern and scale in ecology. *Ecology* 73:1943–1967.

Likens, G. E. 1985. An experimental approach for the study of ecosystems. *Journal of Ecology* 73:381–396.

Mackas, D. L., and C. M. Boyd. 1979. Spectral analysis of zooplankton spatial heterogeneity. *Science* 204:62–64.

MacMahon, T. A. 1973. Size and shape in biology. *Science* 179:1201–1204.

Mandelbrot, B. 1977. *Fractals: Form, Chance, and Dimension.* San Francisco: Freeman.

Mann, K. H., and J. R. N. Lazier. 1991. *Dynamics of Marine Ecosystems.* Boston: Blackwell Science.

May, R. N. 1991. The role of ecological theory in planning re-introduction of endangered species. *Proceedings of the Zoological Society of London* 62:145–163.

McArdle, B. H., J. E. Hewitt, and S. F. Thrush. 1997. Pattern from process: It is not as easy as it looks. *Journal of Experimental Marine Biology and Ecology* 216:229–242.

Mercer, W. B., and A. D. Hall. 1911. The experimental error of field trials. *Journal of Agricultural Science* 4:107–132.

Nixon, S. M. 1988. Physical energy inputs and the comparative ecology of lake and marine ecosystems. *Limnology and Oceanography* 33:1005–1025.

Nixon, S. M. 1997. Prehistoric nutrient inputs and productivity in Narragansett Bay. *Estuaries* 20:253–261.

Officer, C. B., R. B. Biggs, J. L. Taft, L. E. Cronin, M. A. Taylor, and W. R. Boynton. 1984. Chesapeake Bay anoxia: Origin, development, and significance. *Science* 223:22–27.

Olafsson, E. B., C. H. Peterson, and W. G. Ambrose. 1994. Does recruitment limitation structure populations and communities of macro-invertebrates in marine soft sediments: The relative significance of pre- and post-settlement processes. *Oceanography and Marine Biology Annual Review* 32:65–109.

O'Neill, R. V., D. L. DeAngelis, J. B. Waide, and T. F. H. Allen. 1986. *A Hierarchical Concept of Ecosystems.* Princeton: Princeton University Press.

O'Neill, R. V., and A. W. King. 1998. Homage to St. Michael; or, why are there so many books on scale? In D. L. Peterson and V. T. Parker, eds., *Ecological Scale: Theory and Applications,* pp. 3–15. New York: Columbia University Press.

O'Neill R. V., and B. Rust. 1979. Aggregation error in ecological models. *Ecological Modeling* 7:91–105.

Pennycuick, C. J. 1992. *Newton Rules Biology.* Cambridge: Cambridge University Press.

Petersen, J., C.-C. Chen, and W. M. Kemp. 1997. Scaling aquatic primary productivity: Experiments under nutrient- and light-limited conditions. *Limnology and Oceanography* 78:2326–2338.

Platt, T. 1986. Primary production of the ocean water columns as a function of surface light intensity: Algorithms for remote sensing. *Deep-Sea Research* 33:149–163.

Platt, T. R., and K. L. Denman. 1975. Spectral analysis in ecology. *Annual Review of Ecology and Systematics* 6:189–210.

Rastetter, E. B., A. W. King, B. J. Cosby, G. M. Hornberger, R. V. O'Neill, and J. E. Hobbie. 1992. Aggregating fine-scale ecological knowledge to model coarser-scale attributes of ecosystems. *Ecological Applications* 2:55–70.

Reckhow, K. H. 1990. Bayesian inference in non-replicated ecological studies. *Ecology* 71:2053–2059.

Resetarits, W.J., and J. Bernardo (eds.). 1998. *Experimental Ecology Issues and Perspectives*. New York: Oxford University Press.

Ricklefs, R. E. 1987. Community diversity: Relative roles of local and regional processes. *Science* 235:167–171.

Ricklefs, R. E. 1990. Scaling pattern and process in marine ecosystems. In K. Sherman, L. M. Alexander, and B. D. Gold, eds., *Large Marine Ecosystems*, pp. 169–178. Washington, D.C.: American Association for the Advancement of Science.

Ricklefs, R. E., and D. Schluter. 1993. Species diversity: Regional and historical influences. In R. E. Ricklefs and D. Schluter, eds., *Species Diversity in Ecological Communities*, pp. 350–363. Chicago: University of Chicago Press.

Root, T. S, and S. H. Schneider. 1995. Ecology and climate: Research strategies and implications. *Science* 269:334–341.

Rose, G. A., and W. C. Leggett. 1991. The importance of scale to predator-prey correlations: An example of Atlantic fishes. *Ecology* 71:33–43.

Ryder, R. A. 1965. A method for estimating the potential fish production of north-temperate lakes. *Transactions of the American Fisheries Society* 94:214–218.

Schindler, D. W. 1987. Detecting ecosystem response to anthropogenic stress. *Canadian Journal of Fisheries and Aquatic Sciences* 44:6–25.

Schmidt-Nielsen, K. 1984. *Scaling: Why Is Animal Size So Important?* Cambridge: Cambridge University Press.

Schneider, D. C. 1978. Equalisation of prey numbers by migratory shorebirds. *Nature* 271:353–354.

Schneider, D. C. 1994. *Quantitative Ecology: Spatial and Temporal Scaling*. Orlando, Fla.: Academic Press.

Schneider, D. C. 1998. Applied scaling theory. In D. L. Peterson and V. T. Parker, eds., *Ecological Scale: Theory and Applications*, pp. 253–269. New York: Columbia University Press.

Schneider, D. C., T. Bult, R. S. Gregory, D. A. Methven, D. W. Ings, and V. Gotceitas. 1999. Mortality, movement, and body size: Critical scales for Atlantic cod *Gadus morhua* in the Northwest Atlantic. *Canadian Journal of Fisheries and Aquatic Sciences* 56 (Suppl. 1):180–187.

Schneider, D. C., J.-M. Gagnon, and K. D. Gilkinson. 1987. Patchiness of epibenthic megafauna on the outer Grand Banks of Newfoundland. *Marine Ecology Progress Series* 39:1–13.

Schneider, D. C., and R. L. Haedrich. 1989. Prediction limits of allometric equations: A reanalysis of Ryder's morphoedaphic index. *Canadian Journal of Fisheries and Aquatic Sciences* 46:503–508.

Schneider, D. C., and J. F. Piatt. 1986. Scale-dependent correlation of seabirds with schooling fish in a coastal ecosystem. *Marine Ecology Progress Series* 32:237–246.

Schneider, D. C., R. A. Walters, S. F. Thrush, and P. K. Dayton. 1997. Scale-up of ecological experiments: Density variation in the mobile bivalve *Macomona liliana*. *Journal of Experimental Marine Biology and Ecology* 216:129–152.

Shugart, H., ed. 1978. *Time Series and Ecological Processes*. Philadelphia: Society for Industrial and Applied Mathematics.

Skellam, J. G. 1951. Random dispersal in theoretical populations. *Biometrika* 78:196–218.

Smith, P. E. 1978. Biological effects of ocean variability: Time and space scales of biological response. *Journal du Conseil International pour l'Exploration du Mer* 173:117–127.

Sokal, R. R, and F. J. Rohlf. 1995. *Biometry*. 3rd ed. San Francisco: Freeman.

Steele, J. H. 1978. Some comments on plankton patches. In J. H. Steele, ed., *Spatial Pattern in Plankton Communities*, pp. 11–20. New York: Plenum.

Steele, J. H. 1991. Can ecological theory cross the land-sea boundary? *Journal of Theoretical Biology* 153:425–436.

Sutor, G. W. 1996. Abuse of hypothesis testing statistics in ecological risk assessment. *Human and Ecological Risk Assessment* 2:331–347.

Taylor, E. S. 1974. *Dimensional Analysis for Engineers*. Oxford: Clarendon Press.

Thompson, D. W. 1917. *On Growth and Form*. Cambridge: Cambridge University Press (abridged version edited by J. T. Bonner, reissued in 1961).

Thorson, G. 1966. Some factors influencing the recruitment and establishment of marine benthic communities. *Netherlands Journal of Sea Research* 3:267–293.

Thrush, S. F., R. D. Pridmore, V. J. Cummings, P. K. Dayton, R. Ford, J. Grant, J. E. Hewitt, A. H. Hines, S. M. Lawrie, B. H. McArdle, D. C. Schneider, S. J. Turner, R. B. Whitlatch, and M. R. Wilkinson. 1997. Matching the outcome of small-scale density manipulation experiments with larger scale patterns: An example of bivalve adult/juvenile interactions. *Journal of Experimental Marine Biology and Ecology* 216:153–169.

Turner, M. G., and R. H. Gardner, eds. 1991. *Quantitative Methods in Landscape Ecology*. Berlin: Springer-Verlag.

Uhlmann, D. 1985. Scaling of microcosms and the dimensional analysis of lakes. *Internationale Revue der Gesamten Hydrobiologie* 70:47–62.

Underwood, A. J. 1997. *Experiments in Ecology*. Cambridge: Cambridge University Press.

Valentine, J. W. 1973. *The Evolutionary Paleoecology of the Marine Biosphere*. New York: Prentice-Hall.

Weber, L. H., S. Z. El-Sayed, and I. Hampton. 1986. The variance spectra of phytoplankton, krill, and water temperature in the Antarctic Ocean south of Africa. *Deep-Sea Research* 33:1327–1343.

Welsh, A. H., A. T. Peterson, and S. A. Altmann. 1988. The fallacy of averages. *American Naturalist* 132:277–288.

Wessman, C. A. 1992. Spatial scales and global change: Bridging the gap from plots to GCM cells. *Annual Review of Ecology and Systematics* 23:175–200.

West, G. B., J. H. Brown, and B. J. Enquist. 1997. A general model for the origin of allometric scaling laws in biology. *Science* 276:122–126.

Wiens, J. A. 1976. Population responses to patchy environments. *Annual Review of Ecology and Systematics* 7:81–120.

Wiens, J. A. 1989. Spatial scaling in ecology. *Functional Ecology* 3:383–397.

Wiens, J. A., N. C. Stenseth, B. Van Horne, and R. A. Ims. 1993. Ecological mechanisms and landscape ecology. *Oikos* 66:369–380.

Wilson, K. G. 1971. Renormalization group and critical phenomena: (I) Renormalization group and the Kadanoff scaling picture. *Physical Reviews* 4:3174–3184.

Woodin, S. A. 1976. Adult-larval interactions in dense infaunal assemblages: A marine soft-bottom example. *Ecology* 59:274–284.

Young, D. K., and D. C. Rhoads. 1971. Animal-sediment relations in Cape Cod Bay, Massachusetts. I. A transect study. *Marine Biology* 11:242–254.

PART III

SCALING MESOCOSMS
TO NATURE

CHAPTER 5

Getting It Right and Wrong
Extrapolations Across Experimental Scales

Michael L. Pace

EXPERIMENTS IN ECOLOGY ARE CRITICIZED FOR MANY SINS ranging from poor designs (Hurlbert 1984; Underwood 1994) to irrelevance (Peters 1991; Carpenter 1996; Resetarits and Bernardo 1998). These criticisms reflect, in part, a significant concern about relating experimental results to ecological reality. Simply put, how do we use what we learn in experiments? Ideally, experiments reveal interactions, expose underlying mechanisms, and support or refute models. There remains, however, a critical problem about how to relate even the best experiments to the ecological systems of ultimate interest, because results may include scale-dependencies that do not extrapolate.

In considering this problem, two general types of experiments can be distinguished. In one case, a question is tested using a contrived population, community, or ecosystem in a laboratory or quasi-laboratory setting. The Ecotron facility in the Centre for Population Biology at Silwood Park in Ascot, England, is a good example of such a system (Lawton 1996). Here, artificial ecosystems have been created and replicated to carry out controlled experiments that focus primarily on generality, testing key concepts and theoretical models. Extrapolation to a specific field situation is not a direct goal. The hope is for experimental results to be indicative of general interactions and so help illuminate important mechanisms operating in nature, perhaps in a diversity of systems. A second general class of experiments has a more system-specific focus. In these experiments the emphasis is on quasi-realistic conditions. Typically, natural assemblages of organisms are used. These experiments are often referred to as "field experiments." A primary goal of these experiments is extrapolation. The hope is that significant regulatory factors and interactions revealed in the experiment will generalize to the field (i.e., nonexperimental) situation.

A key element in the evaluation of "field experiments" and in the translation of experimental results to understanding is the issue of extrapolation. A number of questions are relevant. How do we extrapolate results? Can we scale up results of experiments conducted at small scales to larger scales of interest? How do treatments imposed in experiments relate to variability observed in natural systems? What ecological interactions create artifacts in experiments, thereby inhibiting extrapolation? These are crucial questions, particularly for what I have labeled "field experiments," because these studies have direct extrapolation as a primary goal. Surprisingly little emphasis, however, is placed on the problem of extrapolation. For example, the general questions posed above about extrapolation of experimental results are not directly treated in recent philosophically oriented texts on ecology (e.g., Peters 1991; Shrader-Frechette and McCoy 1993; Pickett et al. 1994) or in works on experimental ecology (e.g., Hairston 1989; Wilbur 1997). For most research reporting field experiments, consideration of extrapolation is at best relegated to a few paragraphs of discussion. Extrapolation is rarely the subject of a more formal analysis. The neglect of extrapolation, therefore, is a serious blind-spot in experimental ecology. Without understanding the context and limitations of experiments, we cannot build a predictive science.

The goal of this chapter is to explore the concept of extrapolation in the context of experiments. The focus is on aquatic experiments that manipulate nutrient loading and predation and assess responses at the community and ecosystem levels. I begin by defining extrapolation and providing an example of successful extrapolation. This result illustrates the utility of comparative frameworks in extrapolation. I then compare responses of plankton in enclosures and corresponding whole lake experiments to evaluate if enclosure experiments predict responses at the ecosystem level. I evaluate lake ecosystem processes related to success or failure in extrapolation. I argue that experimental research is best done simultaneously with research carried out on the object of interest (i.e., the population, community, or ecosystem of concern) in order to promote ongoing comparisons between experimental results and field observations. Specifying scales of interest and a more careful evaluation of extrapolation should enhance experimental research and facilitate the evaluation of results obtained from multiple approaches.

SUCCESSFUL EXTRAPOLATION: AN EXAMPLE

Extrapolation has a specific meaning in statistics—"to estimate the value of a variable outside its . . . range" and a more informal meaning in common use—"to infer (that which is not known) from that which is known; conjecture" (*Random House Dictionary of the English Language*). Experimentalists think in terms of both meanings. Experimental conclusions are used to infer significance of a process in the field. The first meaning of extrapolation also has significance, however, because an objective may be to project the value of a variable measured in an experiment to the natural situation where the range of variation may be greater than observed in the experimental unit

To explore ecological extrapolation, it is useful to begin with an example. A nutrient fertilization experiment was carried out in mesocosms at the Marine Ecosystems Research Laboratory (MERL) of the University of Rhode Island. The MERL tanks were amended with nitrogen, phosphorus, and silica over a concentration range from no loading to 32 times the background value (Nixon et al. 1984; Oviatt et al. 1986). Hobbie and Cole (1984) measured the responses on planktonic bacteria in these enclosures and found, not surprisingly, that bacterial production increased in correspondence with nutrient loading and primary production. Bacterial production was a strong, linear function of primary production based on means from each enclosure (figure 5.1). Cole et al. (1988) subsequently compared these results with data drawn from the literature for a wide variety of aquatic systems that covered a gradient of trophic conditions. Does this relationship extrapolate either quantitatively or qualitatively to reflect the general pattern of primary and bacterial production? The answer is yes on both counts. The bacterial-primary production relationship derived from the MERL tanks is the same as the relationship derived from field studies in marine and freshwater systems (figure 5.1).

The relationship observed in the MERL tanks extrapolates to the general situation. The MERL tanks, in terms of pelagic bacterial production, are representative of a class of systems observed in nature. What is the basis for this successful extrapolation? First, the microbial system in the 13 m^3 MERL tanks contained at any time $> 10^{13}$ bacteria, and there were on the order of 50 to 200 generations produced over the six-month course of the experiment (Hobbie and Cole 1984). Thus, the time and spatial scales of the experiment were "large," and so, unlike the outcome of many

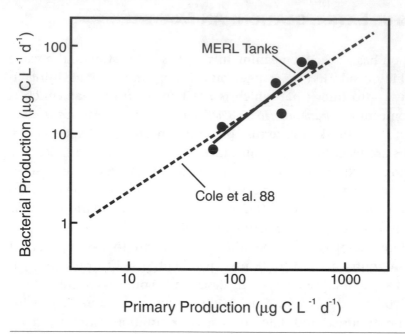

FIGURE 5•1 *Production and Scale*
Relationships between primary and bacterial production in MERL (Marine Ecosystems Research Laboratory) enclosures in comparison with a general relationship derived from the literature by Cole et al. (1988). Each point on the graph is the mean for a tank along a nutrient-loading gradient as measured by Hobbie and Cole (1984). Lines are least square regressions.

experiments (Tilman 1989), the observed response is not a transient. Second, the MERL systems were designed to realistically simulate estuarine systems in a number of ways—including water exchange, temperature, mixing, and light regimes (Oviatt 1981, 1994). These regimes yield realistic patterns of phytoplankton primary production when compared to the adjacent estuary (Keller 1988). Thus, the MERL systems are most similar to the "field experiments" I discussed earlier in that considerable effort was made to create realistic conditions.

A third reason for successful extrapolation was that the experiment created a significant gradient of two major factors regulating bacterial responses—nutrient loading and primary production. Thus, the noise associated with variation in time and space within a particular system was small when compared to the response over the larger scale of comparison created by the experiment. Large manipulations are often most successful at revealing regulatory factors but, unlike the MERL situation, are difficult to create in many small-scale field experiments.

For bacterial production and other processes (Nixon 1992; Kemp et al. 1997), the MERL nutrient fertilization experiment serves as an example for how to assess extrapolation of results. A comparative framework based on a broad empirical relationship (bacterial production; Cole et al. 1988) made the evaluation possible by providing a context for the results.

COMPARATIVE FRAMEWORKS

Comparative analyses provide one foundation for extrapolation. Ecosystem studies can be helpfully organized along gradients of either processes or properties. Limnologists have developed a set of comparative frameworks for this type of analysis. The most significant process that organizes comparisons of lake studies is nutrient loading. For example, productivity of many trophic groups including phytoplankton, bacteria, zooplankton, benthos, and fish are strongly correlated with nutrient loading in lakes (Peters 1986). The lake framework, however, extends well beyond nutrient inputs. Some additional lake gradients include physical and geographic features related to lake size, landscape position, latitude, and shape. In addition, the loading of dissolved organic carbon is emerging as an important gradient for comparisons among lakes (Morris et al. 1995; Williamson et al. 1999). Oxygen dynamics (Nürnberg 1994), sulfate concentrations (Caraco et al. 1989), and fish community structure (Brooks and Dodson 1965) represent other features of lakes that are useful for organizing variation. A similar list could be developed for estuaries as well as other types of ecosystems. In the case of estuaries, the list, at a minimum, would include nutrient loading, estuary size and shape, tidal range, mixing regimes, freshwater input, and human population density in the watershed.

Further development of these comparative frameworks is an important challenge. Comparative frameworks are not only essential for extrapolation; they are also a foundation for theory. They represent major patterns for explanation. Of equal importance, integration of these frameworks provides a means to developing richer and deeper theory (Pickett et al. 1994). Current lake research provides an example of integration as a new focus develops on the interactions of nutrients, food web structure, and dissolved organic carbon (Carpenter and Pace 1997; Carpenter et al. 1998). Comparative analyses of many other ecosystem types lag behind lake research. Again considering the case of estuaries, the lack of comparative analysis is partly due to logistic difficulties, but is also related to a dominance of site-specific studies in estuarine research.

EXTRAPOLATION AND LAKE ENCLOSURE EXPERIMENTS

One way to directly assess extrapolation is to compare similar manipulations done over different experimental scales. I have participated in experiments with colleagues that have considered how nutrient loading and food web structure regulate the abundance and variability of various trophic groups in lakes. Our group has conducted experiments in 45 L plastic buckets (Pace and Funke 1991), 2 m³ plastic bags (Cottingham et al. 1997), and whole lakes (Carpenter and Kitchell 1993; Carpenter et al. 1996; Pace and Cole 1996; Carpenter et al. 1998; Pace et al. 1998). These experiments were all conducted in the same set of lakes located at the University of Notre Dame Environmental Research Center in Gogebic County, Michigan. Here, I compare how well short-term enclosures with fixed manipulations extrapolate to responses observed in whole lake experiments. In addition, results of a longer nutrient loading experiment in enclosures are compared with a similar series of whole lake nutrient additions. In the lakes and experiments reviewed below, the cladoceran, *Daphnia*, was an important grazer and consisted of a mixture of two species, *D. rosea* and *D. pulex*, that were aggregated in counts and biomass estimates. These species are referred to collectively for convenience in this chapter as *Daphnia*.

Short-term Enclosures and Whole Lake Experiments: Responses of Heterotrophic Microbes to Fixed Treatments

Heterotrophic microorganisms including bacteria, flagellates, and ciliates tend to increase strongly in abundance across lake productivity gradients (e.g., Bird and Kalff 1984; Porter et al. 1985; Gasol and Vaqué 1993). Predation by crustacean zooplankton, especially on protozoa, can also be significant and potentially regulate the abundance of these organisms. For example, in one of our study lakes, Pace and Vaqué (1994) measured daily mortality rates of heterotrophic flagellates and ciliates due primarily to *Daphnia* in the range of 0.5–1 and 0.2–0.6 d⁻¹ respectively. These mortality rates are comparable to expected in situ growth rates. Thus, measurements of rate processes indicate that resources and predators are potentially significant in determining the abundance of heterotrophic microbes.

We examined resource and predatory control as well as their interaction in two enclosure experiments (Pace and Funke 1991). These studies were conducted during 1988 and 1989 in two adjacent lakes, Paul and Peter, that are described in Carpenter and Kitchell (1993). The experiments were intentionally short term (4 d) and small scale (45-L enclosures) in order to avoid substantial changes in the manipulated factors (*Daphnia* and nutrients) over time. Bacteria responded positively to nutrients in both experiments—roughly doubling in abundance. *Daphnia* had no effect on abundance and actually increased bacterial productivity in one of the two experiments. Flagellates and ciliates responded positively to nutrients in three out of four cases and negatively to *Daphnia*. Reductions of flagellates by *Daphnia* were stronger in the presence than in the absence of nutrients resulting in significant or marginally significant interactions (Interaction P's = 0.002 and 0.06).

In a series of whole lake experiments beginning in 1991, we manipulated the fish communities of three lakes to create food web structures that either promoted (East and West Long lakes) or suppressed (Peter Lake) the dominant zooplankton grazer, *Daphnia*. A fourth lake (Paul) served as an unmanipulated reference. The three experimental lakes were subsequently fertilized with inorganic nitrogen and phosphorus beginning in 1993. Nutrient loads were at a constant ratio (N:P > 30:1 by atoms), but daily loading rate varied annually to represent a range of enrichments (summarized in Carpenter et al. 1998).

How well did the responses observed in the enclosure studies compare with the whole lake results? Enclosure results provided correct and incorrect predictions (table 5.1). For example, bacterial production increased in response to whole lake nutrient additions, and *Daphnia* had a strong negative effect on the dynamics and average abundance of ciliates. These results were in accord with extrapolations from the enclosures (table 5.1). Bacterial abundance, however, did not increase in the fertilized lakes, a result in contradiction with results from the enclosures (table 5.1). Instead, only specific growth rates (i.e., growth rate per cell) increased with nutrient loading (Pace and Cole 1996). In the enclosure experiments, flagellates increased with nutrient additions in 1 of 2 cases but did not increase in the whole lake experiments. Similarly, ciliates increased in response to nutrients in both enclosure experiments while increases with lake fertilization were observed in only one case (table 5.1). Negative

TABLE 5•1 *Responses of Heterotrophic Microbial Groups to Short-term Enclosure and Whole Lake Manipulations*

Group	Increases in Response to Nutrients		Declines in Response to *Daphnia*	
	Enclosure	Lake	Enclosure	Lake
Bacteria				
Abundance	2/2	0/3	0/2	0/3
Production	2/2	3/3	0/2	1/3
Flagellates	1/2	0/3	2/2	1/3
Ciliates	2/2	1/3	2/2	2/3

NOTE The responses are presented as number of significant increases in response to nutrient addition and significant declines in response to zooplankton over total cases. In the enclosure experiments, responses were judged significant if $P < 0.05$ for nutrient and zooplankton treatments in a factorial analysis of variance (significant interactions noted in text). In the whole lake experiments, responses were judged significant if t-values associated with time series model fits for the variables, phosphorus loading, and *Daphnia* biomass, exceeded 1.96 (approximately, $P < 0.05$).

responses by flagellates and ciliates to *Daphnia* in the enclosures were also evident in the lakes but were not observed as consistently.

In general, lake responses were more variable than those observed in the enclosures. This is not surprising given the limitations on dynamics and potential shifts in community structure imposed by design of the short-term enclosure experiments. Increasing enclosure size and experimental time might lead to better extrapolation as found by Sarnelle (1997) in experiments with *Daphnia galeata mendotae* and microzooplankton, but as discussed next, dynamic responses and potential artifacts can complicate longer-term enclosure experiments.

Long-Term Enclosures and Whole Lake Experiments: Dynamic Responses of Planktonic Communities

Before fertilizing the experimental lakes, we conducted an enclosure experiment to test the efficacy of the projected range of nutrient loads. Phosphorus (P) and nitrogen (N) were added at four levels (ranging from 0–2 µg $PL^{-1}d^{-1}$ at an N:P ratio of 25:1 by atoms). Across this nutrient loading gradient, half the enclosures were sieved weekly with a plankton net to remove large zooplankton. After conducting this enclosure experiment, we fertilized two basins in the same lake where the experiment was performed over the same range of loads. This allowed comparison of the longer-term enclosure results with whole lake results.

In the unsieved enclosures, algal biomass measured as chlorophyll a was similar to whole lake values when loading was < 1 µg PL^{-1}d^{-1} but was very low relative to the whole lake responses at the highest loading (figure 5.2). This suppression can be attributed to the changes in the dominant zooplankton grazers, *Daphnia*. In enclosures, *Daphnia* dominated the zooplankton, and biomass was very high relative to the lake even at low loadings (figure 5.2). In the sieved enclosures, *Daphnia* biomass was reduced at nutrient loadings < 1 µg PL^{-1}d^{-1} relative to the unsieved enclosures and corresponded more closely to lake conditions (figure 5.3). However, at the highest nutrient loading, weekly sieving was not sufficient to suppress *Daphnia*. Chlorophyll a was much lower in this treatment than observed in the lake at the same loading (figure 5.3). Although *Daphnia* was effective at suppressing phytoplankton in the whole lake experiments as documented elsewhere (Carpenter et al. 1996, 1998), the effect of *Daphnia* grazing was exaggerated in the enclosures.

Why did *Daphnia* achieve a higher biomass in the enclosures? Predation pressure was probably much lower in the enclosures as there were no fish and few invertebrate predators. The very high biomass of *Daphnia* even in unfertilized enclosures (figures 5.2, 5.3) supports this contention. Lack of predation, however, is not a complete explanation, because in one of the experimental basins, East Long, planktivory was also very low (Carpenter et al. 1998). Increased food may have promoted higher populations. Enclosures have large plastic surface areas relative to their volume, and these surfaces became enriched with periphyton (Blumenshine et al. 1997) that *Daphnia* is capable of browsing (Horton et al. 1979). In addition, recent studies have emphasized stoichiometric limitations in terms of C:P ratios for zooplankton (Hessen 1992; Sterner et al. 1993; Elser et al. 1996). The large increase in *Daphnia* populations at the lowest phosphorus loading (figures 5.2, 5.3) may partly reflect a release from poor food quality and severe phosphorus limitation. The same process should have occurred in the whole lake experiments but may have been less pronounced given losses of phosphorus from the surface of the stratified lake via sedimentation relative to the greater potential retention of phosphorus within enclosures on surfaces.

These comparisons of responses are based on mean values. A second and perhaps more important issue is variability. How well do enclosure experiments represent the variability at the whole lake scale under similar manipulations? Here, we can compare two metrics (chlorophyll a and bacterial production) measured in similar ways in the whole lake and

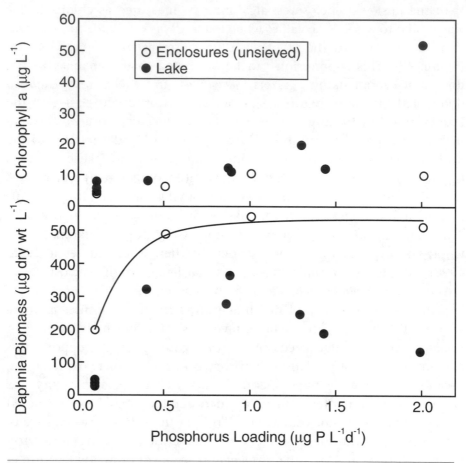

FIGURE 5•2 *Enclosures without Grazer Removal versus Lakes*
Comparison of unsieved enclosures and lake means in Long Lake across a similar range of nutrient loadings. Means for enclosures are derived from triplicates of each loading treatment (Cottingham et al. 1997). Summer season means for Long Lake are for the East and West basins from 1991 to 1995 (Carpenter et al. 1998). The line fit through the enclosure points for *Daphnia* biomass is an exponential model with the form $y = a\,(1-e^{-bx})$.

enclosure experiments. The box plots in figure 5.4 represent major percentiles for data from all the enclosures as well as the Long Lake observations over a comparable range of nutrient loadings (except for lake bacterial production because of missing data at the highest nutrient load). The range of chlorophyll a was 1 to 70 µg L^{-1} in the enclosures versus 3 to 350 µg L^{-1} in the lakes. Enclosures tended to have lower chlorophyll and less extreme blooms relative to the whole lake experiments as suggested by differences in the 75th percentiles of the distributions (figure 5.4) and the differences in the ranges noted above. Limitations on phytoplankton biomass in enclosures are consistent with strong grazing by *Daphnia*. Similarly, the range of variation of bacterial production was lower in the enclosures relative to whole lake experiments (figure 5.4). Grazing by *Daphnia* had important effects on bacteria in the whole lake experiments (Pace and Cole 1996), but bacteria were even more strongly suppressed in the enclosures (Pace and Cole 1994).

The enclosures provided reasonable representations of lake responses in terms of mean chlorophyll at nutrient loads in the range of 0 to 1 µg P $L^{-1}d^{-1}$, but only when *Daphnia* was consistently removed. *Daphnia*

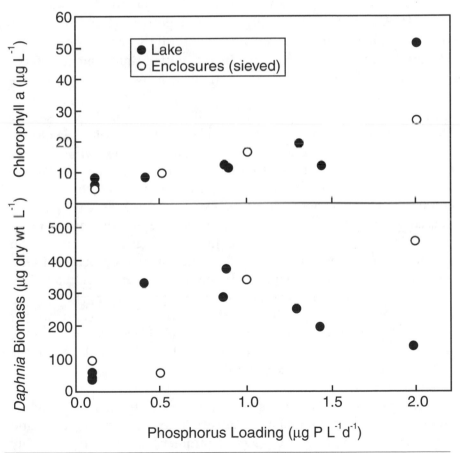

FIGURE 5•3 *Enclosures with Grazer Removal versus Lakes*
Comparison of sieved enclosures and lake means in Long Lake across a similar range of nutrient loadings as in figure 5.2.

populations, however, dominated enclosures, reducing variability and exaggerating grazing effects relative to the whole lake when not removed or when nutrient loading was high. This comparison indicates the difficulty of representing food web effects in enclosures because of potentially unrealistic population dynamics by a critical grazer. The problem is of greater interest than implied simply by the comparison of these experiments. Interpretations of the interactions of nutrient enrichment, grazer effects, and food chain length on algae are derived from patterns in data drawn from both whole lake and enclosure studies (e.g., Sarnelle 1992; Mazumder 1994). If grazing effects are generally magnified in enclosures, analyses should attempt to determine whether enclosures yield different patterns then lake data. In the studies by Sarnelle (1992) and Mazumder (1994), algal responses to nutrient loading and total phosphorus-chlorophyll relationships were similar for enclosures and whole lake experiments, suggesting similar grazer effects, but the specific issue of *Daphnia* abundance in enclosure versus lake experiments was not analyzed.

I have used the whole lake experiments as a standard for assessing extrapolations of the enclosure experiments. The problem could be extended to address whether the whole lake experiments are representative of a larger collection of lakes or if the results are idiosyncratic. Such an analysis is beyond the scope of this chapter. Previous comparisons, however, of the experimental lakes with patterns observed among lakes at a regional scale suggest the experiments accurately represent nutrient loading and grazing effects on phytoplankton (Carpenter et al. 1991; Kitchell and Carpenter 1993). Other studies demonstrate that lake size has a critical effect on many lake processes including mixing, thermocline depth, water clarity, and mercury contamination of fish (e.g., Fee et al. 1996; and studies summarized in Schindler 1998). It may be possible to develop scaling rules to extrapolate results observed in smaller lakes to larger lakes but the evaluation and understanding of differences along lake size gradients is still developing.

Limnologists are well aware of many problems with enclosure studies, including but not limited to uncertainties about mixing and sedimentation (Vanni et al. 1997; Schindler 1998), fish mortality (Threlkeld 1987), excessive predation (O'Brien et al. 1992), and wall growth (Blumenshine et al. 1997). Despite discussions of these problems by the authors noted, potential artifacts from enclosure studies are often glossed over because they complicate papers and undercut inference. This

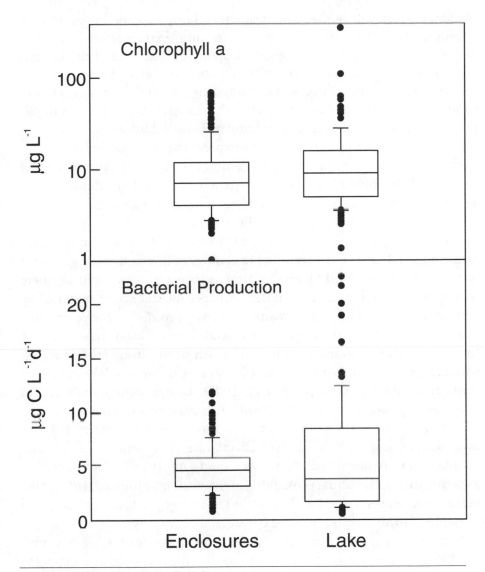

FIGURE 5•4 *Variability in Productivity*

Box plots illustrating the variability in chlorophyll a and bacterial production in the enclosures and lakes over a similar range of nutrient loadings. Upper and lower bars are the 95th and 5th percentiles respectively. Boxes illustrate the 75th (upper line), median (mid-line), and 25th (lower line) percentiles. Points are values less than the 5th percentile or greater than the 95th percentile. Lake data are based on all observations for the years 1992 to 1995 and enclosure data include all observations from the enclosure experiment. Note that the bacterial production box for the "lake" probably underestimates the true variability in this variable, because no comparable data are available from the year with the highest nutrient loading (1995).

tendency to minimize problems leads to a body of experimentation of questionable utility that cannot be readily extrapolated—hardly a desirable situation for a science called upon to make critical predictions about changing ecological systems (e.g., Lubchenco et al. 1991).

Some experiments, of course, do extrapolate well and the point is not to condemn all manipulative, small-scale experiments. For example, phytoplankton are consistently stimulated by additions of inorganic phosphorus and nitrogen whether the experiment is performed in bottles, bags, big tanks, or lakes. This result agrees with knowledge of phytoplankton physiology and the pattern of nutrient limitation in aquatic systems. The general challenge is to translate qualitative and quantitative experimental results to the system of interest.

This challenge becomes more difficult and the potential for misleading results probably rises as the number of ecological interactions (e.g., a food web complex) addressed by an experiment increases. One difficulty here is that manipulations done at different experimental scales may not be equivalent. The whole lake manipulations considered above include numerous interactions between fish and zooplankton that are not included in the enclosures. For example, fish in the study lakes undergo variable recruitment that may result in periods of high or low predation pressure on zooplankton (Post et al. 1997). Under some circumstances (e.g., when planktivores are abundant), fish may also influence nutrient recycling via their migrations between shore and open-water habitats (Schindler et al. 1993). These effects are not simulated by the zooplankton removal methods we used in the larger enclosure experiment. So, the actual ecological interactions manipulated in the lake versus enclosures may be different. Thus, comparisons of the type presented in this chapter must be considered cautiously.

Lakes may represent optimal systems at this point for testing how well small-scale experiments extrapolate to predict the behavior of ecosystems and the shortcomings and advantages of these types of experiments. There is a rich and building literature of whole lake manipulations as well as manipulation experiments done in various kinds and sizes of enclosures. Food web experiments have been especially prominent in lake studies recently and comparisons have been undertaken (e.g., DeMelo et al. 1992; Brett and Goldman 1997). Further analyses, however, are warranted to consider issues such as the comparability of manipulations, the effects of enclosure size, the realism of treatments and responses when compared to whole lake results, and the variability of the same experiment repeated at the whole lake level (e.g., Lodge et al. 1998). One such analysis comes from Schindler (1998) who reviewed his experiences with experiments at numerous size scales (i.e., bottles to whole lakes) over more than thirty years. He found numerous problems arising from studies using bottle and enclosure experiments that could have lead to incorrect scientific conclusions and "erroneous management decisions" (Schindler 1998:331) without additional work at the ecosystem level.

SCALES OF INTEREST, SOFT EXTRAPOLATION, AND CONTEXT

One key to improving experimentation in ecology in general, and to better incorporate research on scaling in particular, is to constantly question how results contribute to solving problems at the ultimate scale of interest. This means first and foremost that the scale of interest must be made explicit. For lakes, one pertinent scale for predictions is the summer season mean and variance of particular lake properties such as nutrient levels, chlorophyll concentrations, nuisance algae, acidity, fish recruitment, and toxin body burdens. Another scale of interest is often regional. Some properties of interest at this scale are distributions of acid lakes, presence or absence of desirable fish species, number of lakes with fish exceeding a toxic advisory limit, and number of lakes occupied by an invader. There are many other scales and many other properties and problems of interest—the point is scientists must be specific and tailor research efforts accordingly.

Not being explicit about ultimate scales of interest leads to "soft extrapolation." By this I mean a vague analysis of how to generalize or hope to generalize results from a particular study. Soft extrapolation is a malady of practice. We tend to rationalize this malady by offering up our scientific contributions as pieces of a larger puzzle, and like many undesirable behaviors, we become habituated to it. For example, a common dodge is to say a particular research result will contribute to models without ever specifying what the models are or how the result might specifically improve a given model. By this means (and others), we avoid confronting the scale of interest by not projecting results against independent measures or frameworks. We avoid the hard job of extrapolation.

Experiments like those done in the Ecotron or in laboratory-like settings must also consider the issue of extrapolation but in a different form. In these types of experiments the goal is generality. The hope is that general mechanisms uncovered in these experiments will apply to a wide variety of systems. Thus, the goal is breadth of extrapolation not specific quantitative prediction. Still, to avoid soft extrapolation, general experiments of this nature should strive to specify the domain over which the results may apply and methods for testing that applicability.

Research that helps evaluate and identify scaling relationships should help with the problem of extrapolation as long as it does not facilitate the soft version. There is a tendency to use the concept of scale in an "I'm okay; you're okay" fashion. Levin (1992) notes that while there is no correct scale for a specific system or problem, that "does not mean that all scales serve equally well." We tend to use mantras like "it's a matter of scale" or "but the scales are different"; these should not serve as a means of perpetuating soft extrapolation. On the contrary, appreciation of scaling-dependencies should promote extrapolation. Again, standards and practice in research are important for improving the science.

Finally, context is critical when considering extrapolation. Soft extrapolation may evolve from an incomplete or poorly defined context for a research program. Obviously, context is related to the scale of interest, but more than just specifying the scale, ecologists need an appreciation for the population, community, or ecosystem of interest. As ecological science becomes more technical and information more extensive, specialization ensues. These developments, while deepening understanding, may also act to limit synthetic perspective.

COMMENTS

Most experiments in ecology are done at scales of time and space far below the scale of interest. This means that experimental results must be extrapolated, but the methods for doing so remain poorly understood. Comparison of enclosure and ecosystem experiments reveals cases of both getting it wrong and getting it right. Thus, experiments can both inform and mislead. The statistical rigor of a well-designed, misleading experiment is not very useful. More careful attention to the problem of extrapolation should improve experimental research and facilitate scaling-up results. The metrics for extrapolation need development and attention. These metrics include comparative frameworks, large-scale experiments, observations from the system of interest, as well as models. Combining these with experiments has been proven and will prove to be effective. A keen sense of ecological systems remains a powerful ally. As technological ability and sophistication of the science increase, we must remember that synthesis and integration are the means to encompassing complexity. This requires scientists capable of seeing beyond the laboratory bench and computer screen to the vistas of the natural world.

ACKNOWLEDGMENTS

I thank members of the Cascade research team for providing data and for the hard work that made the comparisons of enclosure and ecosystem experiments summarized here possible. This research was supported by grants from NSF. I thank S. Carpenter, J. Cole, V. Kennedy, C. Oviatt, J. Petersen, and M. Vanni for thoughtful criticism and suggestions for improving this chapter.

LITERATURE CITED

Bird, D. F., and J. Kalff. 1984. Empirical relationships between bacterial abundance and chlorophyll concentrations in fresh and marine waters. *Canadian Journal of Fisheries and Aquatic Sciences* 41:1015–1023.

Blumenshine, S. C., Y. Vadeboncoeur, D. M. Lodge, K. L. Cottingham, and S. E. Knight. 1997. Benthic-pelagic links: Responses of benthos to water-column nutrient enrichment. *Journal of the North American Benthological Society* 16:466–479.

Brett, M. T., and C. R. Goldman. 1997. Consumer versus resource control in freshwater pelagic food webs. *Science* 275:384–386.

Brooks, J. L., and S. I. Dodson. 1965. Predation, body size, and composition of plankton. *Science* 150:28–35.

Caraco, N. F., J. J. Cole, and G. E. Likens. 1989. Evidence for sulphate-controlled phosphorus release from sediments of aquatic systems. *Nature* 341:316–318.

Carpenter, S. R. 1996. Microcosm experiments have limited relevance for community and ecosystem ecology. *Ecology* 77:677–680.

Carpenter, S. R., J. J. Cole, J. F. Kitchell, and M. L. Pace. 1998. Impact of dissolved organic carbon, phosphorus, and grazing on phytoplankton biomass and production in lakes. *Limnology and Oceanography* 43:73–80.

Carpenter, S. R., T. M. Frost, J. F. Kitchell, T. K. Kratz, D. W. Schindler, J. Shearer, W. G. Sprules, M. J. Vanni, and A. P. Zimmerman. 1991. Patterns of primary production and herbivory in 25 North American lake ecosystems. In J. J. Cole, G. M. Lovett, and S. E. G. Findlay, eds., *Comparative Analyses of Ecosystems: Patterns, Mechanisms, and Theories*, pp. 67–96. New York: Springer-Verlag.

Carpenter, S. R., and J. F. Kitchell, eds. 1993. *The Trophic Cascade in Lakes*. New York: Cambridge University Press.

Carpenter, S. R., J. F. Kitchell, K. L. Cottingham, D. E. Schindler, D. L. Christensen, D. M. Post, and N. Voichick. 1996. Chlorophyll variability, nutrient input, and grazing: Evidence from whole-lake experiments. *Ecology* 77:725–735.

Carpenter, S. R., and M. L. Pace. 1997. Dystrophy and eutrophy in lake ecosystems: Implications of fluctuating inputs. *Oikos* 78:3–14.

Cole, J. J., S. E. G. Findlay, and M. L. Pace. 1988. Bacterial production in fresh and saltwater ecosystems: A cross-system overview. *Marine Ecology Progress Series* 43:1–10.

Cottingham, K. L., S. E. Knight, S. R. Carpenter, J. J. Cole, M. L. Pace, and A. E. Wagner. 1997. Response of phytoplankton and bacteria to nutrients and zooplankton: A mesocosm experiment. *Journal of Plankton Research* 29:995–1010.

DeMelo, R., R. France, and D. J. McQueen. 1992. Biomanipulation: Hit or myth? *Limnology and Oceanography* 37:192–207.

Elser, J. L., D. R. Dobberfuhl, N. A. MacKay, and J. J. Schampel. 1996. Organism size, life history, and N:P stoichiometry. *BioScience* 46:674–684.

Fee, E. J., R. E. Heckey, S. E. M. Kasian, and D. R. Cruikshank. 1996. Effects of lake size, water clarity, and climatic variability on mixing depths in Canadian Shield lakes. *Limnology and Oceanography* 41:912–920.

Gasol, J. M., and D. Vaqué. 1993. Lack of coupling between heterotrophic nanoflagellates and bacteria: A general phenomenon across aquatic systems? *Limnology and Oceanography* 38:657–665.

Hairston, N. G. 1989. *Ecological Experiments: Purpose, Design, Execution*. New York: Cambridge University Press.

Hessen, D. O. 1992. Nutrient element limitation of zooplankton production. *American Naturalist* 140:799–814.

Hobbie, J. E., and J. J. Cole. 1984. Response of a detrital food web to eutrophication. *Bulletin of Marine Science* 35:357–363.

Horton, P. A., M. Rowan, K. E. Webster, and R. H. Peters. 1979. Browsing and grazing by cladoceran filter-feeders. *Canadian Journal of Zoology* 57:206–212.

Hurlbert, S. H. 1984. Pseudoreplication and the design of ecological field experiments. *Ecological Monographs* 54:187–211.

Keller, A. A. 1988. Estimating phytoplankton productivity from light availability and biomass in the MERL mesocosms and Narragansett Bay. *Marine Ecology Progress Series* 45:159–168.

Kemp, W. M., E. M. Smith, M. Marvin-DiPasquale, and W. R. Boynton. 1997. Organic carbon balance and net ecosystem metabolism in Chesapeake Bay. *Marine Ecology Progress Series* 150:229–248.

Kitchell, J. F., and S. R. Carpenter. 1993. Synthesis and new directions. In S. R. Carpenter and J. F. Kitchell, eds., *The Trophic Cascade in Lakes*, pp. 332–350. New York: Cambridge University Press.

Lawton, J. H. 1996. The Ecotron facility at Silwood Park: The value of "big bottle" experiments. *Ecology* 77:665–669.

Levin, S. A. 1992. The problem of pattern and scale in ecology. *Ecology* 73:1943–1967.

Lodge, D. M., S. C. Blumenshine, and Y. Vadeboncoeur. 1998. Insights and applications of large-scale, long-term ecological observations and experiments. In W. J. Resetarits Jr. and J. Bernardo, eds., *Experimental Ecology: Issues and Perspectives*, pp. 202–235. New York: Oxford University Press.

Lubchenco, J., A. M. Olson, L. B. Brubaker, S. R. Carpenter, M. J. Holland, S. P. Hubbell, S. A. Levin, J. A. MacMahon, P. A. Matson, J. M. Melillo, H. A. Mooney, C. H. Peterson, H. R. Pulliam, L. A. Real, P. J. Regal, and P. G. Risser. 1991. The Sustainable Biosphere Initiative: An ecological research agenda. *Ecology* 72:371–412.

Mazumder, A. 1994. Patterns of algal biomass in dominant odd- vs. even-link lake ecosystems. *Ecology* 75:1141–1149.

Morris, D. P., H. Zagarese, C. E. Williamson, H. G. Balseiro, B. R. Hargreaves, B. Modenutti, R. Moeller, and C. Queimalinos. 1995. The attenuation of solar UV radiation in lakes and the role of dissolved organic carbon. *Limnology and Oceanography* 40:1381–1391.

Nixon, S. W. 1992. Quantifying the relationship between nitrogen input and the productivity of marine ecosystems. In M. Takahashi, K. Nakata, and T. R. Parsons, eds., *Proceeding of Advanced Marine Technology Conference*, vol. 5, pp. 57–83. Tokyo: Shimane Prefecture Government.

Nixon, S. W., M. E. Q. Pilson, C. A. Oviatt, P. Donaghay, B. Sullivan, S. Seitzinger, D. Rudnick, and J. Frithsen. 1984. Eutrophication of a coastal marine ecosystem—an experimental study using the MERL microcosms. In M. J. R. Fasham, ed., *Flow of Energy and Material in Marine Ecosystems*, pp. 105–135. New York: Plenum.

Nürnberg, G. K. 1994. Anoxic factor, a quantitative measure of anoxia and fish species richness in Central Ontario lakes. *Transactions of the American Fisheries Society* 124:677–686.

O'Brien, W. J., A. E. Hershey, J. E. Hobbie, M. A. Hullar, G. W. Kipphut, M. C. Miller, B. Moller, and J. R. Vestal. 1992. Control mechanisms of Arctic lake ecosystems: A limnocorral experiment. *Hydrobiologia* 240:143–188.

Oviatt, C. A. 1981. Effects of different mixing schedules on phytoplankton, zooplankton, and nutrients in marine mesocosms. *Marine Ecology Progress Series* 28:57–67.

Oviatt, C. A., D. T. Rudnick, A. A. Keller, P. A. Sampou, and G. T. Almquist. 1986. A comparison of system (O_2 and CO_2) and C-14 measurements of metabolism in estuarine mesocosms. *Marine Ecology Progress Series* 28:57–67.

Pace, M. L., and J. J. Cole. 1994. Comparative and experimental approaches to top-down and bottom-up regulation of bacteria. *Microbial Ecology* 28:181–193.

Pace, M. L., and J. J. Cole. 1996. Regulation of bacteria by resources and predation tested in whole lake experiments. *Limnology and Oceanography* 41:1448–1460.

Pace, M. L., J. J. Cole, and S. R. Carpenter. 1998. Trophic cascades and compensation: Differential responses of microzooplankton in whole lake experiments. *Ecology* 79:138–152.

Pace, M. L., and E. Funke. 1991. Regulation of planktonic microbial communities by nutrients and herbivores. *Ecology* 72:904–914.

Pace, M. L., and D. Vaqué. 1994. The importance of *Daphnia* in determining mortality rates of protozoans and rotifers in lakes. *Limnology and Oceanography* 39:985–996.

Peters, R. H. 1986. The role of prediction in limnology. *Limnology and Oceanography* 31:1143–1159.

Peters, R. H. 1991. *A Critique for Ecology*. New York: Cambridge University Press.

Pickett, S. T. A., J. Kolasa, and C. G. Jones. 1994. *Ecological Understanding*. San Diego: Academic Press.

Porter, K. G., B. F. Sherr, E. B. Sherr, M. L. Pace, and R. W. Sanders. 1985. Protozoans in planktonic food webs. *Journal of Protozoology* 32:409–414.

Post, D. M., S. R. Carpenter, D. L. Christensen, K. L. Cottingham, J. F. Kitchell, D. E. Schindler, and J. R. Hodgson. 1997. Seasonal effects of variable recruitment of a dominant piscivore on pelagic food web structure. *Limnology and Oceanography* 42:722–729.

Resetarits, W. J., and J. Bernardo, eds. 1998. *Experimental Ecology: Issues and Perspectives*. New York: Oxford University Press.

Sarnelle, O. 1992. Nutrient enrichment and grazer effects on phytoplankton in lakes. *Ecology* 73:551–560.

Sarnelle, O. 1997. *Daphnia* effects on microzooplankton: Comparisons of enclosure and whole lake responses. *Ecology* 78:913–928.·

Schindler, D. E., J. F. Kitchell, X. He, S. R. Carpenter, J. R. Hodgson, and K. L. Cottingham. 1993. Food web structure and phosphorus cycling in lakes. *Transactions of the American Fisheries Society* 122:756–772.

Schindler, D. W. 1998. Replication versus realism: The need for ecosystem-scale experiments. *Ecosystems* 1:323–334.

Shrader-Frechette, K. S., and E. D. McCoy. 1993. *Method in Ecology: Strategies for Conservation*. New York: Cambridge University Press.

Sterner, R. W., D. D. Hagemeier, W. L. Smith, and R. F. Smith. 1993. Phytoplankton nutrient limitation and food quality for *Daphnia*. *Limnology and Oceanography* 38:857–871.

Threlkeld, S. T. 1987. Experimental evaluation of trophic-cascade and nutrient-mediated effects of planktivorous fish on phytoplankton community structure. In W. C. Kerfoot and A. Sih, eds., *Predation: Direct and Indirect Impacts on Aquatic Communities*, pp. 161–173. Hanover, N.H.: University Press of New England.

Tilman, D. 1989. Ecological experimentation: Strengths and conceptual problems. In G. E. Likens, ed., *Long-Term Studies in Ecology*, pp. 136–157. New York: Springer-Verlag.

Underwood, A. J. 1994. On beyond BACI: Sampling designs that might reliably detect environmental disturbances. *Ecological Applications* 4:3–15.

Vanni, M. J., C. D. Layne, and S. E. Arnott. 1997. "Top-down" trophic interactions in lakes: Effects of fish and nutrient dynamics. *Ecology* 78:1–20.

Wilbur, H. M. 1997. Experimental ecology of food webs: Complex systems in temporary ponds. *Ecology* 78:2279–2202.

Williamson, C. E., D. P. Morris, M. L. Pace, and O. G. Olson. 1999. Dissolved organic carbon and nutrients as regulators of lake ecosystems: Resurrection of a more integrated paradigm. *Limnology and Oceanography* 44:795–803.

CHAPTER 6

Some Reluctant Ruminations on Scales (and Claws and Teeth) in Marine Mesocosms

Scott Nixon

I AM NOT COMPLETELY SURE WHY THE MENTION OF SCALING MAKES me itch, but it does. And I don't think it has anything to do with the fact that the term brings to mind an unpleasant skin condition. As a working ecologist I have a problem getting a real grip on the subject. On the one hand, many discussions of scale and scaling use terms that sound impressive but lack clear operational definitions. On the other hand, there is a very precise and growing theoretical literature that requires a level of mathematical competence far beyond mine. I suspect that some of my discomfort with the subject must also be due to the fact that I have spent a large part of my scientific effort over the last 35 years on the study of nature as it is manifest in small containers. Those of us who work with microcosms and mesocosms are always running into people who "don't believe anything done in tanks." Then, too, there is the latent danger of space and scale envy that may lie hidden in an ecologist who has lived most of his life in Delaware and Rhode Island. For all of these reasons I tried not to attend the scaling workshop on which this volume is based. But my respect for, and friendship with, the conveners was sufficiently great and of such long standing that I let myself be persuaded to go and present some data from the various mesocosm experiments that have been carried out during the past two decades on the shore of Narragansett Bay. Once there, I itched and scratched through additional meetings of a subgroup charged with discussing scale issues in land-margin ecosystems. The excellent contribution by Walter Boynton et al. in this volume has captured the outcome of those discussions, including some of the frustration many of us felt in dealing with an elusive but intuitively important issue.

THE 1–10 CM DILEMMA

I cannot think of a better way to begin this chapter than to agree with the Boynton et al. report in this volume: "Mesocosms . . . can be used to good effect for some problems, have limited use for others, and cannot be used at all for still others."

The trick, of course, is to know which is the case for any particular problem. The extremes are easy. If our interest is in processes dominated by microorganisms, there is a good chance of getting meaningful results from experiments in containers of several cubic meters. For this reason, mesocosms (and microcosms) have been used relatively frequently and effectively in studies of marine biogeochemical cycling where the high metabolic rates of very small species usually dominate. And no one in his or her right mind would propose mesocosm studies of sea otter–kelp ecosystem dynamics, where large organisms with slow turnover rates and complex life cycles capture our interest. But what about problems and processes in which crabs and shrimp and small or juvenile fish may be important? It is there, with organisms in the approximately 1–10 cm size range, that questions of biological scale really become most difficult for those of us who work with mesocosms.

There is a large and often insightful literature devoted to problems of scale in aquatic mesocosms (see, for example, Balch et al. 1978; Menzel and Steel 1978; Giesy 1980; Grice and Reeve 1982; Lalli 1990; Beyers and Odum 1993; Graney et al. 1994; Carpenter 1996) and my modest hope in this chapter is only to add some personal observations. My first published paper (Nixon 1969) dealt with one aspect of scaling in microcosms and I have continued to struggle with it, and other aspects, for 30 years. Perhaps it is the lack of solution after so long a struggle that really makes me uncomfortable with the subject of scaling.

"AS SIMPLE AS POSSIBLE—BUT NO SIMPLER"

All thoughtful people who develop mesocosms must struggle with the question of what is really important in the larger world in influencing the outcome of the experiment they wish to carry out. But if they really knew what was important, they probably would not have to do the experiment. In the end, every mesocosm is a living hypothesis. If the mesocosm builder is correct, the results of the experiment will have meaning in the larger world; if not, then the results will have meaning only in the mesocosm world—a smaller world that is real and possibly very interesting, but of little relevance to those who want to manage or make predictions about nature on a bigger scale.

It is an under-appreciated benefit of mesocosm design and construction that the process forces us to think in a rigorous and quantitative way about what we think is important in the natural system of interest (Pilson and Nixon 1980). If we want to carry out a mesocosm experiment to find out how the primary production of Narragansett Bay might change with various levels of nitrogen input, do we need to include bottom sediments and benthic infauna? Almost certainly, because we know that sediment-water column interactions are very important in nitrogen cycling. But do we need the macroinfauna or just the sediments and associated meio- and microfauna? Probably, because we know that the macrofauna filter the water column, rework the sediment, and influence pore water chemistry. What about crabs and other epifauna—do we need them? Maybe, if they are important predators on some of the infauna. How many crabs are needed? Data on the field density of crabs are scarce and highly variable. If they average 1 crab 10 m^{-2} and the mesocosm is only 1 m^2, do we put in a crab for 0.1 of each day or 1 day out of 10? Or should we put in a crab that is only 10 percent of the size of the average crab in the field? Or a crab that will exert 10 percent of the feeding pressure of a field crab calculated using some measured or assumed allometric feeding relationship as a function of size? And so on and so forth, until we exhaust our patience and/or our knowledge. Number, size, and time—how and when can we trade among them?

The same exercise must be performed for physical conditions. Some large-scale physical influences, such as tidal water level changes, are easy to include in a mesocosm, whereas smaller scale features, such as turbulent energy dissipation, may be more challenging (e.g., Nixon et al. 1980; Sanford 1997; Petersen et al. 1998). The questions begin again. How important are water level changes, stratification, physical mixing, water residence time, and so on for nitrogen cycling? For crab behavior? For fish feeding? And what about structural or habitat variability within the mesocosm (we know that refuges can be very important in influencing species interactions)? How do we provide ways for organisms to avoid each other in a compressed space? The design of the mesocosm emerges as tentative answers to all these questions and many others accumulate. The final design is, of course, not a perfect replica of larger nature, nor is it intended to be. The mesocosm is a living model—a simplification. The art, the skill, is to make it, as Einstein once said of scientific theory, "as simple as possible, but no simpler."

Some Examples

Our twenty-three-year experience with the Marine Ecosystems Research Laboratory (MERL) mesocosms in Rhode Island as well as with some newer coastal lagoon mesocosms has not necessarily led us to the best answers to these kinds of questions, but it has convinced us that living models can be very useful tools for the study of a variety of ecological processes in coastal waters. The systems have been described frequently in the literature, but a summary may be useful here (table 6.1). The common practice in arguing that mesocosms have been appropriately scaled for any given experiment is to show that the relevant aspects of their behavior correspond with the "real world." Not surprisingly, we have been most successful in this regard with various features of biogeochemical cycling. For example, Pilson (1985) showed that the annual cycles in the concentrations of dissolved inorganic phosphorus (DIP), dissolved inorganic nitrogen (DIN), and silica in unperturbed MERL systems fell within the range of annual variation found in these cycles in Narragansett Bay. Seasonal and annual mass balances for nitrogen and phosphorus have shown that about 20 percent of N input to Narragansett Bay and to the MERL mesocosms is lost to denitrification and that accumulation of N or P within either system is very small (Nowicki and Oviatt 1990; Nixon et al. 1995). Other studies have demonstrated that the behavior of various trace elements and organic pollutants in the mesocosms and the Bay are sufficiently similar that experiments in the tanks can be used to help describe and predict events in larger nature (for example, Hunt and Smith 1982; Santschi et al. 1983; Santschi 1985). A nutrient addition experiment carried out over two annual cycles demonstrated that primary production by phytoplankton in the MERL mesocosms was limited by the rate of supply of nutrients (Nixon et al. 1986). The important point for this discussion, however, is that the relationship between the rate of input of dissolved inorganic nitrogen (DIN) and primary production (^{14}C uptake) in the mesocosms was consistent with field measurements from a range of natural marine ecosystems (figure 6.1).

If we look a bit further up the food chain, it is reassuring that the summer growth rates of at least five species of juvenile finfish are similar in unperturbed control mesocosms and in the field (table 6.2). On the other hand, data for bivalves in the lagoon mesocosms suggest that these systems do not support the more rapid growth observed for these species in the field, perhaps because phytoplankton populations are composed of unsuitable species or abundances are lower in the mesocosms (table 6.3). Unfortunately, the field data we have found for sites near the lagoon mesocosms are all from deeper, phytoplankton-based systems that may be more suitable for filter-feeding shellfish than our shallow macrophyte-dominated lagoon mesocosms. It is not always easy to find the right field data for comparisons with mesocosms.

In 1994, Candace Oviatt wrote a review of some MERL experiments in which she described various examples of important lessons learned from successes and failures of biological scaling. I will use one of each to illustrate her point and to lead into a last, closing rumination (perhaps a poor word choice for thoughts about fish and invertebrates). First, a success. By comparing the biological structure and dynamics of well-mixed water columns with and without benthic organisms, it became

TABLE 6•1 *Marine Mesocosms Currently Operated at the University of Rhode Island Graduate School of Oceanography*

	MERL Mesocosms	**Lagoon Mesocosms**
Number	14	10
Depth (m)	5	1.1
Area (m$^{2)}$	2.6	4
Volume (m$^{3)}$	13.1	4.4
Sediment depth (cm)	~30	~30
Mixing	Vertical plunger or rotary bars	Paddle wheels
Flushing rate	Usually 5% d^{-1}	Usually 5 to 10% d^{-1}; also run at 55% d^{-1}
Primary production	Phytoplankton based	Seagrass (*Zostera marina*) macroalgae, epiphytes, epibenthic microphytes, phytoplankton

NOTE MERL = Marine Ecosystems Research Laboratory

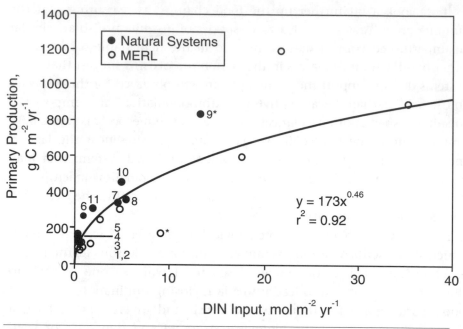

FIGURE 6•1 *Comparing Mesocosms with Natural Systems*
Primary production by phytoplankton as a function of the rate of supply of dissolved inorganic nitrogen. Inorganic phosphorus and silica were also supplied in various amounts approximating the Redfield Ratio with respect to the nitrogen. Open circles are Marine Ecosystem Research Laboratory (MERL) mesocosms during a nutrient addition experiment (Keller 1988). Solid circles are (1) Scotian Shelf, (2) Sargasso Sea, (3) North Sea, (4) Baltic Sea, (5) North Central Pacific, (6) Tomales Bay, Calif., (7) Continental Shelf off New York, (8) Outer Continental Shelf off the southeastern United States, (9) Peru upwelling, (10) Georges Bank, (11) Narragansett Bay (for primary production, see Oviatt et al. 1999; for nitrogen input, Nixon et al. 1995). References and assumptions are given in Nixon et al. (1996) except for the Peru upwelling, which is based on Chavez and Barber (1987). Data for the MERL mesocosm with a star were not included in the regression. The low production in that tank was due to intense filter feeding and depressed chlorophyll levels (Nixon et al. 1986). The Peru upwelling (9) was also excluded because of large uncertainty on the size of the upwelling area (Chavez and Barber 1987). The calculation of nitrogen input per unit area is thus highly uncertain.

very clear from several MERL experiments that whereas the benthic assemblages may enhance pelagic primary production through rapid nutrient regeneration, the individual organisms also prey heavily on the pelagic zooplankton. As a result, "when a benthos is absent or the system is stratified, copepods become abundant and ctenophores, arrow worms, jellyfish, and larval fish appear to prey on an abundant pelagic food supply. These studies suggest benthic dominance of shallow, well-mixed coastal areas and water column dominance of deeper, stratified marine waters" (Oviatt 1994). Oviatt also pointed out that this result from mesocosm observations was consistent with numerous descriptions of field situations, though the ability to completely eliminate the bottom community and to replicate treatments made the mesocosm data especially powerful. The lesson is that by manipulating systems in mesocosms, we can sometimes learn about important features of scale—in this case depth and vertical mixing—that may have remained obscure or ambiguous in field observations.

Now for a failure, or at least a problem. Because the benthos are so important in influencing the biology of the water column in shallow well-mixed systems, anything that has a serious impact on the benthos may also have a major impact on events in the water column. During a long-term (> 2 yr) nutrient addition experiment at MERL, some mesocosms developed very large populations of sand shrimp (*Crangon septemspinosa*) and mud crabs (*Neopanope texana*) that appeared to reduce markedly the abundance of the infauna (Oviatt 1994). Were the epibenthic predators a response to the experimental treatment or just a random event? Should they be

TABLE 6•2 *Summer Growth Rates (mm d^{-1}) of Some Fish in Unperturbed MERL and Coastal Lagoon Mesocosms Compared with Field Estimates for Similar-sized Fish*

Species	Size	Mesocosms	Field
Winter flounder (*Pleuronectes americanus*)	48 mm	0.28 L	0.18[a]
Mummichog (*Fundulus heteroclitus*)	48 mm	0.15 L	0.13[b]
Silversides (*Menidia menidia*)	30 mm	0.44 L	0.35[c]
Stickleback (*Gasterosteus aculeatus*)	31 mm	0.15 L	0.17[d]
Menhaden (*Brevoortia tyrannus*)	30 mm	0.85 M[e]	1.0[f]

[a]C. Powell, R. I. Department of Environmental Management (personal communication, 1988, 1989), Narragansett Bay

[b]Valiela et al. (1977), Cape Cod

[c]Bengtson (1984), Point Judith Lagoon, Rhode Island

[d]Snyder (1991), northern California

[e]Keller et al. (1990), Rhode Island

[f]Reintjes (1969), Rhode Island

NOTE Mean initial length is stated as size. L = lagoon mesocosms, data from B. Buckley, mean for four summers, May–September; M = MERL mesocosms, June–September.

TABLE 6•3 *Summer Growth Rates (μm d⁻¹) of Some Bivalves in Unperturbed MERL and Coastal Lagoon Mesocosms Compared with Field Estimates for Similar-sized Individuals*

Species	Size	Mesocosms	Field
Eastern oyster (*Crassostrea virginica*)	66 mm	0–50 L	140–180[a,b]
Blue mussel (*Mytilus edulis*)	36 mm 20 mm	16–60 L 254 M	70–250[c,d]
Bay scallop (*Argopecten irradians*)	38 mm	5–270 L	60–140[e]
Northern quahog (*Mercenaria mercenaria*)	44 mm 40 mm	50[f] 20–37 M[g]	0–22 L
Nutshell (*Nucula annulata*)	2 mm	3.5[h]	1–1.7[i,j]

[a]Loosanoff and Nomejko (1949), Connecticut
[b]Woodruff (1961), Long Island, New York
[c]Hilbish (1986), Long Island, New York
[d]Incze et al. (1980), Gulf of Maine
[e]Berkman (1986), Charlestown Pond Lagoon, Rhode Island
[f]Pratt and Campbell (1956), Narragansett Bay
[g]P. Doering, personal communication
[h]Craig (1994), Rhode Island
[i]Carey (1962), Long Island, New York, annual mean
[j]Sanders (1956), Long Island, New York, annual mean
NOTE Mean initial length is stated as size. Field data are annual means. L = lagoon mesocosms, data from B. A. Buckley, May–September (numerous years); M = MERL mesocosms.

removed completely from all tanks or reduced to some level? If the latter, to what level? It was in any case hard to census the shrimp and crabs and no one really knew how long they had been present in very high numbers. Trapping was not effective in reducing the shrimp. A fish (*Lophopsetta maculata*; now *Scophthalmus aquosus*) was added to some tanks, but it appeared not to eat the shrimp. It was removed and another species was tried (*Prionotus carolinus*). It, too, failed and was removed. At last, one 15–20 g young cunner (*Tautogolabrus adspersus*) was added to each of the shrimp-infested tanks; the shrimp were reduced to low numbers, and the biomass of infauna increased (Nixon et al. 1986). But there is always the nagging uncertainty of how different some results of the nutrient addition experiment might have been if the fish had been present all the time. On a more positive note is the fact that we learned about the potentially important role of a particular fish species, and the coincidence of dramatic changes in benthic biomass and in water-column biology helped to focus us on the benthic-pelagic links that have turned out to be so important.

HIERARCHY AND SCALE

The first ecology text I ever read (E. P. Odum's 1959 edition of *Fundamentals of Ecology*) introduced the concept of the food chain with what was once a well-known ditty by Jonathan Swift that went:

> Big fleas have little fleas
> upon their backs to bite 'em
> and little fleas have lesser fleas
> and so, ad infinitum.

In spite of its thermodynamic shortcomings, the rhyme stuck in my mind. A few years later I came across L. F. Richardson's (1922) conversion of the fleas into something of more interest to the student of fluid dynamics:

> Big swirls have little swirls
> That prey on their velocity
> And little swirls have lesser swirls
> And so on to viscosity.

From such poetic coincidence comes a conviction that there are certain general principles of nature, and that among them is what H. T. Odum (1983) described as "energy hierarchy":

> Flows of low-quality energy are abundant and widely dispersed, and individual units are small in size. Higher-quality units and their flows, although less in total energy flow, are more concentrated, and each unit is larger in size with a larger territory from which it receives energy and feeds back its actions.

Because the larger, rarer, higher quality units in the hierarchy are very costly in terms of energy to produce and maintain, it seems compelling to assume that the feedback controls they regulate must be very important for system survival. When we design mesocosms we often try to substitute our own work for that of larger natural systems. This probably works well with low-quality energy functions such as light input, temperature control, and perhaps mixing. But I suspect that we are poor substitutes for the high-energy quality components of nature with their claws and beaks and teeth and complex behavior. A Ph.D. in oceanography or zoology doesn't necessarily teach one how to be a good fish.

The hope of the mesocosmologist is that such control feedbacks operate only intermittently, or under special circumstances, and that at other times the activities of the big organisms are sufficiently uncoupled from the smaller that we can make useful observations about nature in their absence (Pilson and Nixon 1980). But even if this hope is realized, we need to remember that "understanding" a phenomenon requires not just a knowledge of how it arises from interactions at the next lower level of organization (traditional scientific reductionism) but also an awareness of the role that the next higher level of organization plays (Nixon 1996). Since mesocosms are deliberately isolated from the next larger scale, the bigger bugs and the large swirls, we must ultimately rely on field observations to scale up the results of mesocosm experiments. Doing so properly requires the best of our highest quality energy of human knowledge and imagination.

ACKNOWLEDGMENTS

Preparation of this note was supported by the Rhode Island Sea Grant College Program. Mike Kemp and Walter Boynton provided helpful suggestions during the review process.

LITERATURE CITED

Balch, N., C. M. Boyd, and M. Mullin. 1978. Large-scale tower tank systems. *Rapports et Procès-Verbaux des Réunions du Conseil International pour l'Exploration de la Mer* 173:13–21.

Bengtson, D. A. 1984. Resource partitioning by *Menidia menidia* and *Menidia beryllina*. *Marine Ecology Progress Series* 18:21–30.

Berkman, P. A. 1986. Ecological relationships between the bay scallop, *Argopecten irradians*, and its epizootic assemblage in Charlestown Pond, Rhode Island. Master's thesis, University of Rhode Island.

Beyers, R. J., and H. T. Odum. 1993. *Ecological Microcosms*. New York: Springer-Verlag.

Carey, A. G. 1962. An ecologic study of two benthic animal populations in Long Island Sound. Ph.D. diss., Yale University.

Carpenter, S. R. 1996. Microcosm experiments have limited relevance for community and ecosystem ecology. *Ecology* 77:677–680.

Chavez, F. P., and R. T. Barber. 1987. An estimate of new production in the equatorial Pacific. *Deep-Sea Research* 34:1229–1243.

Craig, N. I. 1994. Growth of the bivalve *Nucula annulata* in nutrient-enriched environments. *Marine Ecology Progress Series* 104:77–90.

Giesy, J. P., Jr., ed. 1980. *Microcosms in Ecological Research.* DOE Symposium Series 52. Springfield, VA: National Technical Information Service.

Grice, G. D., and M. R. Reeve, eds. 1982. *Marine Mesocosms: Biological and Chemical Research in Experimental Ecosystems.* New York: Springer-Verlag.

Graney, R. L., J. H. Kennedy, and J. H. Rodgers Jr., eds. 1994. *Aquatic Mesocosm Studies in Ecological Risk Assessment.* Boca Raton, Fla.: CRC Press.

Hilbish, T. J. 1986. Growth trajectories of shell and soft tissue in bivalves: Seasonal variation in *Mytilus edulis* L. *Journal of Experimental Marine Biology and Ecology* 96:103–113.

Hunt, C. D., and D. L. Smith. 1982. Controlled marine ecosystems—A tool for studying stable trace metal cycles: Long-term response and variability. In G. D. Grice and M. R. Reeve, eds., *Marine Mesocosms: Biological and Chemical Research in Experimental Ecosystems,* pp. 111–122. New York: Springer-Verlag.

Incze, L. S., R. A. Lutz, and L. Watling. 1980. Relationships between effects of environmental temperature and seston on growth and mortality of *Mytilus edulis* in a temperate northern estuary. *Marine Biology* 57:147–156.

Keller, A. A. 1988. An empirical model of primary productivity (^{14}C) using mesocosm data along a nutrient gradient. *Journal of Plankton Research* 10:813–834.

Keller, A. A., P. H. Doering, S. P. Kelly, and B. K. Sullivan. 1990. Growth of juvenile Atlantic menhaden, *Brevoortia tyrannus* (Pisces: Clupeidae) in MERL mesocosms: Effects of eutrophication. *Limnology and Oceanography* 35:109–122.

Lalli, C. M., ed. 1990. *Enclosed Experimental Marine Ecosystems: A Review and Recommendations. Coastal and Estuarine Studies* (formerly *Lecture Notes on Coastal and Estuarine Studies*). New York: Springer-Verlag.

Loosanoff, V. L., and C. A. Nomejko. 1949. Growth of oysters, *O. virginica,* during different months. *Biological Bulletin* 97:82–94.

Menzel, D. W., and J. H. Steele. 1978. The application of plastic enclosures to the study of pelagic marine biota. *Rapports et Procès-Verbaux des Réunions du Conseil International pour l'Exploration de la Mer* 173:7–12.

Nixon, S. W. 1969. A synthetic microcosm. *Limnology and Oceanography* 14:142–145.

Nixon, S. W. 1996. Regional coastal research: What is it? Why do it? What role should NAML play? *Biological Bulletin* 190:252–259.

Nixon, S. W., D. Alonso, M. E. Q. Pilson, and B. A. Buckley. 1980. Turbulent mixing in aquatic mesocosms. In J. P. Giesy Jr., ed., *Microcosms in Ecological Research,* pp. 818–849. Springfield, Va.: National Technical Information Service.

Nixon, S. W., J. W. Ammerman, L. P. Atkinson, V. M. Berounsky, G. Billen, W. C. Boicourt, W. R. Boynton, T. M. Church, D. M. DiToro, R. Elmgren, J. H. Garber, A. E. Giblin, R. A. Jahnke, N. J. P. Owens, M. E. Q. Pilson, and S. P. Seitzinger. 1996. The fate of nitrogen and phosphorus at the land-sea margin of the North Atlantic Ocean. *Biogeochemistry* 35:141–180.

Nixon, S. W., S. L. Granger, and B. L. Nowicki. 1995. An assessment of the annual mass balance of carbon, nitrogen, and phosphorus in Narragansett Bay. *Biogeochemistry* 31:15–61.

Nixon, S. W., C. A. Oviatt, J. Frithsen, and B. Sullivan. 1986. Nutrients and the productivity of estuarine and coastal marine ecosystems. *Journal of the Limnological Society of South Africa* 12:43–71.

Nowicki, B. L., and C. A. Oviatt. 1990. Are estuaries traps for anthropogenic nutrients? Evidence from estuarine mesocosms. *Marine Ecology Progress Series* 66:131–146.

Odum, E. P. 1959. *Fundamentals of Ecology*. Philadelphia: Saunders.

Odum, H. T. 1983. *Systems Ecology*. New York: Wiley.

Oviatt, C. A. 1994. Biological considerations in marine enclosure experiments: Challenges and revelations. *Oceanography* 7:45–51.

Oviatt, C., A. Keller, and L. Reed. 1999. Primary production patterns in Narragansett Bay during a year with no bay-wide winter-spring diatom bloom. Manuscript.

Petersen, J. E., L. P. Sanford, and W. M. Kemp. 1998. Coastal plankton responses to turbulent mixing in experimental ecosystems. *Marine Ecology Progress Series* 171:23–41.

Pilson, M. E. Q. 1985. Annual cycles of nutrients and chlorophyll in Narragansett Bay, Rhode Island. *Journal of Marine Research* 43:849–873.

Pilson, M. E. Q., and S. W. Nixon. 1980. Marine microcosms in ecological research. In J. P. Giesy Jr., ed., *Microcosms in Ecological Research*, pp. 724–741. Springfield VA: National Technical Information Service.

Pratt, D. M., and D. A. Campbell. 1956. Environmental factors affecting growth in *Venus mercenaria*. *Limnology and Oceanography* 1:2–17.

Reintjes, J. W. 1969. Synopsis of biological data on the Atlantic menhaden, *Brevoortia tyrannus*. *U.S. Fish and Wildlife Services Circular* 320:1–30.

Richardson, L. F. 1922. *Weather Prediction by Numerical Process*. New York: Cambridge University Press.

Sanders, H. L. 1956. Oceanography of Long Island Sound. X. The biology of marine bottom communities. *Bulletin of the Bingham Oceanographic Collection* 15:245–414.

Sanford, L. P. 1997. Turbulent mixing in experimental ecosystem studies. *Marine Ecology Progress Series* 161:265–293.

Santschi, P. H. 1985. The MERL mesocosm approach for studying sediment-water interactions and ecotoxicology. *Environmental Technology Letters* 6:335–350.

Santschi, P. H., D. M. Adler, and M. Amdurer. 1983. The fate of particles and particle-reactive trace metals in coastal waters: Radioisotope studies in microcosms. In C. S. Wong, E. Boyle, K. W. Bruland, J. D. Burton, and E. D. Goldberg, eds., *Trace Metals in Sea Water*, pp. 331–349. New York: Plenum.

Snyder, R. J. 1991. Migration and life histories of the threespine stickleback: Evidence for adaptive variation in growth rate between populations. *Environmental Biology of Fishes* 31:381–388.

Valiela, L., J. E. Wright, J. M. Teal, and S. B. Volkmann. 1977. Growth, production, and energy transformations in the salt-marsh killifish *Fundulus heteroclitus*. *Marine Biology* 40:135–144.

Woodruff, S. 1961. Critical growth and survival studies of the American oyster, *Crassostrea virginica* (Gmelin), in the Long Island area. Master's thesis, University of Rhode Island.

CHAPTER 7

Evaluating and Modeling Foraging Performance of Planktivorous and Piscivorous Fish
Effects of Containment and Issues of Scale

Michael R. Heath and Edward D. Houde

FEEDING BY FISH MAY EXERCISE A "TOP-DOWN" CONTROL OVER LOWER trophic levels in aquatic ecosystems. Both species and size composition of prey may be controlled. The process through which top-down control is exercised is dependent upon feeding behaviors and consumption potential of fish. Evidence for top-down control has been derived primarily from experimentation in artificially enclosed freshwater ecosystems (Threlkeld 1987; Carpenter 1988; Horsted et al. 1988; Carpenter and Kitchell 1992; DeMelo et al. 1992; Abreu et al 1994; Ramcharan et al. 1995, 1996; Proulx et al. 1996; Brett and Goldman 1997; Vanni and Layne 1997). Results of those studies sometimes produced inconsistent and conflicting results, which are difficult to reconcile with respect to behavior of natural ecosystems (DeMelo et al. 1992). Some investigators have argued that artificially enclosing fish to determine effects of predation on plankton community structure is a flawed approach because outcomes are very likely to be dominated by scale-dependent effects of containment (Harte et al. 1980; Frost et al. 1988; Carpenter 1996). Others have argued that there are circumstances under which estimation of predation effects on the size structure of plankton communities and of rate processes in lake ecosystems or experimental enclosures may be meaningful (Neill 1994), particularly when enclosed ecosystems are subjected to large perturbations. Nevertheless, the explicit consequences of containment for the performance of predators and the response of aquatic ecosystems have only occasionally been considered (Englund 1997; MacNally 1997).

Studies on growth and vulnerability of larval fish to predators have provided some of the best examples of the consequences of containment. For example, Paradis et al. (1996) concluded that the vulnerability of fish larvae to predation is strongly and inversely related to enclosure size. There is also clear evidence that growth of fish larvae is inversely related to tank size (Theilacker 1980). In an experiment on growth potential of goby (*Gobiosoma bosc*), Perry et al. (1995) enclosed goby larvae in cylindrical containers of 1 to 150 L and found that survival, growth, and production were strongly depressed in enclosures < 50 L (figure 7.1). The results suggested that reduced foraging by the goby larvae (2–10 mm length) in the smaller containers, which probably was due to increased probability of wall contact, was a principal cause of the reduced production. Similarly, Øiestad (1985, 1990) demonstrated that larval-fish growth rates were higher, and prey-density requirements for survival apparently lower, in experiments conducted in large mesocosms compared to laboratory investigations in smaller tanks. Zooplanktivorous larval fish apparently foraged more successfully on low prey concentrations in the larger enclosures. Based on these observations, Øiestad (1990) concluded that starvation was a less significant source of natural mortality in fish larvae than had been supposed, based upon earlier laboratory experiments. The earlier research, which was mostly conducted in small tanks, lacked appreciation of containment effects and had indicated that conditions suitable for growth and survival of fish larvae were to be found only in dense patches of plankton in the sea.

If containment constrains feeding, it is probable that ingestion rates at a given prey concentration (foraging efficiency) of planktivorous fish will increase as enclosure volumes and dimensions increase. However, for visual-pursuit predators such as piscivorous fish, the situation is less clear and there might be an inverse relationship between rate of ingestion and enclosure dimensions. Predators held under confinement should be able to continuously monitor the available space for the presence of prey, with inevitable consequences, sometimes labeled the "sledgehammer effect," for prey populations (Crowder et al. 1988). In addition, the establishment of density-dependent predation, which may have a strong stabilizing influence on community dynamics in natural ecosystems, must depend critically upon relationships between containment dimensions and search volumes of predators.

In this chapter we address the question "Are there general scaling rules that might allow predator-prey dynamics in natural aquatic ecosystems to be understood from results obtained in experimental systems?" For natural ecosystems such as Chesapeake Bay, where the dominant planktivores are small fish (< 100 mm total length) and jellyfishes that

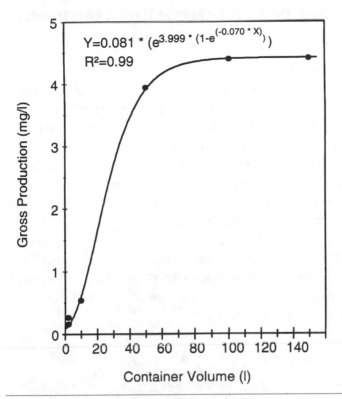

$$Y=0.081 * (e^{3.999 * (1-e^{(-0.070 * X)})})$$
$$R^2=0.99$$

FIGURE 7•1 *Container Effects on Goby Production*
Gross production (mg l^{-1}) of *Gobiosoma bosc* larvae after 20 d in containers ranging from 1 to 150 L in volume. Although not illustrated, equally strong effects of container volume on growth and survival rates were observed. (From Perry et al. 1995)

occur at high mean densities (for example, bay anchovy [*Anchoa mitchilli*] at > 1 m^{-3}), it may be that meaningful experiments to estimate consumption by planktivores and their potential for growth can be carried out in modest-sized experimental ecosystems. Results of recent experiments in the Multiscale Experimental Ecosystem Research Center (MEERC) with planktivorous fish (Houde 1997) support this belief, although they also demonstrated that growth of bay anchovy clearly was inversely related to size of experimental enclosure (Mowitt 1999). Here, we explore issues of spatial scale and implications for foraging efficiency by developing a foraging model to evaluate how efficiencies of planktivorous fish and pursuit piscivores vary in relation to containment dimensions. Based upon modeled results, we recommend enclosure sizes that may be appropriate to contain fish for ecological research.

CONCEPTUAL MODEL OF SCALE-DEPENDENT CONSTRAINTS ON FORAGING

The relationship between the ratios of volume:area and diameter:depth illustrates differences in scale between typical experimental enclosures and natural pelagic ecosystems (figure 7.2). Even the smallest natural lakes have volume:wall area ratios > 50 and diameter:depth ratios > 20, which greatly exceed those ratios in most experimental ecosystems. The latter usually are limited in a logistical sense by available space (or research budgets) and seldom exceed 100 m^3, and correspondingly have low volume:depth and diameter:depth dimensions (Petersen et al. 1999). Large pelagic ecosystems (lakes, estuaries, oceans) lie at far extremes on the extended axes of figure 7.2. The MEERC Pelagic-Benthic experimental ecosystems, which have been used in experiments enclosing small fish (40–60 mm length), range in volume from 1 to 10 m^3. These ecosystems (i.e., mesocosms) are fully represented in the compressed lower-left corner of the response space in figure 7.2.

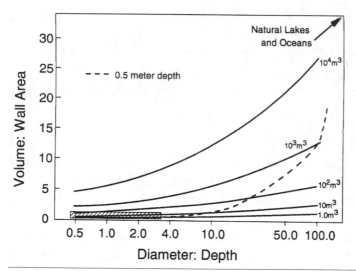

FIGURE 7•2 *Scaling Comparisons of Natural and Experimental Ecosystems*
The relationship between volume:wall area and diameter:depth of cylindrical ecosystems. Lines are plotted for factor of 10 increases in volume. Ecosystems with high volume:wall area and high diameter:depth ratios are the most pelagic. The dashed line represents 0.5 m depth, below which pelagic fish are unlikely to behave "near-normally." The arrow indicates the off-scale directions in which natural lake and marine ecosystems would be represented. The shaded area in the lower left encompasses the dimensions of MEERC pelagic-benthic mesocosms.

Planktivores in natural pelagic ecosystems generally have few immediate physical barriers to impede activities. Environmental barriers such as temperature, salinity, or dissolved oxygen gradients are more likely to constrain foraging. In contrast, encounters with containment walls must alter foraging behavior of such organisms in experimental systems and could distort the functional relationship between consumers and prey. Such enclosure effects should assume increasing significance with decreasing volume:wall area and diameter:depth ratios. Constraints of enclosure probably will affect consumption rates of planktivores by reducing volume clearance rate or by altering swimming speed. In contrast, some pursuit predators such as piscivorous fish might forage most effectively under confinement, at least until enclosure dimensions become too small to allow effective swimming and attack behaviors (figure 7.3).

In our conceptual model, the functional responses of predators in relation to prey are hypothesized to vary with ecosystem size (or the size of predator relative to the enclosed ecosystem) and to differ fundamentally between planktivores and piscivores. The concept is illustrated by the hypothetical

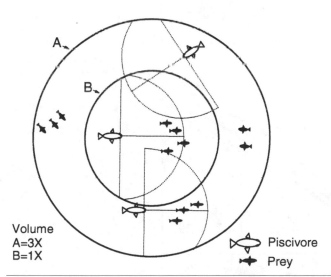

Volume
A=3X
B=1X

Piscivore
Prey

FIGURE 7•3 *Conceptual Model of Enclosed Planktivorous and Piscivorous Fish*
Conceptual model of schooling planktivorous fish and piscivores enclosed in mesocosms (A and B) of surface-area dimensions varying by 3X. Hypothetically, the planktivore will have increased foraging efficiency in the larger enclosure, which is least likely to disrupt schooling and foraging behavior. In contrast, the piscivore is predicted to have greater efficiency in the smaller mesocosm, where it can monitor prey more effectively. Planktivore and piscivore densities are equal in the A and B mesocosms.

functional response curves in figure 7.4. The relative consumption rates at a given prey concentration can be regarded as measures of foraging efficiency. The slope (q) of a functional response curve represents the catchability of the prey, or "predator power." The concepts of foraging efficiency and predator power are analogous to "catch rate" and "catchability" in the fisheries literature (Gulland 1983). Catchability is the fraction of a population taken per unit of fishing effort or, in our analogy, a single unit of predator effort. Fisheries scientists often assume that catchability, or relative vulnerability of a species to a fishing fleet, remains constant as the abundance of fish changes,

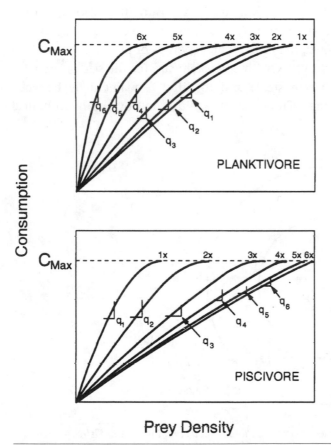

FIGURE 7•4 *Hypothetical Functional Responses of Planktivores and Piscivores in Enclosures* Hypothetical relationships between prey captures and prey densities for top planktivorous fish, and bottom piscivorous fish. The rate of change in foraging effectiveness (planktivore) or "predator power" (piscivore) with prey concentration is indexed by the coefficient q, which is hypothesized to have opposite relationships to ecosystem sizes (1x to 6x) for planktivores and piscivores. C_{max} = maximum consumption or prey capture rate.

and that the catch rate is directly proportional to fish density. But, pelagic fishing fleets can often maintain or even increase catch rates in the face of declining abundances if the habitable range (effective ecosystem size) declines while school sizes remain relatively constant. Under this circumstance, search times by fishing vessels are reduced, encounter rates are elevated, and catchability increases. This situation is analogous to that of a hypothetical piscivore whose consumption rate and "predator power" increase as enclosure size declines (figure 7.4). For planktivores, we hypothesized that there is an opposite trend in which normal foraging behavior of planktivorous fish (schooling and unimpeded swimming) is disrupted by contact with walls in compressed ecosystems, leading to decreased foraging efficiency in small enclosures.

MODELING PLANKTIVORE AND PISCIVORE BEHAVIOR

To test our hypothesis of how containment might affect the ingestion performance of plankton-feeding and pursuit-piscivore fish, we developed individual-based foraging models that simulated interactions between predator and prey, and between predators, under various scenarios of behavioral interactions with containment walls. The objective was to simulate and evaluate the consequences of changes in enclosure (ecosystem) shape and volume with respect to foraging efficiency of planktivores and predatory performance of piscivores.

Model Specification

The individual-based models simulated the swimming paths of fish in a three-dimensional cylindrical tank, with defined rules governing reactions to encounter with the enclosure's wall and interactions between individuals. The planktivorous fish fed by filtering prey from the water. In this case, prey were not explicitly represented and declines in foraging efficiency resulted solely from planktivore behavior with respect to interactions with the enclosure walls. The piscivore fed by preying upon the simulated planktivorous fish. Each fish in the model was assigned a body length, to which mean swimming speed and various detection and reaction distances were scaled accordingly.

The basis for the models was a finite difference scheme to estimate displacement of particles in a three-dimensional, rectilinear coordinate system according to vectors representing the swimming speed and direction of each modeled particle. A swimming vector was assigned to each individual at each time step according to specified rules governing interactions and a scheme to incorporate stochasticity (figure 7.5). Parameters describing characteristics of planktivores and piscivores in the model are listed in table 7.1.

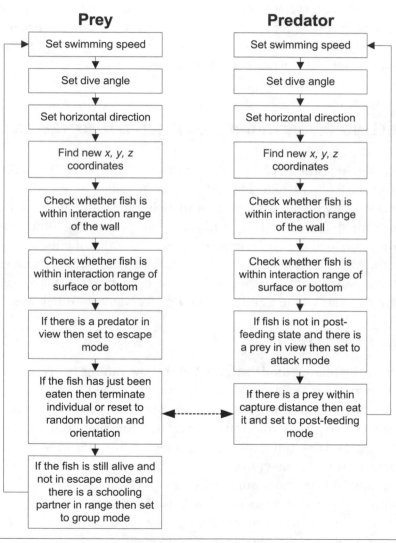

FIGURE 7•5 *Predator-Prey Model Flow Diagram*
Flow diagram of the cycle of operations applied to each individual planktivore and piscivore on each time step of the simulation model.

TABLE 7•1 *Parameters Describing the Behavioral Characteristics of Planktivores and Piscivores in the Simulation Model*

Parameter Description	Value for Planktivores	Value for Piscivores
Body length (cm)	4	40
Maximum separation distance for schooling (body lengths)	3	n/a
Escape reaction distance (prey body lengths)	3	n/a
Prey detection distance (predator body lengths)	n/a	2
Prey proximity for 100% capture probability (predator body lengths)	n/a	0.1
Refractory delay following capture event before initiation of further pursuit (sec)	n/a	30
Maximum turning angle per time step during cruise swimming (degrees/sec)	0.4	0.4
Maximum turning angle at tiller resets (degrees)	45	45
Maximum change in dive angle per time step during cruise swimming (degrees/sec)	0.04	0.04
Maximum change in dive angle at flap resets (degrees)	5	5
Time between tiller and flap resets (sec)	5	5
Maximum stochastic variation in swimming speed (fraction of mean speed)	0.1	0.1
Mean cruise swimming speed (body lengths/sec)	2	2
Ratio of burst speed/cruise speed	5	1
Horizontal angle of detection field (degrees)	180	180
Upward angular limit of detection field (degrees above body axis)	80	80
Downward angular limit of detection field (degrees below body axis)	20	20

NOTE   n/a = not applicable

Stochasticity in individual vectors was introduced at three levels in a manner designed to preserve temporal autocorrelation. First, scalar speed for each individual was drawn at random from a uniform distribution around a fixed value that was scaled to body length. Second, barring interaction by an individual with enclosure walls or with another fish, a "tiller" and "flaps" were reset at prescribed time intervals. The tiller controlled the mean horizontal component of swimming direction until the next reset, and the flaps determined the dive angle. At each reset the new mean orientations were drawn at random from uniform distributions limited by prescribed angular variations around the existing tiller and flap settings. Finally, at each time step the realized horizontal orientation and dive angle were drawn at random from uniform distributions limited by prescribed angular variations around the existing tiller and flap settings. The maximum angular variation associated with each time step was typically one order of magnitude smaller than that applied at tiller and flap resets, and variations in dive angle were about one order of magnitude less than those in horizontal orientation.

Two alternative scenarios were defined for probable behaviors of individuals at or near the enclosure wall. Scenarios were selected to resemble fish behaviors that had been observed in MEERC mesocosms. In the first and simplest scenario (W_A), fish were simply reflected back into the tank as would be a ray of light. This scenario was referred to as "ray reflection" and planktivores were considered incapable of foraging during a time step in which they were in contact with the wall. In the second scenario (W_B), whenever individuals were within 3 body lengths of the wall their tiller was reset tangential to the wall. When between 3 and 1 body lengths of the wall a small stochastic component, equivalent to that applied on each time step, also was applied to the tiller. Within 1 body length of the wall a larger stochastic component, equivalent to that applied on normal tiller reset, was added. Upon contact with the wall, individuals behaved as in the ray reflection case. In scenario W_B, planktivores were considered incapable of foraging when within 1 body length of the wall. The piscivores, when near or in contact with the enclosure walls, were permitted to capture planktivores.

Fish interacting with either the surface or bottom of the enclosure were capable of foraging. In the event of surface or bottom contact, the flaps were reset to ensure horizontal swimming until the next reset. Any stochastically assigned vectors that implied displacement through the bottom or surface of the tank were defaulted to maintain the fish at its present depth.

Alternative rules governing probable interactions between planktivores also were specified (Pitcher and Parrish 1993). In each alternative, fish adjusted course to swim parallel to their nearest neighbor within a prescribed radius of body lengths. A group of individuals so aligned was referred to as a school. Stochastic components of speed and direction continued to be applied at each time step to each fish, but were subsequently reset for members of a school according to one of two alternatives that were judged to represent possible behaviors. In the first scenario (S_A), each member adopted the mean speed and orientation of all members of the school, while in the second scenario (S_B) school members adopted the speed and direction of a lead fish, defined as being the member nearest the enclosure wall. Individuals were free to leave schools at any time if wall interactions caused them to move outside the prescribed radius of existing school members.

Interactions between piscivores and planktivores were governed primarily by reaction and detection distances that were linearly scaled to body length, but also by the visual field of both the predator and prey (Miller et al. 1992, 1993). Fish typically have a visual field that approximates one quarter of a spherical volume (Dunbrack and Dill 1984). In the horizontal plane, the visual field typically extends forward from a line approximately perpendicular to the axis, while in the vertical plane vision is typically limited to approximately 80° upward from the horizontal and 20° downward.

There were three components to piscivore-planktivore interactions in the model. First, on each time step the planktivores reacted to the nearest piscivore within their visual field (if one was present) by resetting their tiller to swim directly away from the threat with a burst speed that was a linear multiple of their normal cruising speed (table 7.1). Second, a piscivore attacked the nearest planktivore within its visual field (if one was present) by resetting its tiller toward the prey. Finally, if a planktivore came within a prescribed fraction of the piscivore's body length then it was assumed to have been attacked and captured. Following capture of a prey, the predator experienced a refractory period of prescribed duration during which further encounters with prey failed to elicit an attack.

Simulation Performance Indices

The average foraging efficiency of each individual planktivore over the course of a simulation was calculated as the proportion of time *not* spent interacting with the tank wall or evading a predator:

$$\varepsilon_t = \frac{1}{1 + \textit{(Average interactions/second)}}$$

Time-step dependency was eliminated by rescaling:

$$\varepsilon_{min} = \frac{1}{1 + \textit{(Time steps/second)}}$$

$$\varepsilon = \frac{\varepsilon_t - \varepsilon_{min}}{1 - \varepsilon_{min}}$$

In the piscivore simulations, each captured prey was immediately replaced with a new, randomly located individual, such that prey density remained constant despite the predation. Two indices of predator performance were defined. The mortality rate of the prey is

$$M = \frac{\textit{(prey eaten/unit time)}}{\textit{(prey numbers)}}$$

"Predator power" (analogous to catchability) is

$$C = \frac{\textit{(prey eaten/unit time)} \cdot \textit{(tank volume)}}{\textit{(prey numbers)} \cdot \textit{(predators numbers)}}$$

Simulation Strategies

Two categories of simulations were specified. The first simulated the foraging performance of planktivores in relation to containment scale and was conducted without predators present. The second category examined the performance of piscivores in relation to containment scale, and included an evaluation of the consequences of predation pressure on foraging efficiency of the planktivores. In the first category of simulations, enclosure depth was irrelevant because, in the model, proximity to the bottom or surface did not affect foraging efficiency or horizontal swimming of the planktivores. Consequently, scaling relationships between planktivore body size and enclosure size could be expressed relative to enclosure diameter alone. However, for the second category of simulations, depth was potentially very important because the visual field of piscivores and the predator-detection field of planktivores were three-dimensional volumes that clearly could impinge upon the bottom or surface of an enclosure.

In the tests that included both planktivores and piscivores, simulations were conducted for combinations of enclosure volume and shape (diameter:depth ratio) for several scenarios of fish behavior and concentration. In all cases the number of piscivores was constant at one per enclosure. These simulations were run in two sets. In the first set, enclosure volume was held constant and the concentration of planktivores varied between 0.2 m^{-3} and 51 m^{-3} for each of three different enclosure shapes (diameter:depth). In the second set, the concentration of planktivores was held constant (5 m^{-3}) and enclosure volume varied between 0.2 m^3 and 102 m^3 for each of six enclosure shapes. In all cases, behavior of the piscivore at the enclosure wall was configured for ray reflection (scenario W_A), and sets of runs were repeated with different planktivore behaviors.

All simulations were run for 8,000 time steps with a step duration of 0.2 s. This value was selected as a compromise between the time required to run the model, and the number of predator-prey interactions in each run. Typically, 20 to 40 (max 51) capture events were simulated in 8,000 time steps with the configurations specified here, and a typical run required about 5 min on a 133MHz Pentium processor PC, but this depended on the number of predators and prey involved. Sensitivity analysis indicated that, at the predator and prey configurations specified, results were free of numerical instabilities at this step duration and were independent of initial conditions beyond approximately 2,000 steps. The model code was written in a compiled version of BASIC.

Model Results

SIMULATED TRAJECTORIES. In the absence of schooling behavior, simulated trajectories of fish indicated that individuals tracked along the enclosure walls when wall behavior scenario W_B (turn parallel to the wall) was specified and spent more time in the central part of the enclosure when scenario W_A (ray reflection) was operative (figure 7.6). When schooling, the planktivore fish tended to circle the cylindrical enclosure, with individuals closest to the wall dictating behavior of the school. Behavior at the wall and the schooling behavior option (S_A or S_B) combined to determine the proportion of time that each individual spent in a school. The proportion was highest for wall scenario W_A combined with schooling scenario S_A (adopt group average velocity), and lowest for wall scenario W_B combined with schooling scenario S_B (adopt leader's velocity) (figure 7.7).

PLANKTIVORE FORAGING EFFICIENCY. In the absence of schooling behavior, planktivore foraging efficiency varied in an orderly manner with respect to enclosure (= tank) diameter and body length. Foraging efficiency

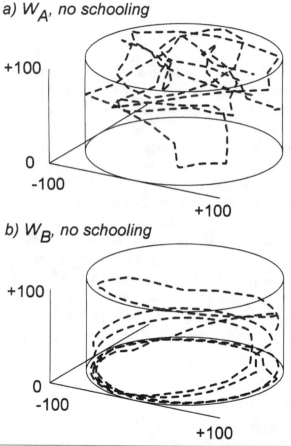

a) W_A, no schooling

+100

0
-100

+100

b) W_B, no schooling

+100

0
-100

+100

FIGURE 7•6 *Modeled Trajectories of Nonschooling Fish in Enclosures*
Examples of the simulated trajectories of individual nonschooling planktivores (4 cm body length) in a tank of 200 cm diameter and 100 cm depth with different wall behavior scenarios. In each case the trajectory shown corresponds to one individual and has a duration of 400 s. (a) Wall behavior scenario W_A (ray reflection). The individual shown had a foraging efficiency of 0.948. (b) Wall behavior scenario W_B (turn parallel). The individual shown had a foraging efficiency of 0.959.

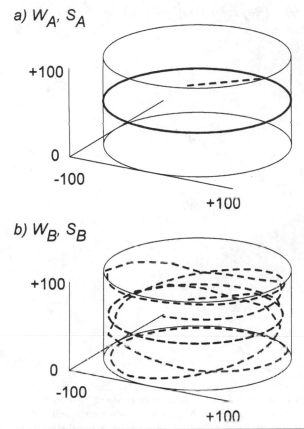

FIGURE 7•7 *Modeled Trajectories of Schooling Fish in Enclosures*
Examples of the simulated trajectories of individual schooling planktivores (4 cm body length) in a tank of 200 cm diameter and 100 cm depth with different wall behavior scenarios. In each case the trajectory shown corresponds to one individual out of a group of 6 in the simulation and has a duration of 400 s. (a) Wall behavior scenario W_A (ray reflection) and schooling scenario S_A (group average). The individual shown had a foraging efficiency of 0.427 and spent 100 percent of the time in a school. (b) Wall behavior scenario W_B (turn parallel) and schooling scenario S_B (follow the leader). The individual shown had a foraging efficiency of 0.867 and spent 41 percent of the time in schools.

increased with enclosure diameter and decreased with body length (figure 7.8). Variations in body length, which determined swimming speeds (i.e., 2 body lengths sec^{-1}), were exactly accounted for by relating foraging efficiency to the ratio of mean swimming speed/tank diameter. This relationship was broadly similar for the different wall-behavior scenarios, but with quite different sensitivity (figure 7.9). As the ratio of swimming speed/tank diameter increased (i.e., decreasing enclosure diameter for a given size of fish), foraging efficiency declined, and it declined more

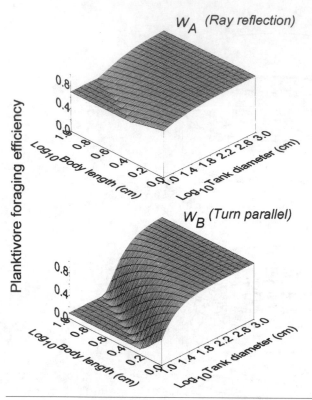

FIGURE 7•8 *Effects of Enclosure Size and Body Size on Planktivore Foraging Efficiency*
Response surface of planktivore foraging efficiency with respect to body length and tank diameter for different wall behavior scenarios (ray reflection; turn parallel). In both cases the mean swimming speed was 2 body lengths per second.

rapidly when planktivores within 3 body lengths of the wall were programmed to turn parallel to the wall (scenario W_B) than it did with ray reflection (scenario W_A).

Schooling behavior modified the relationship between foraging efficiency and the ratio of swimming speed/tank diameter under some combinations of scenarios. For group average schooling (S_A), the efficiency decreased rapidly with increasing ratio values to a threshold and then remained relatively constant (figure 7.10). This response was observed because, in the model, foraging efficiency was inversely related to the proportion of time spent in schools by an individual planktivore, a consequence of the trajectories of schools being driven by proximity to the enclosure walls. Above swimming speed/tank diameter ratios of approximately 0.06 s^{-1}, the constraints of the enclosure walls forced individuals permanently into schooling formations. For planktivore fish

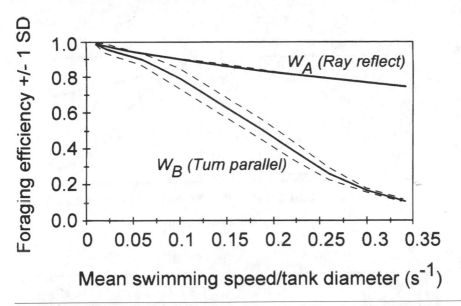

FIGURE 7•9 *Enclosure Scale, Wall-avoidance Behavior, and Foraging Efficiency of Non-schooling Fish*
Generic relationships between planktivore foraging efficiency and the ratio of mean swimming speed to tank diameter for different wall behavior scenarios (ray reflection; turn parallel). Dashed lines indicate the variability (±1 SD) among individuals within a simulation run.

of 4 or 8 cm length under scenario S_A, enclosure diameters at which threshold foraging efficiency became constant were 133 and 266 cm diameter, respectively. Efficiencies increased rapidly in larger enclosures (figure 7.10).

Under scenario S_B, in which the outermost individual in a school effectively protected others in the group from interaction with the wall by dictating the group swimming speed and direction, the mean foraging performances (figure 7.10) differed little from those of the nonschooling planktivores (figure 7.9). With scenario S_B in effect for schooling planktivores of 4 or 8 cm length, enclosure diameters of 133 and 266 cm, respectively, were associated with foraging efficiencies of approximately 90 percent. Efficiencies decreased steadily as enclosure diameters decreased (figure 7.10).

PISCIVORE PERFORMANCE WITH CONSTANT ENCLOSURE VOLUME AND VARIABLE PREY CONCENTRATION. In the first set of runs to simulate piscivory, a single piscivore was added to enclosures in which planktivore concentrations varied between 0.2 and 51 m^{-3} in cylindrical enclosures of different shape

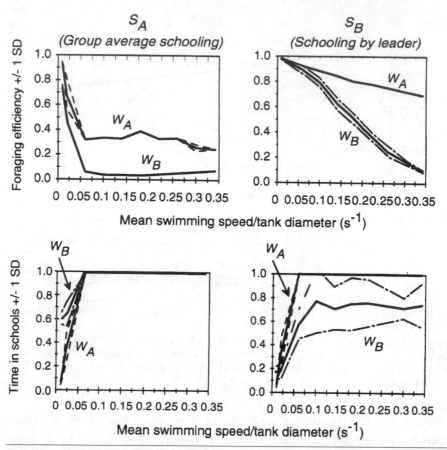

FIGURE 7•10 *Enclosure Scale, Schooling Behavior, and Foraging Efficiency of Schooling Fish* *Upper panel*: Relationship between planktivore foraging efficiency and the ratio swimming speed:tank diameter for different wall behavior and schooling scenarios. *Lower panel*: Relationship between the proportion of time spent in schools and the ratio swimming speed:tank diameter for different wall behavior and schooling scenarios. W_A = ray reflection; W_B = turn parallel. Dashed lines indicate the variability (±1 SD) among individuals within a simulation run.

but equal volume. Planktivore (the prey) behavior was configured for the simplest scenario (W_A, no schooling). Relative predation rate conformed to a saturating function, with a half saturation prey concentration of approximately 0.5 m^{-3}, and was independent of tank shape (figure 7.11). In these runs, predator power varied in precisely the same manner as prey mortality rate because the concentration of predators was constant. Predator power and planktivore mortality rate were inversely related to planktivore concentrations (figure 7.12). Both decreased as a power function of prey concentration, except at low prey concentrations (< 0.4

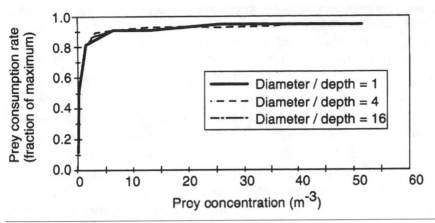

FIGURE 7•11 *Effects of Enclosure Dimensions on Piscivore Consumption*
Consumption rate of planktivores by a single piscivore in relation to planktivore concentration in simulations with constant enclosure volume (4 m³), but variable shape expressed by the ratio of tank diameter:depth.

m^{-3}) in enclosures of diameter:depth < 4 (figure 7.12). The decline of predator power in narrow enclosures at the lowest prey concentrations was observed because the predator's (a 40 cm fish) searching efficiency was impaired under those conditions.

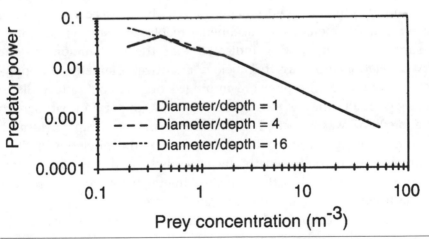

FIGURE 7•12 *Predator Power (Catchability) of Piscivores in Enclosures*
"Predator power" in relation to prey concentration in simulations with constant enclosure volume (4 m³) but variable shape expressed by the ratio of tank diameter:depth.

PISCIVORE PERFORMANCE WITH CONSTANT PREY CONCENTRATION BUT VARIABLE ENCLOSURE SHAPE AND VOLUME. Simulations were run at a constant saturating concentration of planktivore prey ($5 \ m^{-3}$). Under that specification, the numbers of prey varied between 1 and 512 for the range of enclosure sizes investigated. Sets of runs were repeated with two combinations of planktivore behavioral scenarios (W_A, no schooling; and W_A, S_A). In the simulations with no schooling by planktivores, their mortality rate decreased as a power function of enclosure volume for enclosures of diameter:depth ≥ 4 across the range of enclosure volumes tested (figure 7.13). Enclosures with a smaller diameter:depth ratio also conformed to this relationship if enclosure volumes were $\geq 3 \ m^3$. However, mortality rates were lower than rates indicated by the general relationship in enclosures $< 3 \ m^3$ when enclosure diameter:depth was < 4 (i.e., narrow enclosures). The power function decline in prey mortality rate was generated because piscivore numbers remained constant in these simulations while prey numbers increased with enclosure volume. Under the simulated conditions, predator power declined rapidly in narrow enclosures (diameter:depth ratios ≤ 4) as enclosure volumes decreased, but was essentially independent of enclosure volume or shape when diameter:depth was > 4 (figure 7.13).

With nonschooling prey, predator power in an enclosure of given volume was related to enclosure diameter, but not depth, by a general relationship in which the performance of a piscivore was constant above some critical diameter and declined steeply in smaller diameter enclosures. Sensitivity testing indicated that the explanation for the observed relationship was that below a critical diameter the prey-detection volume of the piscivore impinged on the enclosure walls to some degree at all times. In effect, the volume swept per unit path length of the predator was impaired, especially in narrow deep enclosures, making the predator a less efficient forager. For the piscivore in these simulations (40-cm body length, mean cruise speed 80 cm s^{-1}, prey detection distance 80 cm), the critical diameter below which predator power declined rapidly was approximately 200 cm.

In an exact repetition of the above simulations, except that group-averaged schooling behavior was designated for the planktivore prey (scenario W_A, S_A), results differed dramatically. With schooling planktivore prey in enclosures of diameter:depth ≥ 4, predator power of the piscivore decreased approximately exponentially with increasing enclosure volume up to approximately $50 \ m^3$ and was relatively constant

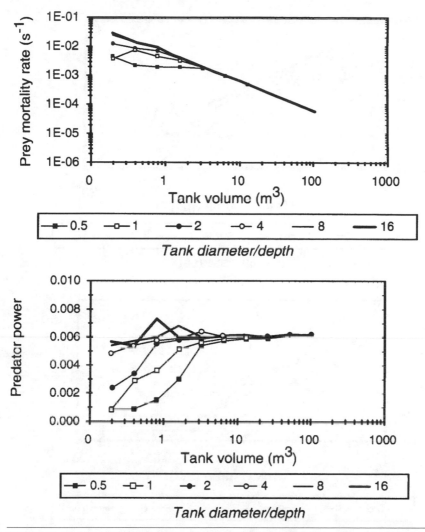

FIGURE 7•13 *Mortality of Nonschooling Prey and Predator Power of Enclosed Piscivores*
Upper panel: Mortality rate of nonschooling planktivores in simulations with constant planktivore concentration (5 m^{-3}) but variable enclosure volume and shape, and a single predator in each case. *Lower panel*: "Predator power" in simulations with a constant concentration of nonschooling planktivore prey (5 m^{-3}) but variable enclosure volume and shape.

thereafter (figure 7.14). In tanks smaller than 50 m^3, predator power was sensitive to tank shape, with low diameter:depth ratios leading to the greatest reduction in the piscivore's efficiency. The lowest diameter:depth ratio (0.5) resulted in lowest values of predator power (≤ 0.002) across the range of volumes investigated. We conclude that large volume and/or tall-narrow enclosures provided refuge from predation for the schooling

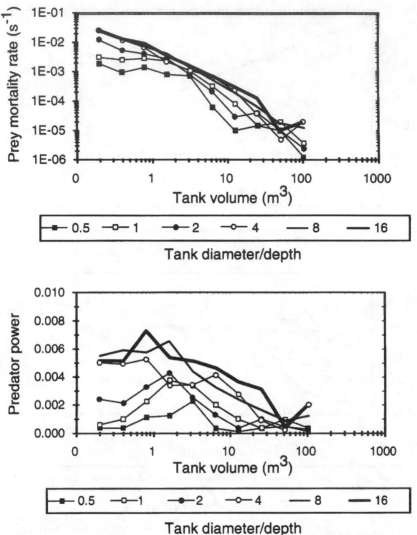

FIGURE 7•14 *Mortality of Schooling Prey and Predator Power of Enclosed Piscivores*
Upper panel: Mortality rate of schooling planktivores in simulations with constant planktivore concentration (5 m^{-3}) but variable enclosure volume and shape, and a single predator in each case. *Lower panel*: "Predator power" in simulations with a constant concentration of schooling planktivore prey (5 m^{-3}) but variable enclosure volume and shape.

planktivores. Wide-shallow (diameter:depth $\geq$ 4) enclosures did not provide refuge for schooling prey. Prey mortality rates and predator power both were highest in enclosures of those dimensions. The simulation results under the prey-schooling scenario (figure 7.14) strongly supported our conceptual model of predatory potential (figures

7.3 and 7.4). However, results for nonschooling prey, while partly consistent with our hypothesized relationships, did not show the hypothesized decline in predator power with increasing enclosure dimensions (figure 7.13).

EFFECTS OF PREDATION PRESSURE ON THE FORAGING EFFICIENCY OF PLANKTIVORES. In the simulations, foraging efficiency of planktivores was expected to decline with increasing predation pressure because, in the model, planktivores did not feed during escape reactions. Results from model runs with a constant planktivore concentration and a single predator indicated that foraging efficiency by the planktivores was markedly lower than in the absence of a predator (figure 7.15). In the first set of piscivore-planktivore runs (planktivore behavior W_A, no schooling) the largest effect occurred in simulations combining the smallest enclosure volumes (i.e., high predator concentration relative to prey concentration) and highest diameter:depth

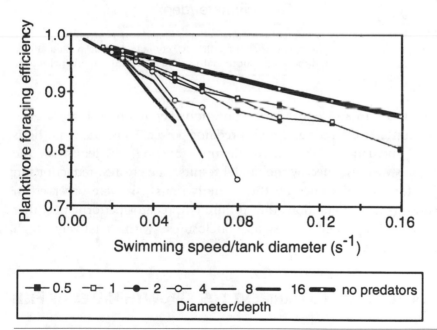

FIGURE 7•15 *Enclosure Scale, Predator Presence, and Planktivore Foraging*
Foraging efficiency of nonschooling planktivores in simulations with predation by a single piscivore and constant planktivore concentration (5 m^{-3}), but variable enclosure volume and shape. Also plotted is the generic relationship obtained in the absence of predation (see figure 7.9).

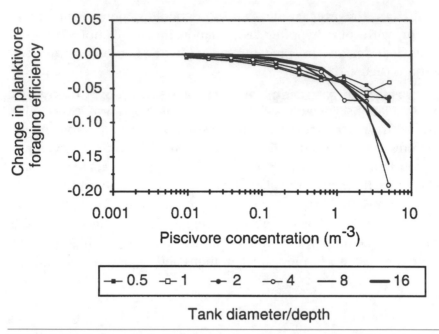

FIGURE 7•16 *Enclosure Scale, Predator Concentration, and Planktivore Foraging*
Differences in nonschooling planktivore foraging efficiency compared between simulations with predation (figure 7.15) and those without predators present (figure 7.9), in enclosures of different volume and shape.

ratios (figures 7.15 and 7.16). These conditions favored both high rates of predator contacts per prey and high predator power. Thus, the hypothesis that high predation risk should disrupt foraging efficiency of the planktivores was sustained by the model results. The greatest reductions in modeled foraging efficiency by the planktivorous fish were observed in enclosures with configurations that promoted predator performance, that is, nonschooling prey in small enclosures with relatively high diameter:depth ratios.

IMPLICATIONS OF CONTAINMENT FOR GROWTH RATES OF FISH

The individual-based models simulated foraging behaviors of fish and demonstrated that foraging efficiencies and consumption rates were sensitive to enclosure dimensions. Interpreting the consequences with respect to effects on growth rates of fish required a further stage of modeling. We were able to address the issue only for the planktivores by applying a bioenergetics model (Hewett and Johnson 1987, 1992) that is parameterized for bay anchovy (*Anchoa mitchilli*) (Houde and Madon 1995). Application of this model indicated that decreases in foraging efficiency and associated reductions in consumption of 10 percent lead to > 50 percent declines in weight-specific growth rates (table 7.2). Furthermore, the model predicted that containment of 4-cm planktivorous fish, if they behaved according to behavior scenario W_A, S_B (figure 7.10) and were contained in cylindrical enclosures of 200-cm diameter, would result in a growth-rate reduction of 25 percent from that expected by bay anchovy in the sea at the same prey concentration. The modeling also predicted that these fish would not grow in enclosures < 50-cm diameter (table 7.2). If the fish behaved according to other modeled scenarios (figure 7.10), expected reductions in foraging efficiencies would be greater still, and corresponding reductions in resultant consumption and growth would be very much reduced.

TABLE 7•2 *Predicted Foraging Efficiencies, Weight-specific Consumption, and Weight-specific Growth Rates of 4 and 6 cm Length Bay Anchovy* Anchoa mitchilli *in Enclosures of Different Diameters*

Fish length (cm)	Enclosure diameter (cm)	Foraging efficiency	Consumption $(g\ g^{-1}d^{-1})$	Weight-specific growth (d^{-1})
4	—	1	0.334	0.040
	400	0.98	0.326	0.034
	200	0.96	0.320	0.03
	100	0.92	0.308	0.022
	50	0.86	0.287	0.008
	25	0.76	0.254	−0.014
6	—	1	0.309	0.040
	400	0.97	0.299	0.033
	200	0.94	0.290	0.027
	100	0.89	0.275	0.017
	50	0.8	0.248	−0.001
	25	0.7	0.216	−0.022

NOTE Foraging efficiencies at each enclosure diameter were derived from the foraging model herein. Consumption and growth rates were derived from a bioenergetics model for bay anchovy (Houde and Madon 1995). Foraging model results are based on scenario W_A, S_B (see text and figure 7.10 for explanations).

Results of MEERC experiments to date have supported the modeled results. Bay anchovy in MEERC enclosures of 244-cm and 357-cm diameters have grown at rates similar to those reported in Chesapeake Bay (Zastrow et al. 1991; Newberger and Houde 1995), and significantly faster than growth rates in smaller 113-cm diameter enclosures (Mowitt 1999). Zooplankton populations were correspondingly lower in the 244-cm and 357-cm enclosures, indicating higher consumption rates by the anchovies under the less confining conditions (Mowitt 1999).

COMMENTS

Detecting top-down effects of fish on ecosystem-level processes is difficult under any circumstance and especially so in enclosure experiments (Carpenter and Kitchell 1992). Some ecologists would argue that enclosure experiments generally are insufficient to address complex issues in community and ecosystem ecology (Carpenter 1996). One major fault in applying enclosure approaches is the lack of appreciation or consideration of scale-dependence of outcomes in the design of enclosure experiments (Petersen et al. 1999). Modeling that we have undertaken here and experiments being conducted in MEERC (Mowitt 1999) have demonstrated that dimensions and sizes of enclosed ecosystems can have a profound effect on estimates of consumption by contained fish and on their potential to exercise control over lower trophic levels in pelagic communities. More research is needed but models and experimental results both suggest that some generalized scaling rules could be developed to understand how fish foraging efficiency and associated growth and production may govern ecosystem processes at lower trophic levels. The modeled results reported here are a first step in developing such scaling rules.

Most effects of containment that were modeled and discussed here are the consequences of altered fish behavior under confinement and can be considered experimental artefacts. It is conceivable that similar scale-dependent controls on foraging efficiency might occur in natural ecosystems under some circumstances, for example where vertical stratification or other precipitous boundaries might constrain activity and feeding behavior of planktivorous fish. Although the models that we have developed are quite complex, they provide only a crude approximation of the behavior of living fish. Based upon our observations of fish in enclosures of 1 to 10 m^3 volumes (1.1 to 3.6 m diameter), we selected some possible behaviors for modeling purposes. These resemble reported behaviors of fish in cylindrical tanks of differing sizes (Takagi et al. 1993). However, for modeling purposes many alternative behaviors are possible and added sophistication can be envisaged including, for example, incorporation of adaptive behavior in which individuals "learn" to better negotiate the walls of a container or perhaps to feed opportunistically near the walls.

Adding detail to these models, while conceptually possible, could render the results less understandable. In their present configuration, it is already difficult to unequivocally identify the functional properties contributing to major features of results. Further elaboration and complexity, without a better understanding of planktivorous and piscivorous fish behaviors, risks producing a model system as impenetrable as real life. At the present level of complexity, it is clear that alternative scenarios designating possible behaviors of fish upon encountering each other and the walls of a container may significantly alter the performance of individuals as consumers. Much more needs to be known about fish behavior before models could be parameterized to accurately scale up enclosure results to natural ecosystems. Nevertheless, some generic conclusions regarding scale-dependency can be drawn from the simulation experiments.

One clear conclusion is that foraging efficiency of planktivorous fish does scale in an orderly manner to enclosure diameter and relative swimming speed (and size) of fish. The precise details will be sensitive to several factors, for example, how fish react to enclosure walls, if they form schools, whether they are able to feed (either particle-selecting or filtering) close to walls, and will be dependent upon zooplankton prey distributions and abundances in the enclosures. There is some evidence, based upon observations in freshwater enclosures, that mesozooplankton concentrate near the walls of large, but not small, enclosures (Stephenson et al. 1984). If this were generally true, predators also might aggregate near the walls to elevate their consumption rates.

The introduction of schooling rules into the behavioral repertoire of the modeled planktivores can substantially alter the scale-dependent effects of enclosure size, but details are critically dependent upon individual behavior within schools and in relation to the walls. Despite present limitations, the model supports observations of several authors who have demonstrated that feeding and growth rates of planktivorous fish and fish larvae are impaired by small-scale containment (Theilacker 1980; MacKenzie et al. 1990; Perry et al. 1995; Mowitt 1999).

The simulations of piscivore performance in relation to containment yielded interesting results that have provided new insights into factors that can control the predation process under experimental conditions. The efficiency of the predator and the vulnerability of the prey were both sensitive to enclosure volume and shape. If the prey did not school, the performance of the predator was impaired by reductions in enclosure diameter due to interaction between the walls and the visual field of the predator, thereby reducing the volume searched per unit length of swimming path. However, when schooling was included in the prey behavioral pattern, volume (and depth) of the enclosure assumed greater importance, because they provided potential refuge for schooling prey.

Modeled results provided information on appropriate dimensions and sizes of experimental units that may be minimum requirements for experimental research on fish consumption and growth. For planktivores, only small reductions in foraging efficiencies were experienced when the ratio of swimming speed:enclosure diameter was in the range 0.05 to 0.10 (figures 7.9 and 7.10). If fish swim at 2 body lengths sec^{-1}, then planktivores of 10 cm and 5 cm lengths would require enclosures of 200 to 400 cm and 100 to 200 cm diameter, respectively, to forage at near maximal efficiency. For piscivores, the situation is more complex because enclosure depth, diameter, volume, and prey behavior, in addition to piscivore size, all interact to control predator power. Results from model runs suggested that enclosure diameter:depth ratio should be > 4 to support efficient piscivory and also suggested that swimming speed:enclosure diameter ratios should not exceed an approximate range of 0.32 to 0.40 to allow piscivores to function with reasonably high efficiency. These scaling generalities are helpful to design experiments that predict consumption rates and growth potential of planktivore or piscivore fish, but alone they are insufficient to predict ecosystem responses to top-down control.

Overall, the modeling exercise provided a quantitative demonstration of how scale-dependent factors must be considered in any attempt to interpret aquatic ecosystem dynamics from enclosure-based trophic or behavioral experiments. Although too little is known now about the behavior of aquatic predators under containment to allow specific guidance on experimental designs or to reach conclusions about top-down control by fish in aquatic ecosystems, some general principles can be established. In particular, results of the simulations lead us to anticipate the existence of ordered scaling principles relating foraging efficiencies to ratios of body size or swimming speed and to dimensions of enclosures. Results of most experiments reported in the literature apparently were obtained with little or no regard to scale-dependency (Petersen et al. 1999) or to evaluation of artifacts from containment. As a consequence, parameterizing such relationships is generally not possible at present. Investigation of containment effects to estimate consumption or predation rates should be an integral part of experimental designs (e.g., de Lafontaine and Leggett 1987) as is evident in the review by Paradis et al. (1996) of larval fish vulnerability to predators in experimental systems. Better experimental designs will be dependent, in part, on enhanced knowledge of fish behavior, which can be gained only by focused research on predators in contained ecosystems.

ACKNOWLEDGMENTS

Development of the individual-based models described in this chapter was facilitated by a Yonge Travel fellowship awarded to MRH by the Scottish Association for Marine Science and the University of Maryland. Support for EDH was provided by U.S. EPA funding to the Multiscale Experimental Ecosystem Research Center (MEERC).

LITERATURE CITED

Abreu, P. C., E. Graneli, C. Odebrecht, D. Kitzmann, L. A. Proenca, and C. Resgalla Jr. 1994. Effect of fish and mesozooplankton manipulation on the phytoplankton community in the Patos Lagoon Estuary, southern Brazil. *Estuaries* 17:575–584.

Brett, M. T., and C. G. Goldman. 1997. Consumer versus resource control in freshwater pelagic food webs. *Science* 275:384–386.

Carpenter, S. R. 1988. Transmission of variance through lake food webs. In S. R. Carpenter, ed., *Complex Interactions in Lake Communities*, pp. 119–135. New York: Springer-Verlag.

Carpenter, S. R. 1996. Microcosm experiments have limited relevance for community and ecosystem ecology. *Ecology* 77:677–680.

Carpenter, S. R., and J. F. Kitchell. 1992. Trophic cascade and biomanipulation: Interface of research and management—a reply to the comments by DeMelo et al. *Limnology and Oceanography* 37:208–213.

Crowder, L. B., R. W. Drenner, W. C. Kerfoot, D. J. McQueen, E. L. Mills, U. Sommer, C. N. Spencer, and M. J. Vanni. 1988. Food web interactions in lakes. In S. R. Carpenter, ed., *Complex Interactions in Lake Communities*, pp. 141–160. New York: Springer-Verlag.

de Lafontaine, Y., and W. C. Leggett. 1987. Effects of container size on estimates of mortality and predation rates in experiments with macrozooplankton and larval fish. *Canadian Journal of Fisheries and Aquatic Science* 44:1534–1543.

DeMelo, R., R. France, and D. J. McQueen. 1992. Biomanipulation: Hit or myth? *Limnology and Oceanography* 37:192–207.

Dunbrack, R. L., and L. M. Dill. 1984. Three-dimensional prey reaction field of the juvenile coho salmon (*Oncorhynchus kisutch*). *Canadian Journal of Fisheries and Aquatic Science* 41:1176–1182.

Englund, G. 1997. Importance of spatial scale and prey movements in predator caging experiments. *Ecology* 78:2316–2325.

Frost, T. M., D. L. DeAngelis, S. M. Bartell, D. J. Hall, and S. H. Hurlbert. 1988. Scale in the design and interpretation of aquatic community research. In S. R. Carpenter, ed., *Complex Interactions in Lake Communities*, pp. 229–258. New York: Springer-Verlag.

Gulland, J. A. 1983. *Fish Stock Assessment: A Manual of Basic Methods*. Chichester, U.K.: Wiley.

Harte, J., D. Levy, J. Rees, and E. Saegebarth. 1980. Making microcosms an effective assessment tool. In J. P. Giesy Jr., ed., *Microcosms in Ecological Research*, pp. 105–137. Springfield, Va.: National Technical Information Service.

Hewett, S. W., and B. J. Johnson. 1987. *A Generalized Bioenergetics Model of Fish Growth for Microcomputers*. Sea Grant Technical Report WIS-SG-87-245. Madison: University of Wisconsin, Wisconsin Sea Grant Program.

Hewett, S. W., and B. J. Johnson. 1992. *An Upgrade of a Generalized Bioenergetics Model of Fish Growth for Microcomputers*. Sea Grant Technical Report WIS-SG-92-250. Madison: University of Wisconsin, Wisconsin Sea Grant Program.

Horsted, S. J., T. G. Nielsen, B. Riemann, J. Pock-Steen, and P. K. Bjornsen. 1988. Regulation of zooplankton by suspension-feeding bivalves and fish in estuarine enclosures. *Marine Ecology Progress Series* 48:217–224.

Houde, E. D. 1997. Trophic complexity and scaling issues: Fish predators in MEERC. In V. S. Kennedy, ed., *Multiscale Experimental Ecosystem Research Center*, pp. 29–38. Annual Report, 1 October 1996–30 September 1997. Cambridge: University of Maryland System, Center for Environmental Science.

Houde, E. D., and S. P. Madon. 1995. Fish in MEERC: Top-down controls— Bioenergetics models and experimental protocols. Annual Progress Report to MEERC. Cambridge: University of Maryland, Center for Environmental Science.

MacKenzie, B. R., W. C. Leggett, and R. H. Peters. 1990. Estimating larval fish ingestion rates: Can laboratory-derived values be reliably extrapolated to the wild? *Marine Ecology Progress Series* 67:209–225.

MacNally, R. 1997. Scaling artefacts in confinement experiments: A simulation model. *Ecological Modelling* 99:229–245.

Miller, T. J., L. B. Crowder, and J. A. Rice. 1993. Ontogenetic changes in behavioural and histological measures of visual acuity in three species of fish. *Environmental Biology of Fishes* 37:1–8.

Miller T. J., L. B. Crowder, J. A. Rice, and F. P. Binkowski. 1992. Body size and the ontogeny of the functional response in fishes. *Canadian Journal of Fisheries and Aquatic Science* 49:805–812.

Mowitt, W. P. 1999. An analysis of top-down control by bay anchovy (*Anchoa mitchilli*) and its scale-dependence in estuarine mesocosms. Master's thesis, University of Maryland.

Neill, W. E. 1994. Spatial and temporal scaling and the organization of limnetic communities. In P. S. Giller, A. G. Hildrew, and D. G. Raffaelli, eds., *Aquatic Ecology: Scale, Pattern and Process*, pp. 189–231. London: Blackwell Science.

Newberger, T. A., and E. D. Houde. 1995. Population biology of bay anchovy *Anchoa mitchilli* in the mid-Chesapeake Bay. *Marine Ecology Progress Series* 116:25–37.

Øiestad, V. 1985. Predation on fish larvae as a regulatory force, illustrated in mesocosm studies with large groups of larvae. *NAFO Scientific Council Studies* 8:25–32.

Øiestad, V. 1990. Specific application of meso- and macrocosms for solving problems in fisheries research. In C. M. Lalli, ed., *Enclosed Experimental Marine Ecosystems: A Review and Recommendations*, pp. 136–154. Coastal and Estuarine Studies 37. New York: Springer-Verlag.

Paradis, A. R., P. Pepin, and J. A. Brown. 1996. Vulnerability of fish eggs and larvae to predation: Review of the influence of the relative size of prey and predator. *Canadian Journal of Fisheries and Aquatic Science* 53:1226–1235.

Perry, L. G., E. D. Houde, and S. D. Leach. 1995. Effects of container volume on the survival, growth, and production of larval *Gobiosoma bosc* (Lacepede). Unpublished report, University of Maryland Center for Environmental Science, Chesapeake Biological Laboratory, Solomons.

Petersen, J. E., J. Cornwell, and W. M. Kemp. 1999. Implicit scaling in the design of experimental aquatic ecosystems. *Oikos* 85:3–18.

Pitcher, T. J., and J. K. Parrish. 1993. Functions of schooling behaviour in teleosts. In T. J. Pitcher, ed., *The Behaviour of Teleost Fishes,* 2nd ed., pp. 363–439. London: Chapman and Hall.

Proulx, M., F. R. Pick, A. Mazumder, P. B. Hamilton, and D. S. Lean. 1996. Effects of nutrients and planktivorous fish on the phytoplankton of shallow and deep aquatic systems. *Ecology* 77:1556–1572.

Ramcharan, C. W., R. L. France, and D. J. McQueen. 1996. Multiple effects of planktivorous fish on algae through a pelagic trophic cascade. *Canadian Journal of Fisheries and Aquatic Science* 53:2819–2828.

Ramcharan, C. W., D. J. McQueen, E. Demers, S. A. Popiel, A. M. Rocchi, N. D. Yan, A. H. Wong, and K. D. Hughes. 1995. A comparative approach to determining the role of fish predation in structuring limnetic ecosystems. *Archives of Hydrobiology* 133:389–416.

Stephenson, G. L., P. Hamilton, N. K. Kaushik, J. B. Robinson, and K. R. Solomon. 1984. Spatial distribution of plankton in enclosures of three sizes. *Canadian Journal of Fisheries and Aquatic Science* 41:1048–1054.

Takagi, T., K. Nashimoto, K. Yamamoto, and T. Hiraishi. 1993. Fish schooling behavior in water tanks of different shapes and sizes. *Nippon Suisan Gakkaishi* 59:1279–1287.

Theilacker, G. H. 1980. Rearing container size affects morphology and nutritional condition of larval jack mackerel, *Trachurus symmetricus*. *Fishery Bulletin* 78:789–791.

Threlkeld, S. T. 1987. Experimental evaluation of trophic-cascade and nutrient-mediated effects of planktivorous fish on plankton community structure. In W. C. Kerfoot and A. Sih, eds., *Predation: Direct and Indirect Impacts on Aquatic Communities*, pp. 161–173. Hanover, N.H.: University Press of New England.

Vanni, M. J., and C. D. Layne. 1997. Nutrient recycling and herbivory as mechanisms in the "top-down" effect of fish on algae in lakes. *Ecology* 78:21–40.

Zastrow, C. E., E. D. Houde, and L. G. Morin. 1991. Spawning, fecundity, hatch-date frequency and young-of-the-year growth of bay anchovy *Anchoa mitchilli* in mid-Chesapeake Bay. *Marine Ecology Progress Series* 73:161–171.

CHAPTER 8

Experimental Validity and Ecological Scale as Criteria for Evaluating Research Programs

Shahid Naeem

EXPERIMENTAL ECOLOGY LIES WITHIN THE SPECTRUM OF ECOLOGICAL methodology, a continuum that runs from passive observation to pure thought (i.e., theoretical abstractions of nature, either mathematical, verbal, or graphical). In experimental ecology, experiments are constructed to test hypotheses derived from either observation or theory. Several authors have reviewed the design, interpretation, and role of experiments in ecology (Hurlbert 1984; Hairston 1989; Peters 1991; Scheiner and Gurevitch 1993; Hilborn and Mangel 1997; Underwood 1997). Because patterns derived from passive observation cannot identify mechanisms whereas theories employing different mechanisms can describe the same pattern, an important objective in experimental ecology is to conduct a series of experiments that together provide sufficient information to link ecological theory with observations of nature.

There are two factors that emerge as critical factors in designing a series of experiments that will link theory with observation. These are ecological scale (spatial, temporal, and biotic scales in ecological systems) and experimental validity. Ecological experiments are therefore characterized not only by the scales they investigate, but also by the balance they strike between internal and external validity. The importance of ecological scale is well known (O'Neill et al. 1986; Weins 1989; Schneider 1994), but its importance has only fairly recently been appreciated in ecology as is reflected in this volume. In contrast, the importance of experimental validity is less well appreciated. Manly (1992) offers the following definition of validity, which is divided into "internal" and "external" validity:

> *Internal validity* concerns whether the apparent effects or lack of effects shown by the experimental results are due to the factor

being studied, rather than some alternative factor. *External validity* concerns the extent to which the results of an experiment can be generalized to some wider population of interest.

For example, the experimental addition of fertilizer to replicate sites to test if resource limitation regulates weed invasions has low internal validity but high external validity. Internal validity is low because one cannot know if fertilizer addition changed other factors (e.g., changes in soil pH, herbivory, or other factors not associated with resource limitation), which encouraged or discouraged weed invasion. External validity is high, however, because we have confidence that all patches similar to the replicates will show the same response to fertilizer addition irrespective of the mechanism. If this experiment were conducted in a greenhouse using a pot with sterile, homogenous soil and equal biomass of plants, and if weed seeds were added to all replicates to test for invasion success, internal validity of this experiment would be high while its external validity would be low. Because all factors were fully controlled, there is little doubt that the single factor manipulated (nutrient levels) is responsible for an observed increase or decrease in success of germinating weeds. Because no patch in nature would resemble such a pot, however, there is little confidence one can generalize the results derived from a pot to the wider population of interest.

By these definitions, internal validity is likely to be highest when all factors are under the experimenter's control; such a condition is most likely when the experimenter conducts laboratory experiments where conditions are assumed to be well documented and controlled. External validity is highest when there is certainty that the experimental system corresponds to the natural system under investigation in all critical aspects, assuming the system under investigation is well defined. These latter conditions are most likely to occur when the experimenter conducts a field experiment in which the ecosystem and community are minimally perturbed and only one or a few factors are manipulated. Of course, there are no guarantees. Results from a field experiment conducted at a unique site may not reflect what would happen at any other site. Likewise, a growth chamber may exhibit more uncontrolled variability than is commonly observed in nature.

There is a tradeoff between external and internal validity. Maximizing external validity (i.e., minimizing the number of factors controlled by the experimenter) compromises internal validity because it is often impossible to disentangle causation from a web of correlations and interactions among the unmanipulated factors. Similarly, maximizing internal validity (i.e., maximizing the number of factors under the experimenter's control) compromises external validity because restricting the freedom of unmanipulated factors to vary, covary, or exhibit nonlinear or stochastic behavior (e.g., holding unmanipulated factors constant) may remove important sources of indirect, higher order, or complex interactions critical to the pattern being examined. Although this tradeoff might seem obvious, it is worth reflection. It is not uncommon for investigators to assume erroneously that there is some ideal point along the continuum at which all experimental approaches should reside.

The interpretation, extrapolation, and inference from ecological experiments bounded by the limits imposed both by scale and validity are examined in this chapter. The main purpose of this review is to demonstrate that research aimed at evaluating the relative merits of conflicting ecological theories requires conducting empirical studies that span not only spatial and temporal scales but also employ a range of experimental methods from those maximizing internal validity to those maximizing external validity. As will become apparent, for practical reasons such a full program of research is seldom feasible. A second goal is to show how using these two axes of experimental design, ecological scale and experimental validity, can provide a framework for evaluating how thoroughly a body of experimental research has explored a particular issue.

A recent set of ecological experiments that have tested hypotheses concerning the relationship between biodiversity and ecosystem functioning will provide an example of the importance of considering both scale and experimental validity in interpreting experimental results. These experiments are used not only because of my familiarity with them, but also because they illustrate how apparent conflicts in empirical findings may be better understood when scale and experimental validity are considered. Against the scale-validity framework, this program of research will be shown to have covered a very narrow range of experimental designs. I will argue that current controversies in biodiversity-ecosystem functioning are premature and cannot be resolved with the information to hand.

SCALE, VALIDITY, AND ECOLOGICAL EXPERIMENTS

The central thesis of this chapter is that both scale and experimental validity are important dimensions of ecological research. A brief discussion of scale, in particular biotic scale, and experimental validity are provided here before developing the scale-validity framework for critiquing programs of experimental ecological research.

Biotic versus Temporal and Spatial Scales

Biotic scales, unlike temporal and spatial scales, are often difficult to define and measure precisely. Biotic scales can be defined along different axes such as genetic (e.g., subspecies, species, genus, and higher taxonomic groupings), population (e.g., from a population on a single patch to an entire metapopulation), or ecosystem properties (e.g., high or low levels of expression of biogeochemical processes such as carbon sequestration or nitrogen fixation).

Different scales are measured in different units. Temporal and spatial scales are measured in standard units (e.g., days and meters, respectively), even if the operational units (e.g., generations, seasons, or tidal cycles for temporal scales or microhabitat, habitat, or landscape for spatial scales) vary among studies. In contrast, biotic scales not only lack standard units for measurement but their operational units may vary from one study to the next. Operational units might be population-based (e.g., individual, population, local metapopulation, species), ecosystem-based (e.g., local contribution to ecosystem process by species, regional contributions, contribution to global cycles), or genetically based (e.g., alleles within an individual, alleles within a local population, alleles within the gene pool of species).

"Species" is often used as the unit of choice for measuring biotic scales. Species however is not as cleanly defined as "meter" or "day." For example, the pattern of grouping of species by a genetically based biotic scale shows no necessary relationship with the pattern of grouping made according to an ecosystem-based biotic scale. *Vicia vilosa* and *Astragalus canadensis*, for example, by a genetically based biotic scale, are distinct species of herbacious plants. A community containing exclusively one or the other would be considered lower on a genetically based biotic scale than a community that contained them both. By an ecosystem-based classification, however, they are both "nitrogen fixers" because of their symbiotic relationship with *Rhizobium* sp. In an ecosystem-based biotic scale, patches that contained one or both species would not be considered different because both patches fix atmospheric nitrogen. The importance and ramifications of this problem, the lack of correspondence among different types of biotic scales (Allen and Hoekstra 1992), will become more evident in the examples drawn from biodiversity experiments reviewed below.

This brief overview of scale and experimental validity points to the importance of these factors in the design and interpretation of ecological experiments. More important, they describe a multidimensional space (i.e., internal and external validity and spatial, temporal, and biotic scales) within which most experimental studies may be defined. As discussed below, there are three widely used classes of ecological experiments that describe regions within this space, though they are defined primarily along the lines of internal versus external validity.

Ecological Experiments

For the purposes of this discussion, ecological experiments can be generally grouped into three classes: (1) field, (2) model-ecosystem (macro-, meso-, and microcosm), and (3) simulation experiments. These classes represent places along a continuum defined by the gradient between internal and external validity (figure 8.1). One end represents experiments that test theory-generated hypotheses while the other end represents experiments that test observation-generated hypotheses. The three classes of experiments vary in their relationship to validity and scale. Note that some field experiments attempt to test theory whereas some laboratory experiments attempt to test patterns observed in nature, but under the definitions provided above, such experiments generally lack the appropriate internal and external validity to do so.

Field experiments attempt to manipulate or perturb one or more factors within a defined system (e.g., nutrient levels, CO_2 levels, presence or absence of predators) in a series of replicate plots while using matching, but unmanipulated, plots as "controls" or "references." Very little internal validity is achievable in such experimental designs because many processes, whether stochastic, known, or unknown, interfere with identifying causation. Whole lake experiments (Schindler 1990; Carpenter and Kitchell 1993) are examples of such field experiments that have been successful in providing insights into the ecology of freshwater ecosystems.

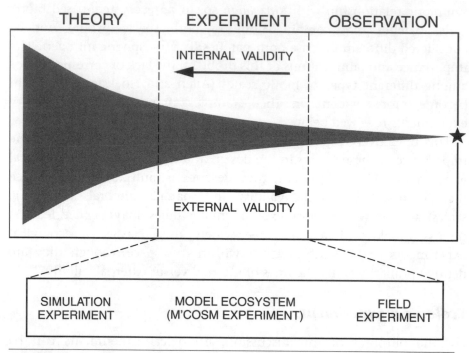

THEORY   EXPERIMENT   OBSERVATION

INTERNAL VALIDITY

EXTERNAL VALIDITY

SIMULATION
EXPERIMENT

MODEL ECOSYSTEM
(M'COSM EXPERIMENT)

FIELD
EXPERIMENT

FIGURE 8•1 *Theory-Observation Continuum*
Experimental ecology within the theory-observation continuum. In this chapter, I classify all ecological experiments as simulation, model-ecosystem, or field experiment. The classification is designed to distinguish experiments that more appropriately validate theory (simulation experiments) and those that more appropriately validate patterns observed in nature. The star represents a precisely documented pattern in nature. The shaded area flaring out as one moves left represents the region in which ecologists work, the width of the band representing the range of possible studies. The figure proposes that constraints are fewer on theory than on accurate observation. Experiments manipulate nature, either reconstructing or simulating it, or carrying out *in situ* manipulations of one or a few factors. Because there is a tradeoff between internal and external validity, experiments are most valuable when conducted across all three classes to provide continuous linkage of specific theory and observations. "M'cosm" experiment refers to micro-, meso-, and macrocosm experiments (see text for further discussion).

In some instances, an experiment done outdoors is termed a "field experiment." If the experimental ecosystem was constructed rather than naturally occurring, however, although it may share similarities with field experiments (e.g., natural variation in temperature or solar period), such an experiment falls into the next category of model ecosystem experiments.

Model ecosystem experiments are designed to replicate natural systems. These system-experiments may be further grouped according to physical scales as micro-, meso-, and macrocosm experiments (m'cosm experiments). Model ecosystem experiments are what Lawton (1995) calls "halfway houses" in which hypothesis testing is used to bridge nature and theory. Such experiments involve constructing replicate models of nature in which all ecological processes important to the theory being examined are present and factors relating to the hypotheses being tested are experimentally varied.

In these experiments, scale should ideally correspond to the spatial, temporal, and biotic scale of the processes under investigation. The prefixes "macro," "meso," and "micro," however, seldom relate to ecological scale and may be of little use. When the model system is physically large it is generally called a "macrocosm." The absolute dimensions are usually not specified. There are few examples of macrocosms, although Biosphere 2 (Nelson et al. 1993) and Bios-3 (Salisbury et al. 1997) are often considered macrocosm experiments. When the experimental system is physically intermediate in size, it is referred to as a "mesocosm," such as the Ecotron (Lawton et al. 1993; Lawton 1995, 1996). When the experimental system is physically small, it is termed a "microcosm" such as microbial microcosms and soil faunal microcosms (Drake et al. 1996). Microcosms, primarily for reasons of tractability, are much more common in research.

As noted above, the distinction between macro-, meso-, and micro- is somewhat arbitrary and based principally on physical size rather than scale. Because size is often determined by a mixture of pragmatic (e.g., cost, feasibility, engineering limitations) and biological considerations (e.g., the size necessary to sustain viable populations of the organisms of interest), it is often difficult to attach significance to the size of the experimental unit. That is, there are no necessary reasons to consider results obtained from macrocosms to be of greater merit than those obtained from microcosms. Experimenters may also select the size of their experimental unit based on the assumption that physical scale correlates with ecological complexity (Allen and Hoekstra 1992), what Silbernagel (1997) refers to as "size scale" and "ecological scale," respectively. Thus, a mesocosm might be chosen over a microcosm because the researcher assumes that the mesocosm will contain and sustain more ecological complexity (e.g., more biotic interactions, more possibilities for environmental heterogeneity, more possibilities for complex dynamics) than a microcosm and therefore be a closer model of the ecosystem being studied. This assumption, however, that size and complexity are related, is seldom tested. Furthermore, it is unclear if the additional complexities that physically larger experimental units may contain are the same as the complexities observed in nature.

Another difficulty with the prefixes "micro-," "meso-," and "macro-" is the fact that the physical scale of the experimental unit may not be related to the complexity of the experimental system as a whole (Allen and Hoekstra 1992). For example, Drake et al. (1993) examined long-term (i.e., many generations) processes such as community assembly and its interaction with landscape level processes such as immigration and emigration, but their experimental units were 1 L bottles. Although these 1 L microcosms are small, a series of 300 L cattle tanks, typical of what might be considered a mesocosm, may be more simple than Drake et al.'s (1993) microcosms if these mesocosms are stocked with only a few long-lived species and are closed to immigration and emigration.

Note that a model ecosystem experiment is not an m'cosm experiment if its design is based on a theory rather than on patterns derived from natural ecosystems, communities, or observations of nature. For example, studies of spatial heterogeneity on predator-prey systems (Huffaker 1958; Holyoak and Lawler 1996; Burkey 1997) may be better treated as simulations or biological analogs of theory, the next class of experiments.

Simulation experiments construct biological analogs of verbal, graphical, or mathematical theory. Although the theory being tested is generally derived from observations of nature, the experimental systems in this class focus on the theory rather than the observations that motivated the theory. The central assumption is that experimental support or refutation of the theory being tested will provide insights into the theory's ability to accurately describe natural phenomena. Such experimental systems serve to examine the importance of assumptions in theory, the sensitivity of parameters or variables in theory, to test predictions of theory, and to explore theory that is analytically intractable or not feasible to test under field conditions. The scale of the simulation experiment typically reflects the implicit scales in the theory being examined (e.g., number of generations, number of patches, strengths of interactions). Although the theories themselves are ultimately derived from thoughts about nature or experience, the experimental systems are explicitly constructed as analogs of theory, even if construction yields systems that can be found nowhere in nature. Internal validity is likely to be highest in these experiments because the investigator controls most of the variables. External validity is lowest because the experimental system focuses on the model being tested rather than systems in nature. Good examples include Gause's (1934) and

Vandermeer's (1969) classic experiments that used microbial analogs of Lotka-Volterra models to gain insights into Lotka-Volterra theory. These studies used microbial communities (e.g., protozoa, yeast, and bacteria) in small containers under laboratory conditions. In nature, however, such simple microbial communities are not likely to exist nor are environmental conditions likely to be so homogeneous or stable. Rather, the conditions of simple community composition and homogeneous environmental conditions were chosen to meet the assumptions of the Lotka-Volterra models under examination.

Simulation and model ecosystem experiments are often considered to be the same, but as they are defined above, they are quite different. Simulation experiments are sometimes incorrectly regarded as flawed because they do not mimic nature closely enough. For example, Davis et al. (1998) examined shifts in species distributions among communities in response to global warming using *Drosophila* sp. assemblages. To some, this type of experiment may seem flawed because it does not model nature in sufficient detail. That is, how can fruit flies in boxes provide an adequate means for investigating such large-scale phenomena as global warming and shifts in species geographic ranges when the experimental arena is a small box? Davis et al. (1998), however, explicitly addressed predictions that species shifts will correspond to preferred temperatures of species, not whether geographical location of natural populations, communities, ecosystems, or biomes would shift in response to temperature change. Their simulation experiment only needed a set of species whose preferred temperatures could be determined both in isolation and in combination, and an experimental system that would permit species movement along a temperature gradient. Their experiment met these criteria. As a simulation experiment, Davis et al.'s (1998) approach is correct. This experiment, however, does not meet the criteria of a model ecosystem (m'cosm) experiment because its design was not modeled after any system in nature.

A SCALE-VALIDITY FRAMEWORK FOR EXPERIMENTAL ECOLOGY

A framework for these various types of experiments as they relate to scale and validity may be represented as a matrix of possible experiments (figure 8.2). This matrix includes experimental designs that range from short-term, microspatial, genetic-level, simulation experiments to long-term, macrospatial, ecosystem-level field experiments. Even at this level of classification, experiments can be grouped into 54 (2 temporal x 3 spatial x 3 biotic scale levels x 3 classes of experiments) different kinds.

This exercise of constructing a matrix of possible experimental types is not meant to imply that all 54 types of experiments must be conducted before a solid understanding of nature is achievable. Rather, it is meant to provide a framework within which existing (or potential) experimental

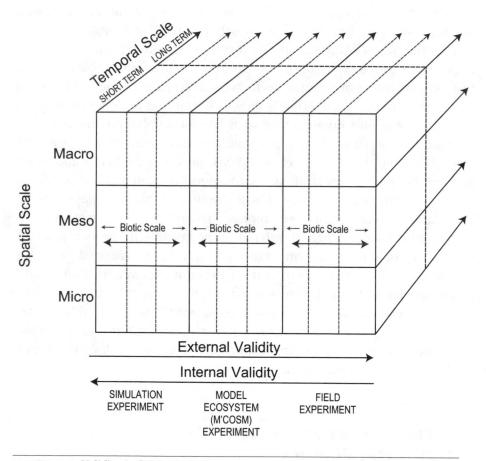

FIGURE 8•2 *Validity-Scale Matrix for Experiments*
Experimental types as classified by experimental validity and ecological scale. Each cell in this multidimensional matrix represents a possible experiment addressing a single hypothesis. The biotic scale (see text for discussion) is arbitrarily divided into three levels (dashed lines). "M'cosm" refers to micro-, meso-, or macrocosm experiments.

work might be placed to identify the boundaries and limits of the information our experiments might provide us. By examining where there are gaps in the matrix of possibilities, we can identify future directions for additional experimental research. As discussed below, we may also assess how thorough an experimental research program has been before evaluating the relative merits of conflicting theories in ecology. Realistically, it may be difficult to place experiments precisely in this framework because the boundaries between experimental types are often diffuse. For example, an experiment that includes microbial decomposers with generation times on the scale of hours and plants with generation times on the scale of years may be long term for microbes yet short term for plants. Locating such an experiment in the matrix would be difficult. Scaling rules (Schneider 1994) may help to simplify the process of experimental classification.

Implications for Scope of Inference

Choice of an experimental design, as described above, affects three aspects of inferential capabilities. First, a given ecological scale constrains inference to ecological phenomena that occur at that scale. Second, a given choice determines if inference will be more applicable to theory or to nature. As discussed above, choosing to conduct a field, model ecosystem, or simulation experiment provides different strengths and weaknesses in making inferences about nature or theory. Third, choice determines the scope of inference (i.e., the ability to extrapolate findings to systems other than the system directly examined in the experiment). Graphically, this third point means that once an experiment is located in the matrix of experimental designs, its inferential space is also determined (figure 8.2). For example, a short-term, microspatial, population-level simulation experiment has high internal validity and is most appropriate for testing theory concerning such phenomena as short-term, spatially *localized* population response to environmental perturbations. If, in fact, one makes inferences about theory concerning short-term, *global* population response to environmental perturbations, then the inferential boundaries of the chosen experimental design are violated. Violating inference boundaries defined by the scale of an experiment has been referred to as "scale confusion" (Belsky 1986; Peters 1991), but, clearly, proper restriction of inference to boundaries set by validity are also important.

BIODIVERSITY AND ECOSYSTEM-FUNCTIONING EXPERIMENTS AS ILLUSTRATION

In this section, I review a recent set of experiments devoted to testing and evaluating the relative merits of the different hypotheses proposed for the relationship between biodiversity and ecosystem functioning. The purpose of this review is to provide an example of how considerations of ecological scale and experimental validity can aid in evaluating a program of experimental research. I will review the issues, the experiments, and the results from these experiments, and show what insights can be gained by placing a set of experiments within the context of scale and validity.

Introduction to the Issue

The biota of an ecosystem is viewed as being a significant contributor to the magnitudes and rates of biogeochemical processes and properties of ecosystems (Odum 1953; Vitousek and Hooper 1993). Using the word "function" in the sense of activity and not purpose or design, one may ask, to what extent does the biodiversity of an ecosystem contribute to its ecosystem functioning? Given the dramatic declines in biotic diversity (Soulé 1991; Jenkins 1992; Tilman et al. 1994; Pimm et al. 1995; MacPhee and Flemming 1997; Malakoff 1997; Nee and May 1997; Stork 1997), and the dependence of global environmental conditions on biogeochemical processes (Butcher et al. 1992; Schlesinger 1997), a sense of urgency motivated an international conference convened in Bayreuth, Germany, to address this question (Schulze and Mooney 1993). After reviewing relevant ecological studies, the ecologists in attendance provided a variety of ideas, which in turn has stimulated many others (figure 8.3). Because these ideas, referred to here as biodiversity-ecosystem-functioning (BD-EF) hypotheses, have been reviewed elsewhere in more detail (Vitousek and Hooper 1993; Naeem et al. 1995a; Johnson et al. 1996; Peterson et al. 1998), only their essential differences will be discussed.

Although each hypothesis traces a different trajectory across the bivariate space of biodiversity and ecosystem functioning (asymptotic, nonlinear, or exponential), they differ primarily in the degree to which they are each sensitive to variation in biodiversity. Breaking up each hypothetical trajectory into those portions that predict insensitivity to variation in biodiversity (dashed lines) or sensitivity to variation in

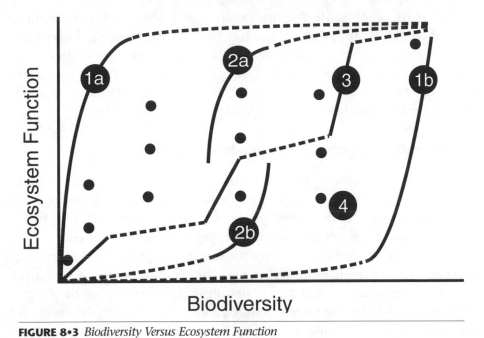

FIGURE 8•3 *Biodiversity Versus Ecosystem Function*
Representative biodiversity/ecosystem functioning hypotheses. Numbers refer to the
following published biodiversity/ecosystem functioning hypotheses: Compensating/
keystone hypothesis (1a, b; Sala et al. 1996); Nonlinear hypothesis (2 a, b; Pickett and
Carpenter 1996); Rivet (3; Ehrlich and Ehrlich 1981); and Idiosyncratic hypothesis (4, the
series of unconnected circles; Naeem et al. 1995b). Dashed lines indicate portion of curve
that reflects species equivalency. Solid lines indicate portions of curves that reflect species
singularity.

biodiversity (solid lines), figure 8.3 illustrates how most BD-EF theories
vary primarily in their treatments of sensitivity and insensitivity to
variation in diversity.

Ecosystem functions have been found to be either sensitive or
insensitive to variation in biodiversity. Ecological redundancy, or the
ability for one species to substitute for another in ecosystem functioning,
is typically regarded as the basis for the lack of ecosystem sensitivity
(Walker 1992, 1995; Lawton and Brown 1993; Beare et al. 1995; Gitay et
al. 1996; Jaksic et al. 1996; Naeem 1998). "Keystone" species, species with
high "community importance values" (Power et al. 1996), species that
function as "ecosystem engineers" (Jones et al. 1994), or other species
that have dramatic or disproportionate effects on community or
ecosystem functioning form the bases for ecosystem sensitivity to
biodiversity loss. A separate line of support for ecosystem sensitivity to
biodiversity loss is based on population (MacArthur 1955; Gardner and

Ashby 1970; May 1974; Dennis et al. 1997) and food web theory (Hairston et al. 1960; Pimm and Lawton 1977; Abrams 1993; Sterner et al. 1997). This body of theory suggests that community properties and ecosystem properties (De Angelis 1992; Loreau 1994; Zheng et al. 1997) are likely to be sensitive to variation in species composition because changes in composition lead to changes in dynamic stability.

Biodiversity-Ecosystem-Functioning Experiments

Observation and theory provided conflicting hypotheses concerning ecosystem response to variation in biodiversity (figure 8.3). Such a situation, where neither observation nor theory speak clearly to a scientific concern, is a clear case where experimentation is needed.

To test BD-EF hypotheses requires the direct manipulation of biological diversity as an experimental factor and the measuring of one or more ecosystem responses to this manipulation in order to determine which hypothesis best fits the experimental results. Biodiversity, however, is difficult to define (Solbrig 1991; Groombridge 1992; Harper and Hawksworth 1995; Gaston 1996) and difficult to manipulate (Naeem et al. 1994). Biodiversity includes genetic, population, and ecosystem diversity, and manipulating any one of these in communities that typically contain dozens to hundreds of species of metazoa and metaphyta is generally not feasible. The principal experimental approach thus far has been to construct replicates of a diverse community modeled after a natural ecosystem and simultaneously construct replicate, depauperate versions of this community for comparisons. Differences between the diverse "control" and the depauperate "treatments" are attributed to differences in biodiversity. Recently, considerable controversy has arisen concerning the interpretation of studies that employ this approach (Grime 1997; Huston 1997; Hodgson et al. 1998; Lawton et al. 1998) and the relative merits of different experimental designs used in such studies (Allison 1999), but the focus of this chapter is on issues concerning scale, so I will not review the controversy here.

In spite of the difficulty of defining and manipulating biodiversity and the controversy over the appropriate experimental design, several experimental tests of BD-EF hypotheses have been conducted since 1992 (table 8.1). Although a wide range of BD-EF experiments has been conducted (table 8.1), from varying plant species and functional group diversity among replicate plots under field conditions (e.g., Tilman et al.

1996; Tilman, Knops et al. 1997; Hooper and Vitousek 1997, 1998) to varying multitrophic-level assemblages of microbes in laboratory microcosm experiments (e.g., McGrady-Steed et al. 1997; Naeem and Li 1997), several general unifying features of these experiments can be observed when they are organized by the framework described in figure 8.2. First, BD-EF experiments consist primarily of model ecosystem experiments (table 8.1). That is, the majority of experiments used replicates of control (full diversity) and treatment ecosystems (depauperate versions of the control) modeled after systems observed in nature. Soil composition, climatic conditions, community composition, and relative abundances of species in the Ecotron were based on what was typically found in weedy fields in the south of England. This experiment began with sterile soil, then added plants, microbes, soil fauna, and insect herbivores and predators to re-create the weedy field ecosystem (Naeem et al. 1994). Grassland plants and pot experiments similarly begin with homogenized soil to which seeds are added to produce model communities that resemble natural grasslands (Hooper and Vitousek 1997, 1998; Tilman, Knops et al. 1997; Hooper 1998; Symstad et al. 1998; Naeem, Byers et al. 2000). Wardle, Bonner et al. (1997) manipulated plant litter diversity and found no relationship between litter diversity and litter decomposition, but they did not manipulate living biodiversity, thus I do not include this study in my review.

Microbial microcosm experiments differed from other experiments by focusing primarily on manipulating diversity rather than re-creating communities that closely matched natural systems. Although these experiments also began with homogeneous conditions (containers of constant volume, constant environmental conditions, and sterile media) and were modeled loosely on natural communities (i.e., freshwater ponds) (McGrady-Steed et al. 1997; Naeem and Li 1997, 1998), the emphasis was on obtaining large numbers of replicates, running experiments for multiple generations, and including multiple trophic levels. These experiments provided important direct tests of BD-EF theory (figure 8.3), but extrapolation to natural systems that are larger in scale, open to immigration and emigration, and more environmentally heterogeneous is difficult.

Second, although BD-EF experiments differed in size (e.g., ranging from 1 L to 50 ml and < 0.1 m^2 to > 100 m^2), spatial and temporal scale was never varied and used as an experimental factor. A multinational experiment involving eight European sites (Hector et al. 1999) did replicate

TABLE 8•1 *Experimental Approaches with Examples from Biodiversity-Ecosystem Function (BD-EF) Research*

Class	Internal Validity	External Validity	Scale	Method	Example BD-EF Experiments
I. Field experiments	Low	High	Match with scale of hypothetical pattern observed in nature	Manipulate one or more factors	None
II. Model ecosystem (m 'cosm)	Intermediate	Intermediate	Match with scale of hypothetical pattern observed in nature or hypothesis derived from theory	Reconstruct nature varying one or more factors among replicates	Ectron[ab] Grassland plots[cdefgh] Pot experiments[ijkl]
III. Simulation	High	Low	Match with scale of theory for simulation	Biological analog of theory	Microbial microcosm[mno]

[a]Naeem et al. (1994)
[b]Naeem et al. (1995b)
[c]Tilman (1996)
[d]Tilman et al. (1996)
[e]Tilman, Knops et al. (1997)
[f]Hooper and Vitousek (1997)
[g]Hooper and Vitousek (1998)
[h]Hector et al. (1999)
[i]Naeem et al. (1996)
[j]Symstad et al. (1998)
[k]Van der Heijden et al. (1998)
[l]Naeem, Byers et al. (2000)
[m]Naeem, Hahn et al. (2000)
[n]Naeem and Li (1997)
[o]McGrady-Steed et al. (1997)

NOTE Scale matching refers to the necessity of matching scale of inference with the appropriate scale of the hypothesis being tested. The Ecotron is an integrated series of controlled environmental chambers (Lawton et al. 1993).

a BD-EF experiment across different sites, with each site in a different country. By pooling data across sites, one can compare large scale, cross-site patterns with small scale, within-site patterns, but this experiment did not explicitly manipulate spatial scale (e.g., size of plots) as an experimental variable. Operationally, temporal scales were short among the majority of experiments (< 3 generations for plants in the Ecotron, 1 to 3 generations in the experimental grassland plots, 1 generation in pot experiments). Microbial microcosms were observed over relatively long time courses (> 25 generations). In no experiment were long-term experimental results compared with replicated short-term results.

Third, unlike spatial and temporal scales, biotic scale was manipulated in some BD-EF experiments. For example, in grassland plot and pot experiments, only one trophically defined functional group (i.e., autotrophs or plants) was used, but the numbers of biogeochemically defined functional groups within plants (e.g., C_3, C_4, legume, and nonleguminous forbs, early and late season perennials) were varied experimentally (Hooper and Vitousek 1997, 1998; Tilman, Knops et al. 1997; Hooper 1998; Symstad et al. 1998). Microbial microcosm experiments also manipulated biotic scale by varying species richness both within and across several trophically defined functional groups (McGrady-Steed et al. 1997; Naeem and Li 1997).

BD-EF Experimental Results and Observations

Results from these BD-EF experiments are not consistent with correlations between diversity and ecosystem functions observed in nature. All BD-EF experiments to date have confirmed that linear or asymptotic associations exist between ecosystem functioning and variation in diversity whereas observational studies show either no association or the inverse of what has been observed in experiments (figure 8.4).

Scale differences between the two sets of studies, however, may explain why observational and experimental results do not agree. Observational studies compared ecosystems at the level of the landscape, examining similar but geographically separate ecosystems that varied not only in diversity but also in age or other environmental conditions. Experimental studies, however, compared replicate ecosystems under identical environmental conditions that essentially examined the relationship between diversity and ecosystem functioning *within* an ecosystem. Thus, given the differences in scales, it is not surprising that observation and experiment disagree. Results derived from within-system studies can readily disagree completely with those derived from among-system studies. This is similar to what statisticians call Simpson's Paradox where pooling

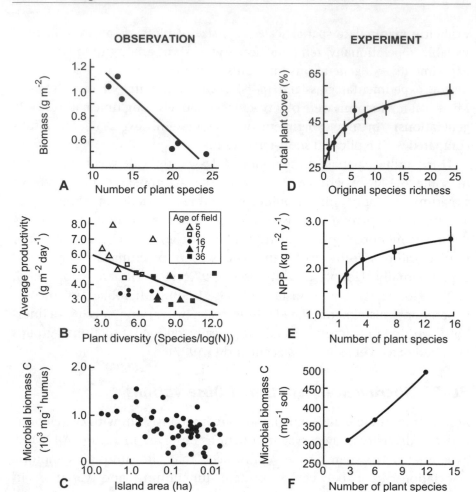

FIGURE 8•4 *Biodiversity Versus Function, Observation Versus Experiment*
Comparison among experimental and observational studies of the relationship between
biodiversity and ecosystem functioning. Left plots represent results from observational
studies whereas right plots represent results from experimental studies. Results are drawn
from the following studies: (a) plant production and plant diversity in savannas (Bulla
1996); (b) plant production and plant diversity in old fields (McNaughton 1993); (c)
microbial production and plant diversity (inversely correlated with island size) (Wardle et al.
1997a); (d) plant production and plant diversity in prairie grassland plots (Tilman et al.
1996); (e) plant production and plant diversity in the Ecotron (Chapin et al. 1998); and (f)
microbial production and plant diversity in Mediterranean grassland monoliths collected by
J. Roy and colleagues (Chapin et al. 1998).

data across systems can yield proportions that are inverse to those obtained
within systems (Manly 1992). This is the same phenomenon that led
researchers to propose that fluoridation of water increases cancer rate, an

observational result derived from cross-site comparisons that did not agree with numerous experiments (Manly 1992). The resolution to such seemingly paradoxical findings is to either identify the mechanism or conduct experiments controlling for confounding factors. For example, in the case of the mistaken fluoride-cancer connection, laboratory experiments showed no connection between fluoride and cancer while reexamination of observational studies found confounding demographic factors (e.g., cities with fluoridated water had larger numbers of older people who exhibited higher incidences of cancer). Conflicting results should motivate further research, not discourage it.

Evaluation of BD-EF Experimental Studies Using the Scale-Validity Framework

Using the scale-validity framework (figure 8.3) to evaluate BD-EF experimental research (table 8.1) reveals that research has been concentrated in a very small portion of the realm of possible experimental designs. The majority of the experiments fall into the micro-spatial, micro-temporal, model-ecosystem cell in the matrix. Although two different biotic scales have been examined (i.e., species richness, functional group, or both), because these scales are unrelated (species being a genetic-based scale and functional group being an ecosystem-based scale), no experiment was actually conducted at different levels of the same biotic scale. The microbial microcosm experiments occupy the micro-spatial, long-term (i.e., multigenerational), simulation experiment cell. Thus, only 2 cells of 54 possible are occupied in the scale-validity framework.

DISCUSSION

The realm of possible experimental designs in ecology is characterized most simply by scale and validity. Inference from such studies is limited by choice of experimental design. Although ecologists agree that the strongest inferences are when experiment, observation, and theory agree, there is a tendency to express preferences when knowledge is incomplete. For example, in a recent exchange, Tilman, Naeem et al. (1997) stated that, "Direct experimental tests of the effects of plant diversity on ecosystem properties may give more valid results than would correlations that are uncorrected for collinear variables," to which Wardle et al. (1997a) replied, "We believe that if outcomes of short term experimental and theoretical studies do not concur with patterns and processes observed in nature, then it is the experiments and theory that should be queried." Similar opposing preferences have been expressed concerning microcosm versus field experiments (Carpenter 1996, 1999; Drake et al. 1996; Drenner and Mazumder 1999; Huston 1999), even though ecologists recognize that both kinds of experiments serve useful purposes in research. The principle of experimental validity, however, should remind us that such preferences for one approach over another are unwarranted. Theory, experiment, and patterns observed in nature represent an inseparable triad of ecological methodologies each mutually dependent on the findings of the other to achieve linkage between observations of nature, theory, and mechanisms.

The general, contemporary method by which ecologists make observations about nature analyze these observations for patterns, construct theories to explain these patterns, generate hypotheses from these theories, and test these hypotheses by experimentation (Hairston 1989; Scheiner and Gurevitch 1993; Hilborn and Mangel 1997; Underwood 1997) provides much information about nature, but as the number of experiments increases over time, a collected appraisal of the studies becomes necessary. These appraisals can assist in identifying areas that may have been neglected, identify possible future directions for further experimental work, and expose the strengths and weaknesses of research programs. As suggested here, one qualitative way to evaluate such a program of research is to examine to what extent scale and validity have been dealt with by the studies. This is not an exact or quantitative appraisal, especially because both ecological scale and experimental validity are complex concepts and often neglected in experimental design. Nevertheless, a matrix can be constructed (figure 8.2) and a body of experimental research may be examined by placing it within this framework to determine where gaps might exist. In the absence of this exercise, results from different experiments may appear to conflict with one another when, in fact, they do not, and debates about interpretations of different findings may ensue when debate is unnecessary.

Experiments on biodiversity and ecosystem functioning experiments serve as a good example of evaluating a program of research with respect to scale and experimental validity. Observations of nature have provided a variety of conflicting hypotheses concerning the relationship between biodiversity and ecosystem functioning (figure 8.3). The need to clarify this relationship motivated the experiments that have been done to date. Exploring scale and experimental validity in the designs of this set of experiments revealed the relatively narrow approach of the experiments. Examining these experiments with respect to scale revealed that observations were made at the level of the landscape whereas experiments were very small in size. With respect to validity, BD-EF experiments have been primarily model ecosystem experiments that are intermediate in internal and external validity. Given this small area explored in the realm of possible experimental designs, the current debate (André et al. 1994; Givnish 1994; Garnier et al. 1997; Grime 1997; Huston 1997; Tilman 1997; Tilman, Naeem et al. 1997; Wardle et al. 1997b; Hodgson et al. 1998; Lawton et al. 1998) seems premature and most likely a reflection of the need for further research.

The matrix presented (figure 8.2) allows us to see what issues have not been addressed by this body of experimental work and to provide suggestions for where best to place future effort. Quite clearly, current BD-EF experiments have not addressed issues of temporal or spatial scale adequately (though one might argue this is common to much of experimental ecology). Current experiments have mostly chosen small spatial scales and relatively short temporal scales (a few generations). Only one experiment known as BIODEPTH, consisting of BD-EF experiments replicated in several European countries, has recently provided unique insights concerning within- versus cross-site patterns of association between production and ecosystem functioning (Hector et al. 1999). The apparent conflict shown in figure 8.4 was readily addressed by this study where grouping across all sites revealed no apparent relationship between production and plant species richness while analyses within sites revealed a positive associations.

Those few BD-EF experiments that did examine more than one scale provided valuable insights into the roles of biodiversity and ecosystem functioning that support the utility of expanding the kinds of experiments researchers conduct. For example, experiments that manipulated species (genetically based biotic scale) and functional groups (ecosystem-based biotic scale) uncovered potential reasons for why such a wide variety of BD-EF hypotheses have been derived from theory and observation (Hooper and Vitousek 1997, 1998; Tilman, Knops et al. 1997; Hooper 1998). Sensitivity to variation in biodiversity occurs when either trophically defined or biogeochemically defined functional groups are added or lost, whereas ecosystem functioning is relatively insensitive to redundant species within functional groups. The within- versus across-site BD-EF experiments of BIODEPTH showed how the spatial scale of such an investigation is critical to interpreting the relationship between plant diversity and production. Further, BD-EF experiments that manipulated numbers of species, numbers of trophic groups, and numbers of species within trophic groups uncovered a potential relationship between biodiversity and ecosystem predictability or reliability (the probability that a system will perform in a specific way; for example, the probability that a certain amount of production will occur given community composition and specific environmental conditions) (McGrady-Steed et al. 1997; Naeem and Li 1997; Naeem, Hahn et al. 2000).

As ecology continues to contribute to environmental problem solving, the issues of ecological scale and experimental validity will be increasingly important. The timeframe for finding solutions to many environmental problems will often be short and experimentalists will have to choose from a large number of possible experimental types. Ideally, a set of experiments will be selected that may serve most effectively and efficiently to bridge theory and associated mechanisms with patterns observed in nature. The concepts of experimental validity and ecological scales provide one framework by which we can design such programs of experimental research.

ACKNOWLEDGMENTS

I thank P. Abrams, R. Gardner, S. Tjossem, and an anonymous reviewer for critically evaluating the manuscript. This effort was supported in part by NSF DEB 9709559.

LITERATURE CITED

Abrams, P. A. 1993. Effect of increased productivity on the abundance of trophic levels. *American Naturalist* 141:351–371.

Allen, T. F. H., and T. W. Hoekstra. 1992. *Toward a Unified Ecology*. New York: Columbia University Press.

Allison, G. W. 1999. The implications of experimental design for biodiversity manipulations. *American Naturalist* 153:26–45.

André, M., F. Bréchignac, and P. Thibault. 1994. Biodiversity in model ecosystems. *Nature* 371:565.

Beare, M. H., D. C. Coleman, D. A. Crossley, P. F. Hendrix, and E. P. Odum. 1995. A hierarchical approach to evaluating the significance of soil biodiversity to biogeochemical cycling. *Plant and Soil* 170:5–122.

Belsky, A. J. 1986. Does herbivory benefit plants? A review of the evidence. *American Naturalist* 127:870–892.

Bulla, L. 1996. Relationships between biotic diversity and primary productivity in savanna grasslands. In O. T. Solbrig, E. Medina, and J. F. Silva, eds., *Biodiversity and Savanna Ecosystem Processes*, pp. 97–120. Berlin: Springer-Verlag.

Burkey, T. V. 1997. Metapopulation extinction in fragmented landscapes: Using bacteria and protozoa communities as model ecosystems. *American Naturalist* 150:568–591.

Butcher, S. S., R. J. Charlson, G. H. Orians, and G. V. Wolfe. 1992. *Global Biogeochemical Cycles*. London: Academic Press.

Carpenter, S. R. 1996. Microcosm experiments have limited relevance for community and ecosystem ecology. *Ecology* 77:677–680.

Carpenter, S. R. 1999. Microcosm experiments have limited relevance for community and ecosystem ecology: Reply. *Ecology* 80:1085–1087

Carpenter, S. R., and J. F. Kitchell, eds. 1993. *The Trophic Cascade in Lakes*. Cambridge: Cambridge University Press.

Chapin, F. S., III, O. E. Sala, I. C. Burke, J. P. Grime, D. U. Hooper, W. K. Lauenroth, L. A. Lombard, H. A. Mooney, A. R. Mosier, S. Naeem, S. W. Pacala, J. Roy, W. L. Steffen, and D. Tilman. 1998. Ecosystem consequences of changing biodiversity. *BioScience* 48:45–52.

Davis, A. J., L. S. Jenkinson, J. H. Lawton, B. Shorrocks, and S. Wood. 1998. Making mistakes when predicting shifts in species range in response to global warming. *Nature* 391:783–786.

De Angelis, D. L. 1992. *Dynamics of Nutrient Cycling and Food Webs*. London: Chapman and Hall.

Dennis, B., R. A. Desharnais, J. M. Cushing, and R. F. Costantino. 1997. Transitions in population dynamics: Equilibria to periodic cycles to aperiodic cycles. *Journal of Animal Ecology* 66:704–729.

Drake, J. A., T. E. Flum, G. J. Witteman, T. Voskuil, A. M. Hoylman, C. Creson, D. A. Kenney, G. R. Huxel, C. S. LaRue, and J. R. Duncan. 1993. The construction and assembly of an ecological landscape. *Journal of Animal Ecology* 62:117–130.

Drake, J. A., G. R. Huxel, and C. I. Hewitt. 1996. Microcosms as models for generating and testing community theory. *Ecology* 77:670–677.

Drenner, R. W., and A. Mazumder. 1999. Microcosm experiments have limited relevance for community and ecosystem ecology: Comment. *Ecology* 80:1081–1084.

Ehrlich, P. R., and A. H. Ehrlich. 1981. *Extinction: The Causes and Consequences of the Disappearance of Species*. New York: Random House.

Gardner, M. R., and W. R. Ashby. 1970. Connectance of large dynamical (cybernetic) systems: Critical values of stability. *Nature* 288:784.

Garnier, E., M.-L. Navas, M. P. Austin, J. M. Lilley, and R. M. Gifford. 1997. A problem for biodiversity-productivity studies: How to compare the productivity of multispecific plant mixtures to that of monocultures? *Acta Oecologica* 18:657–670.

Gaston, K., ed. 1996. *Biodiversity: A Biology of Numbers and Differences*. Oxford: Blackwell Science.

Gause, G. F. 1934. *The Struggle for Existence*. New York: Williams and Wilkins.

Gitay, H., J. B. Wilson, and W. G. Lee. 1996. Species redundancy: A redundant concept? *Journal of Ecology* 84:121–124.

Givnish, T. J. 1994. Does biodiversity beget stability? *Nature* 371:113–114.

Grime, J. P. 1997. Biodiversity and ecosystem function: The debate deepens. *Science* 277:1260–1261.

Groombridge, B., ed. 1992. *Global Biodiversity*. London: Chapman and Hall.

Hairston, N. G. S. 1989. *Ecological Experiments*. Cambridge: Cambridge University Press.

Hairston, N. G. S., F. E. Smith, and L. B. Slobodkin. 1960. Community structure, population control, and competition. *American Naturalist* 106:249–257.

Harper, J. L., and D. L. Hawksworth. 1995. Preface. In D. L. Hawksworth, ed., *Biodiversity Measurement and Estimation*, pp. 5–23. London: Chapman and Hall.

Hector, A., B. Schmid, C. Beierkuhnlein, M. C. Caldiera, M. Diemer, P. G. Dimitrakopoulos, J. A. Finn, H. Freitas, P. S. Giller, J. Good, R. Harris, P. Higberg, K. Huss-Danell, J. Joshi, A. Jumpponen, C. Korner, P. W. Leadly, M. Loreau, A. Minns, C. P. H. Mulder, G. O. O'Donovan, S. J. Otway, J. S. Pereira, A. Prinz, D. J. Read, M. Scherer-Lorenzen, E.-D. Schulze, A.-S. Siamantziouras, D., E. M. Spehn, A. C. Terry, A. Y. Troumbis, F. I. Woodward, S. Yachi, and J. H. Lawton. 1999. Plant diversity and productivity experiments in European grasslands. *Science* 286:1123–1127.

Hilborn, R., and M. Mangel. 1997. *The Ecological Detective*. Princeton: Princeton University Press.

Hodgson, J. G., K. Thompson, A. Bogaard, and P. J. Wilson. 1998. Does biodiversity determine ecosystem function? The Ecotron experiment reconsidered. *Functional Ecology* 12:843–848.

Holyoak, M., and S. P. Lawler. 1996. The role of dispersal in predator-prey metapopulation dynamics. *Journal of Animal Ecology* 65:640–652.

Hooper, D. U. 1998. The role of complementarity and competition in ecosystem responses to variation in plant diversity. *Ecology* 79:704–719.

Hooper, D. U., and P. M. Vitousek. 1997. The effects of plant composition and diversity on ecosystem processes. *Science* 277:1302–1305.

Hooper, D. U., and P. M. Vitousek. 1998. Effects of plant composition and diversity on nturient cycling. *Ecological Monographs* 68:121–149.

Huffaker, C. B. 1958. Experimental studies of predation: Dispersion factors and preadator-prey oscillations. *Hilgardia* 27:343–383.

Hurlbert, S. H. 1984. Pseudoreplication and the design of ecological field experiments. *Ecological Monographs* 54:187–211.

Huston, M. A. 1997. Hidden treatments in ecological experiments: Re-evaluating the ecosystem function of biodiversity. *Oecologia* 110:449–460.

Huston, M. A. 1999. Microcosm experiments have limited relevance for community and ecosystem ecology: Synthesis of comments. *Ecology* 80:1088–1089.

Jaksic, F. M., P. Feinsinger, and J. E. Jiménez. 1996. Ecological redundancy and long-term dynamics of vertebrate predators in semiarid Chile. *Conservation Biology* 10:252–262.

Jenkins, M. 1992. Species extinction. In B. Groombridge, ed., *Global Biodiversity*, pp. 192–205. London: World Conservation Monitoring Centre.

Johnson, K. H., K. A. Vogt, H. J. Clark, O. J. Schmitz, and D. J. Vogt. 1996. Biodiversity and the productivity and stability of ecosystems. *Trends in Ecology and Evolution* 11:372–377.

Jones, C. G., J. H. Lawton, and M. Shachak. 1994. Organisms as ecosystem engineers. *Oikos* 69:373–386.

Lawton, J. H. 1995. Ecological experiments with model systems. *Science* 269:328–331.

Lawton, J. H. 1996. The Ecotron facility at Silwood Park: The value of "big bottle" experiments. *Ecology* 77:665–669.

Lawton, J. H., and V. K. Brown. 1993. Redundancy in ecosystems. In E. D. Schulze and H. A. Mooney, eds., *Biodiversity and Ecosystem Function*, pp. 255–270. New York: Springer-Verlag.

Lawton, J. H., S. Naeem, L. J. Thompson, A. Hector, and M. J. Crawley. 1998. Biodiversity and ecosystem functioning: Getting the Ecotron experiment in its correct context. *Functional Ecology* 12:843–856.

Lawton, J. H., S. Naeem, R. M. Woodfin, V. K. Brown, A. Gange, H. C. J. Godfray, P. A. Heads, S. P. Lawler, D. Magda, C. D. Thomas, L. J. Thompson, and S. Young. 1993. The Ecotron: A controlled environmental facility for the investigation of populations and ecosystem processes. *Philosophical Transactions of the Royal Society of London, B* 341:181–194.

Loreau, M. 1994. Material cycling and stability in ecosystems. *American Naturalist* 143:508–513.

MacArthur, R. 1955. Fluctuations of animal populations and a measure of community stability. *Ecology* 36:533–536.

MacPhee, R., and C. Flemming. 1997. Brown-eyed, milk-giving, and extinct. *Natural History* 106:84–88.

Malakoff, D. 1997. Extinction on the high seas. *Science* 277:486–488.

Manly, B. F. J. 1992. *The Design and Analysis of Research Studies*. Cambridge: Cambridge University Press.

May, R. M. 1974. *Stability and Complexity in Model Ecosystems*. Princeton: Princeton University Press.

McGrady-Steed, J., P. M. Harris, and P. J. Morin. 1997. Biodiversity regulates ecosystem predictability. *Nature* 390:162–165.

McNaughton, S. J. 1993. Biodiversity and function of grazing ecosystems. In E. D. Schulze and H. A. Mooney, eds., *Biodiversity and Ecosystem Function*, pp. 361–384. New York: Springer-Verlag.

Naeem, S. 1998. Species redundancy and ecosystem reliability. *Conservation Biology* 12:39–45.

Naeem, S., D. Byers, S. F. Tjossem, C. Bristow, and S. Li. 2000. Plant neighborhood diversity and production. *Ecoscience* (in press).

Naeem, S., D. Hahn, and G. Schuurman. 2000. Producer-decomposer codependency modulates biodiversity effects. *Nature* 403:762-767.

Naeem, S., and S. Li. 1997. Biodiversity enhances ecosystem reliability. *Nature* 390:507–509.

Naeem, S., and S. Li. 1998. Consumer species richness and autotrophic biomass. *Ecology* 79:2603–2615.

Naeem, S., L. J. Thompson, T. H. Jones, J. H. Lawton, S. P. Lawler, and R. M. Woodfin. 1996. Changing community composition and elevated CO_2. In C. Körner and F. A. Bazzaz, eds., *Carbon Dioxide, Populations, and Communities*, pp. 93–99. San Diego: Academic Press.

Naeem, S., L. J. Thompson, S. P. Lawler, J. H. Lawton, and R. M. Woodfin. 1994. Declining biodiversity can alter the performance of ecosystems. *Nature* 368:734–737.

Naeem, S., L. J. Thompson, S. P. Lawler, J. H. Lawton, and R. M. Woodfin. 1995a. Biodiversity and ecosystem functioning: Empirical evidence from experimental microcosms. *Endeavour* 19:58–63.

Naeem, S., L. J. Thompson, S. P. Lawler, J. H. Lawton, and R. M. Woodfin. 1995b. Empirical evidence that declining species diversity may alter the performance of terrestrial ecosystems. *Philosophical Transactions of the Royal Society of London, B* 347:249–262.

Nee, S., and R. M. May. 1997. Extinction and the loss of evolutionary history. *Science* 278:692–694.

Nelson, M., T. L. Burgess, A. Alling, N. Alvarez-Romo, W. F. Dempster, R. L. Wallford, and J. P. Allen. 1993. Using a closed ecological system to study Earth's biosphere. *BioScience* 43:225–236.

Odum, E. P. 1953. *Fundamentals of Ecology*. Philadelphia: Saunders.

O'Neill, R. V., D. L. De Angelis, J. B. Waide, and T. F. H. Allen. 1986. *A Hierarchical Concept of Ecosystems*. Princeton: Princeton University Press.

Peters, R. H. 1991. *A Critique for Ecology*. Cambridge: Cambridge University Press.

Peterson, G., C. R. Allen, and C. S. Holling. 1998. Ecological resilience, biodiversity, and scale. *Ecosystems* 1:6–18.

Pickett, S. T. A., and S. R. Carpenter. 1996. Drivers and dynamics of changes in biodiversity. In UNEP, ed., *Global Biodiversity Assessment*, pp. 311–318. Cambridge: Cambridge University Press.

Pimm, S. L., and J. H. Lawton. 1977. Number of trophic levels in ecological communities. *Nature* 268:329–331.

Pimm, S. L., G. J. Russel, J. L. Gittleman, and T. M. Brooks. 1995. The future of biodiversity. *Science* 269:347–350.

Power, M. E., D. Tilman, J. A. Estes, B. A. Menge, W. J. Bond, S. Mills, G. Daily, J. C. Castilla, J. Lubchenco, and R. T. Paine. 1996. Challenges in the quest for keystones. *BioScience* 46:609–620.

Sala, O. E., W. K. Lauenroth, S. J. McNaughton, G. Rusch, and X. Zhang. 1996. Biodiversity and ecosystem function in grasslands. In H. A. Mooney, J. H.

Cushman, E. Medina, O. E. Sala, and E. D. Schulze, eds., *Functional Role of Biodiversity: A Global Perspective*, pp. 129–150. New York: Wiley.

Salisbury, F. B., J. I. Gitelson, and G. M. Lisovsky. 1997. Bios-3: Siberian experiments in bioregenerative life support. *BioScience* 47:575–585.

Scheiner, S. M., and J. Gurevitch, eds. 1993. *Design and Analysis of Ecological Experiments*. London: Chapman and Hall.

Schindler, D. W. 1990. Experimental perturbations of whole lakes as tests of hypotheses concerning ecosystem structure and function. *Oikos* 57:25–41.

Schlesinger, W. H. 1997. *Biogeochemistry*. 2nd ed. San Diego: Academic Press.

Schneider, D. C. 1994. *Quantitative Ecology: Spatial and Temporal Scaling*. San Diego: Academic Press.

Schulze, E. D., and H. A. Mooney, eds. 1993. *Biodiversity and Ecosystem Function*. New York: Springer-Verlag.

Silbernagel, J. 1997. Scale perception—From cartography to ecology. *Bulletin of the Ecological Society of America* 78:166–169.

Solbrig, O. T. 1991. *From Genes to Ecosystems: A Research Agenda for Biodiversity*. Cambridge, Mass.: IUBS.

Soulé, M. E. 1991. Conservation: Tactics for a constant crisis. *Science* 253:744–750.

Sterner, R. W., A. Bajpai, and T. Adams. 1997. The enigma of food chain length: Absence of theoretical evidence for dynamic constraints. *Ecology* 78:2258–2262.

Stork, N. 1997. Measuring global biodiversity and its decline. In M. L. Reaka-Kudla, D. E. Wilson, and E. O. Wilson, eds., *Biodiversity II*, pp. 41–68. Washington, DC: Island Press.

Symstad, A. J., D. Tilman, J. Wilson, and J. Knops. 1998. Species loss and ecosystem functioning: Effects of species identity and community composition. *Oikos* 81:389–397.

Tilman, D. 1996. Biodiversity: Population versus ecosystem stability. *Ecology* 77:350–363.

Tilman, D. 1997. Distinguishing the effects of species diversity and species composition. *Oikos* 80:185.

Tilman, D., J. Knops, D. Wedin, P. Reich, M. Ritchie, and E. Sieman. 1997. The influence of functional diversity and composition on ecosystem processes. *Science* 277:1300–1302.

Tilman, D., R. M. May, C. L. Lehman, and M. A. Nowak. 1994. Habitat destruction and the extinction debt. *Nature* 371:65–66.

Tilman, D., S. Naeem, J. Knops, P. Reich, E. Siemann, D. Wedin, M. Ritchie, and J. Lawton. 1997. Biodiversity and ecosystem properties. *Science* 278:1866–1867.

Tilman, D., D. Wedin, and J. Knops. 1996. Productivity and sustainability influenced by biodiversity in grassland ecosystems. *Nature* 379:718–720.

Underwood, A. J. 1997. *Experiments in Ecology*. Cambridge: Cambridge University Press.

Vandermeer, J. H. 1969. The competitive structure of communities: an experimental approach with protozoa. *Ecology* 50:175–179.

Vitousek, P. M., and D. U. Hooper. 1993. Biological diversity and terrestrial ecosystem biogeochemistry. In E. D. Schulze and H. A. Mooney, eds., *Biodiversity and Ecosystem Function*, pp. 3–14. New York: Springer-Verlag.

Walker, B. 1995. Conserving biological diversity through ecosystem resilience. *Conservation Biology* 9:747–752.

Walker, B. H. 1992. Biological diversity and ecological redundancy. *Conservation Biology* 6:18–23.

Wardle, D. A., K. I. Bonner, and K. S. Nicholson. 1997. Biodiversity and plant litter: Experimental evidence which does not support the view that enhanced species richness improves ecosystem function. *Oikos* 79:247–258.

Wardle, D. A., O. Zackrisson, G. Hörnberg, and C. Gallet. 1997a. The influence of island area on ecosystem properties. *Science* 277:1296–1299.

Wardle, D. A., O. Zackrisson, G. Hörnberg, and C. Gallet. 1997b. Biodiversity and ecosystem properties. *Science* 278:1867–1869.

Weins, J. A. 1989. Spatial scaling in ecology. *Functional Ecology* 3:385–397.

Zheng, D. W., J. Bengtsson, and G. I. Ågren. 1997. Soil food webs and ecosystem processes: Decomposition in donor-control and Lotka-Volterra systems. *American Naturalist* 149:125–148.

PART IV

SCALE AND EXPERIMENT IN DIFFERENT ECOSYSTEMS

CHAPTER 9

Scaling Issues in Experimental Ecology
Freshwater Ecosystems

Thomas M. Frost, Robert E. Ulanowicz, Steve C. Blumenshine,
Timothy F. H. Allen, Frieda Taub, and John H. Rodgers Jr.

EXPERIMENTS HAVE CONTRIBUTED SUBSTANTIALLY TO OUR UNDER-
standing of ecological processes in aquatic and terrestrial ecosystems (Hairston 1989; Resetarits and Bernardo 1998). Their use has
been particularly effective in studies of freshwater ecosystems (Lodge,
Blumenshine et al. 1998), in part because many key processes in aquatic
habitats operate at scales that lend themselves to experimentation.
Informative experiments have been conducted at scales ranging from
test tubes and liter-sized containers to whole lakes and streams. Whole-
ecosystem manipulations have had substantial advantages for
experimental work (Carpenter et al. 1995), but such large scale
experiments also have some significant drawbacks and can not always
be used to investigate ecological phenomena. Effectively tailoring the
scale of freshwater experiments to the questions and habitats being
addressed is a critical component of advancing the understanding of
aquatic ecosystems.

Microcosms play a critical role in applying an experimental approach
to a full range of freshwater habitats and ecosystems (Daehler and Strong
1996). Microcosms cannot, however, incorporate the complete range of
organisms, habitats, or processes that are important in freshwater
ecosystems. Ecologists must be conservative and clever in extrapolating
the results of experiments in containers to natural ecosystems. In some
cases researchers can realistically simulate natural phenomena in their
experiments to gain an understanding of ecological processes. Such
realistic simulations are not the only situations that can help to explain
ecological phenomena, however. Important insights can be gained in
experiments that examine gradients of scale that represent critical

ecosystem features or that generate realistic results in situations where such results might not be expected.

Our goal in this chapter is to present an overview of the varied approaches to gaining information from freshwater experiments. We discuss important scaling issues that must be considered in the design and interpretation of experiments that simulate natural ecological processes. We also consider examples of how driving factors have been evaluated without experiments using environmental gradients or using unrealistic experiments.

Experimental studies of freshwater systems regularly incorporate considerations of spatial and temporal scales (Frost et al. 1988). In some cases, the scales of observation used realistically simulate processes operating in natural systems. For example, the spatial scales that are critical to some aquatic organisms can be represented realistically in test tubes or containers of a few liters in volume. Others require mesocosms 1 to 5 m in diameter and several meters in depth, a size that is manageable in many natural habitats. Some of the important interactions involving phytoplankton, for example, appear to be incorporated realistically in containers 1 m^3 in volume or less (Reynolds 1997). Of course, not all freshwater ecosystem organisms or processes can be realistically represented in practical container experiments.

Time scales representing many aquatic ecosystem processes can also be incorporated into mesocosm experiments. Experiments to evaluate a wide range of freshwater processes can be carried out over less than one week. The direct effects of zooplankton on phytoplankton can be assessed over the course of a few days, a period over which in-container conditions are not expected to differ markedly from in-lake conditions. As with spatial considerations, though, there are important freshwater ecosystem processes that take a substantial period of time to play out, a length of time that is difficult to handle experimentally. Some experiments certainly involve time periods that are too long or too short and containers that are too small, but there are many cases where microcosm experiments can be conducted over temporal and spatial scales that are realistic for freshwater ecosystems.

Not all of the interacting elements of aquatic ecosystems can be expected to be fully represented even in very large mesocosm experiments, however. Under-represented elements include important components of aquatic communities (e.g., fishes and macrophytes), major physical features of aquatic habitats (e.g., mixing patterns and thermal structure), and the complete range of interacting habitats in aquatic ecosystems (e.g., pelagic and benthic regions). That is not to say that worthwhile insights can only be gained from reasonably scaled experiments, however. We will discuss cases where useful insights can be derived from unrealistic conditions.

As suggested previously, the potential to manipulate entire natural ecosystems directly is a substantial advantage for experimental work in freshwater habitats (Carpenter et al. 1995). Several highly successful experiments have been conducted on whole lakes or streams to evaluate fundamental ecosystem processes or responses to stress (e.g., Peterson et al. 1985; Schindler 1990; Brezonik et al. 1993; Carpenter and Kitchell 1993). These experiments have provided important insights into basic system characteristics that can be extended to ecosystems that have not been manipulated.

Whole-ecosystem experiments have their own limitations, too, in terms of their replication (e.g., Carpenter et al. 1989; Stewart-Oaten et al. 1992), in their ability to control for natural variability in important system drivers, and in their capacity to include processes occurring in their surrounding watersheds. It may also be difficult to conduct large-scale experiments over a period of time necessary for the influence of an important factor to propagate through an ecosystem. For example, fish populations were only reduced after several years in a lake acidification experiment because only young fish were affected by the pH depression (Schindler et al. 1985). It must also be considered that, due to the complex nature of interactions within aquatic ecosystems, what is intended as a single stress, such as acidification, may actually generate a number of stresses operating on the manipulated ecosystem. The specific factors generating a particular ecosystem response may not be simple to determine (Frost et al. 1999).

Experimental manipulations in containers can be particularly effective in conjunction with whole-ecosystem experiments to help understand the mechanisms operating in aquatic ecosystems (Frost et al. 1988). Also, because only a limited number of entire aquatic ecosystems can, or should, be manipulated experimentally, systematic considerations of scale are essential for extending information from the few experimental manipulations that are practical to developing a full understanding of natural ecosystem function.

SCALE CONSIDERATIONS WHEN CONDUCTING FRESHWATER EXPERIMENTS

The complexity of freshwater ecosystems limits how effectively experimental conditions can be expected to represent natural situations. There are many critical factors that must be considered when attempting to draw inferences about natural phenomena from experiments, particularly those in mesocosms. In this section we present an overview of some of the factors that must be considered in the design of experimental systems.

There are some variables for which scaling issues are not an obvious factor in the ways that they are imposed in experiments. Examples include temperature or the concentration of a chemical constituent like N, P, or H. Such variables have been termed "intensive" (e.g., Denbigh 1964) and contrast sharply with distinctly scaling variables such as space or time, which are termed "extensive." Intensive variables should be incorporated extrinsically into the design of an experiment but they function quite differently from extensive variables.

Space and time are two obvious scaling factors that must be considered in the design and interpretation of any experiment. Experiments that are conducted over too long a time or in too small a container have obvious limitations in their ability to simulate natural conditions. Understanding how to define "too," however, can often be a problem. Concerns about space and time may seem so obvious as to be trivial, but one does not have to delve very deeply into the ecological literature to see how comfortable some investigators seem to be with drawing inferences from artificially scaled experiments (e.g., Wilbur 1997). It is valuable to consider the ways that spatial and temporal factors that constrain an experimental system can have major influences over the inferences that can be drawn about ecological systems.

Freshwater experiments can be conducted over time periods that are either too short or too long. Leibold et al. (1997) attribute different patterns of experimental versus naturally occurring phytoplankton-zooplankton relationships to the short duration of experiments that have investigated these relationships, regardless of the size of the unit in which experiments were carried out. They attribute these differences to a reduced scope of possible outcomes in enclosure experiments due to the reduced potential for changes in community composition, and thus ecosystem function. Most freshwater experiments have been conducted with a relatively short duration. In a recent review, Lodge, Blumenshine et al. (1998) found a geometric mean of experiment durations of 29 days in 327 studies. In short experiments, an absence of longer-term processes and indirect effects can accentuate the role of direct interactions (Ives 1995; Diehl and Kornijow 1997; Leibold et al. 1997; Sarnelle 1997). Although some experiments have attempted to evaluate the influence of experimental duration (Lodge, Blumenshine et al. 1998) or the generation times of study organisms (Ives et al. 1996), the relationships of these variables with physical and biological complexity remain unclear.

Some experiments are also too long. When enclosure effects increase over time, such as the growth of periphyton on chamber walls, the differences between a mesocosm and a natural system will magnify as an experiment proceeds. Temporal factors such as the life spans of keystone organisms and the growth rates of important system components are critical experimental considerations.

The size of enclosures can influence experiments in a variety of ways. Enclosures can operate through direct effects on physical conditions and by constraints on the movement and behavior of larger organisms. The walls of an enclosure can have dramatic impacts on experimental conditions by exaggerating the influence of attached organisms relative to natural situations in lakes and streams (Bloesch et al. 1988; Lyche et al. 1996). Enclosure effects can also exclude processes from freshwater experiments that are important in natural settings. For example, mixing patterns and thus the horizontal and vertical diffusion of heat and material, turbulence, light penetration, nutrient sources, and nutrient concentration can be substantially altered by enclosure effects (Bloesch et al. 1988). Other factors, such as seasonal weather cycles, physical disturbances, or colonization will be very different in enclosures than in lakes or streams (Polis et al. 1996). Many organisms, particularly large fishes, may not survive in enclosures and, even if they survive, their behavior can be substantially influenced by container conditions. Community responses to treatments will be constrained to those processes that can operate within the size of an experiment's enclosure. If large-scale processes operating in natural systems outweigh those that occur at smaller scales, some fundamental community and ecosystem patterns, along with responses to perturbations, will differ markedly between enclosures and natural settings.

Enclosures may limit the complexity of biological patterns and processes that can operate in experiments. Direct interactions may be accentuated by spatial constraints in enclosures relative to natural settings. For example, a reduction of spatial heterogeneity favors observing direct effects of predation and competition because potential refuges and alternative food resources are diminished in simplified enclosure habitats (Sarnelle 1997). The isolation of lake habitats, communities, and ecosystem processes can be major artifacts of experimental enclosure design. Such components of lake and stream ecosystems do not function independently of each other. Although benthic and pelagic communities are often studied in isolation, even in natural settings, they are linked through competition for light and nutrients (Sand-Jensen and Borum 1991). Interactions of separate freshwater habitats must be a fundamental consideration of any experimental design, and of its evaluation.

Experiments dependent on including a full range of producer communities are particularly subject to difficulties when all habitats are not fully represented. For example, the contribution of benthic algae to whole-lake algal biomass and production may be substantial. Lodge, Blumenshine et al. (1998) suggest that the percent of whole-lake primary production due to benthic algae is higher in shallow lakes and may range from 35 to 95 percent of lake primary production for lakes with mean depth < 6m. The high surface area:volume ratios in most enclosure designs relative to lake habitats will exaggerate the importance of benthic communities. Studies that explicitly focus on pelagic organisms and processes may employ short experimental duration (days) or manual removal of periphyton on enclosure walls to compensate for wall effects (e.g., Bloesch et al. 1988; Hansson and Carpenter 1993; Lyche et al. 1996; Chen et al. 1997). In some instances, growth of periphyton on enclosure walls has been quantified to evaluate compensatory patterns of benthic and pelagic algal responses to nutrient additions (Hansson 1988; Mazumder et al. 1989; Blumenshine et al. 1997). Increases in near-surface phytoplankton with nutrient enrichment can also reduce light penetration and concentrate primary production in surface waters (Hansson 1988). Limitations on the spatial distribution of producer communities in an experiment must be a fundamental consideration in its interpretation.

The distribution of nutrients and the importance of different nutrient sources, particularly from distinct freshwater habitats, must be considered in any experiments where the level of primary production is an important factor. The full range of natural mechanisms of nutrient delivery and uptake can be difficult to simulate in many enclosure designs. For example, nutrient inputs from allocthonous sources are important in many smaller lakes, but whether nutrients are derived from surface runoff, subsurface inputs, or other routes will affect spatial distribution of algal communities (e.g., Cole et al. 1990). In well-lit, lake-bottom communities, epipelic algal mats affect sediment oxidation patterns and the release of phosphorus to the overlying water column (Carlton and Wetzel 1988). At higher nutrient-loading rates, increases in phytoplankton may retard production of light-limited benthic algae. Thus, in nutrient manipulation experiments, water-column nutrient-addition rates, duration, and benthic algal substrate type and area will interact to affect how and where algae accumulate in enclosures (Blumenshine et al. 1997). Investigations of the effects of nutrient additions, or other factors that act on lake production, need to consider whether enclosures are intended to simulate only open water communities in isolation or whether the effects of benthic communities are being considered as influencing phytoplankton responses to nutrient manipulations.

Enclosure can limit the potential for the inclusion of ecosystem feedback processes in experiments. Feedback processes are essential features of an ecosystem's behavior that account for much of its basic structure and function (DeAngelis et al. 1986). Feedback can be strongly influenced by scaling factors in experimental manipulations. For example, Forrester (1987) characterized major shifts in the behavior of living systems as signaling the displacement of control from one feedback configuration to that of another. Ulanowicz (1991) described at least eight properties of autocatalytic feedback that are nonmechanical in nature. Feedback can be viewed as an emergent property of ecosystems related to size, in the sense that, with an increase in the scale of observation, comes a greater likelihood that all the components of a feedback loop will be included. Failing but one member, a feedback loop loses its identity, along with associated properties, and becomes but a linear concatenation of a process. An enclosure that does not include all feedback loop components is incapable of incorporating realistic elements of ecosystem behavior.

Ulanowicz's (1991) considerations on feedback have special relevance for ecological experiments in enclosures. It is usually impossible to include representatives of all the elements of an ecosystem in a contained space. Furthermore, the limited spatial extent of enclosures can affect certain processes, such as feeding and predator avoidance, so as to preclude these interactions from the dynamics. Thus, with ever smaller mesocosms comes the growing likelihood that species components or processes that are elements of a feedback loop will be excluded from the community. As a result, the associated feedback control will be excluded from the dynamics of the experiment and, following Forrester's (1987) observation, the behavior of the enclosed system might come to depart significantly from that of the natural system it is intended to simulate. The influence of starting conditions that are specific to experimental enclosures may also result in different trajectories of experimental and natural systems, suggesting different controlling factors and therefore raising the potential for errors with extrapolations to natural situations (Bloesch et al. 1988).

Experiments intended to examine particular ecosystem processes may have special limitations. In experiments to examine how grazing by herbivores modifies algal assemblages (reviews in Brett and Goldman 1997; Leibold et al. 1997), herbivore grazing pressure is commonly manipulated through initial stocking, sieving grazers from enclosures, or indirectly by manipulation of plantivorous fish. Sieving may unintentionally remove some nontarget plankton, such as larger algae (Bloesch et al. 1988), and result in weaker contrasts in grazer assemblages between treatments especially with respect to body size (Cottingham et al. 1997; Persson 1997). Sieving may also represent an important export of nutrients from nutrient-poor systems. In some enclosure experiments where grazing pressure has been modified by the manipulation of planktivorous fish, results have typically compared well with patterns in natural lake communities (e.g., Mazumder et al. 1989; Hansson and Carpenter 1993; Vanni et al. 1997). This may occur because of the inclusion of fish effects on nutrient cycling, such effects being absent in experiments that rely on sieving. Grazer-manipulation experiments can be conducted in a variety of ways and must be interpreted with specific cautions that are in some ways similar to those in other freshwater experiments.

The use of fish in enclosures is another experimental design that can be difficult to extrapolate to natural settings. Lakes differ substantially in the composition or abundance of their fish assemblages much more than simply in terms of fish presence or absence. The effects of fish predation on prey communities in lakes may operate over a gradient of fish-predation pressure (Blumenshine et al. 2000). Interannual differences in weather conditions can affect over-winter survival or spring recruitment of fish and their prey. A limited range of seasonal effects in enclosure experiments could, for example, overemphasize the importance of predation for prey communities, thus masking the importance of other processes.

An additional consideration for enclosures involves the spatial distributions of fish that are often dictated by behavioral interactions of predator and prey species. Some aggregative or refuge-use patterns in response to the presence or absence of a predator may be observed in relatively large enclosures, but patterns observed in natural settings may operate on larger spatial and temporal scales than can be observed in enclosures (Eklov 1995). Other organisms may be concentrated in particular lake habitats such that predator-prey interactions and their resulting community patterns may be expected to extrapolate well to enclosure experiments. Lodge, Stein et al. (1998) were able to realistically compare snail, crayfish, and pumpkinseed sunfish distributions in littoral communities from laboratory aquaria and field cages to distributions within lakes. Predator and prey variability patterns on a whole-lake scale are linked with factors on spatial and temporal scales that would not be incorporated in most enclosure experiments and support the suggestion that enclosures enhance direct interactions relative to indirect factors that occur in natural systems in equilibrium (Diehl and Kornijow 1997; Englund 1997; Sarnelle 1997). A primary challenge for freshwater experiments is not only to determine which processes control the patterns of interest, but to resolve the spatial and temporal boundaries of each process and to determine why these change with scale (Fisher 1994).

ASSESSING RESPONSES IN FRESHWATER EXPERIMENTS

The responses in an experiment can be assessed with variables that reflect very different scales of resolution. The degree to which organisms in a community are identified separately, or the degree of taxonomic aggregation (e.g., to species, genera, families, etc.), are critical scale considerations when assessing how an ecosystem responds during experimental manipulations (Frost et al. 1988). In some ways this can be considered as parallel to space and time in the design of an experiment but taxonomic aggregation is fundamentally different from spatial or temporal scales of an observation. The extent of taxonomic aggregation is frequently driven by an observer's experiences and perspectives. Its extent has important consequences for the ways in which system processes are perceived.

Two examples of different kinds of taxonomic aggregation illustrate how levels of aggregation can provide very different kinds of information about an ecosystem. When evaluating the primary-producer community in a lake, an investigator can assess the biomass present for each of the 30 or more species that are likely to occur in the algal community present in a liter of water. Alternatively, the photosynthetic potential of the same community can be evaluated by quantifying the total amount of chlorophyll present in the same liter. Obviously, knowing all about the species present in the producer community provides much more detailed information than does quantifying total chlorophyll. The effort necessary to obtain this detailed information, however, is much greater than that needed to measure total chlorophyll and there are many assessments of producer communities in fresh water for which the chlorophyll concentration provides all of the information that is needed. Moreover, it appears that predictable and understandable responses by phytoplankton, even to fairly major nutrient additions, are those that are closer to the total chlorophyll measurements than to assessments of individual phytoplankton species (Cottingham and Carpenter 1998).

Responses to nutrient additions contrast with the effects of acidification for zooplankton (Frost et al. 1995) and for phytoplankton communities (Schindler et al. 1985). In response to acid additions, individual species provide much stronger responses than aggregated variables like total biomass or production. For example, in a paired-basin, whole-lake acidification experiment, some individual zooplankton species showed dramatic declines with acidification (figure 9.1). At the same time other zooplankton species increased substantially as pH decreased (figure 9.2). Increases by some species were sufficiently large that they masked declines in the total biomass of major groups of zooplankton or in the entire zooplankton community (Frost et al. 1995). Different responses at different levels of aggregation were quite evident in these responses. An integrated measure of zooplankton community composition provided by a similarity calculation comparing the relative biomass of the members of the zooplankton communities in the treatment and reference basins showed an early response to acid effects (figure 9.3). In contrast, the difference in total zooplankton biomass between the treatment and reference basins responded at a later stage of the experiment and never exhibited as substantial a difference as the similarity index (figure 9.3). Phytoplankton species showed a similarly higher sensitivity to acid effects than total algal biomass or production (Schindler et al. 1985).

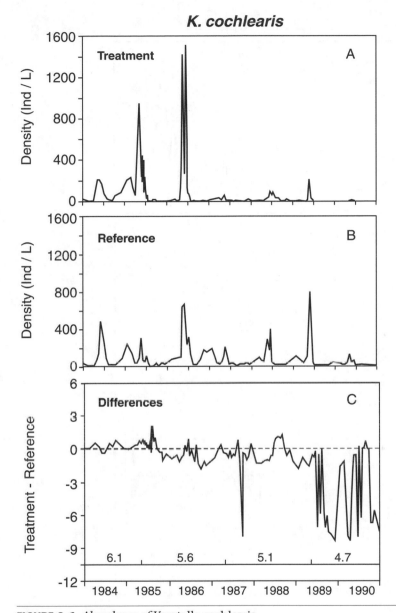

FIGURE 9•1 *Abundance of* Keratella cochlearis
Abundance of *Keratella cochlearis* in Little Rock Lake from 1984 to 1990 in (A) the
treatment basin and (B) reference basin, and (C) differences between the log10 (treatment
abundance + 1) and the log10 (reference abundance + 1). (From Gonzalez and Frost 1994)
Ind = individuals.

This contrast between responses to two different forms of ecosystem
stress illustrates important ways that ecosystem responses can be

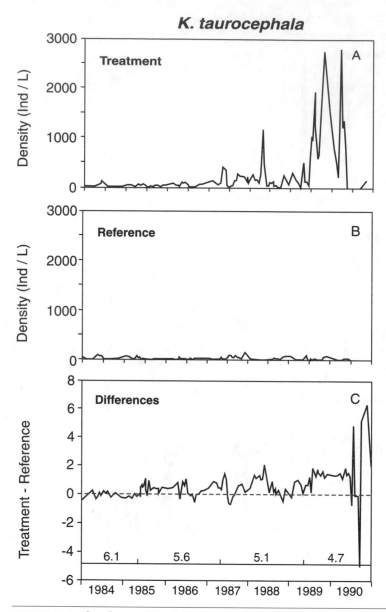

FIGURE 9•2 *Abundance of* Keratella taurocephala
Abundance of *Keratella taurocephala* in Little Rock Lake from 1984 to 1990 in (A) the treatment basin and (B) reference basin, and (C) differences between the log10(treatment abundance + 1) and the log10(reference abundance + 1). (From Gonzalez and Frost 1994) Ind = individuals.

perceived quite differently. In attempting to optimize the amount of information that can be obtained in evaluating ecosystem processes in experiments of any scale, there are important choices that must be made in terms of how to best evaluate a community's response.

A need to understanding human effects on ecosystems is a driving factor in much present-day environmental research. In assessing human effects, the choice of an appropriate aggregation scale has strong potential for maximizing understanding, particularly in optimizing the information obtained from sampling efforts that are limited by financial resources. In assessing ecosystem stress, investigators have to consider which level of aggregation provides the best way to assess system condition, options that range from individual species to the total mass of organisms performing an ecosystem function. In choosing an effective indicator, sensitivity to stress must be weighed against the variability that a variable would exhibit in a habitat if no stress were present (Frost et al. 1992). Both sensitivity and variability will differ among variables that aggregate different numbers of taxa. Functional compensation limits responses to stress by aggregated zooplankton variables (Frost et al. 1995). Not all species are sensitive to a stress, however, and a variable at an intermediate level of

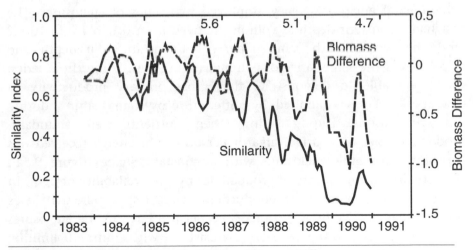

FIGURE 9•3 *Zooplankton Species Similarity and Biomass Difference*
Similarity in species composition and differences in total zooplankton biomass between the treatment and reference basins of Little Rock Lake calculated for all sampling dates between 1983 and 1991. Both were smoothed using five-point moving averages. Similarity was calculated following Inouye and Tilman (1988). The differences in biomass were calculated following Stewart-Oaten et al. (1986) as log (Treatment Biomass) - log (Reference Biomass). Original biomass units were µg/L. (From Frost et al. 1995)

aggregation might provide a more generally detectable response to a variety of stresses than would individual species.The consideration of different levels of aggregation itself can generate important insights into basic ecosystem properties (Frost et al. 1995).

EXPLICIT TESTS OF SCALING GRADIENTS

Some evaluations have been undertaken to investigate directly the effects of different scales of experimental treatments on perceptions of fundamental ecosystem processes. In some cases, these studies have involved direct experimental manipulations that have compared driving factors along distinct gradients to reveal important ecological principles. Some investigations have also used mathematical models to evaluate gradients of experimental conditions. Using either approach, researchers have developed understandings of the roles of key environmental factors.

A clear example of using a gradient of different forms of enclosures comes from estuarine studies. Petersen et al. (1997) conducted experimental tests on the effects of the size and the configuration of experimental enclosures. They employed two series of enclosures. The first had a constant depth as volume changed from 0.1 to 1.0 to 10 m^3 while the second had the same shape, with a constant radius:depth ratio of 5.6 across the same range of volumes (figure 9.4). Using these series, they were able to demonstrate that gross primary productivity in Chesapeake Bay was limited by different ecosystem features during different seasons. During spring when nutrients were abundant, productivity per unit surface area or per unit light energy received was constant across the enclosures with a constant shape (figure 9.5a), consistent with the control of productivity by the availability of light. In contrast, during summer when nutrients were low, productivity was constant per unit volume and increased per unit area in the enclosures with greater depth (figure 9.5b), consistent with nutrient limiting conditions. The summer pattern in the enclosures was shifted toward spring conditions when nutrients were added to enclosures. During fall, enclosures exhibited patterns with elements of those seen during both spring and summer suggesting a combination of light and nutrient limitation (figure 9.5c). The different controlling factors in Chesapeake Bay were revealed from these evaluations of gradients of small enclosures. Similar patterns are likely in freshwater studies.

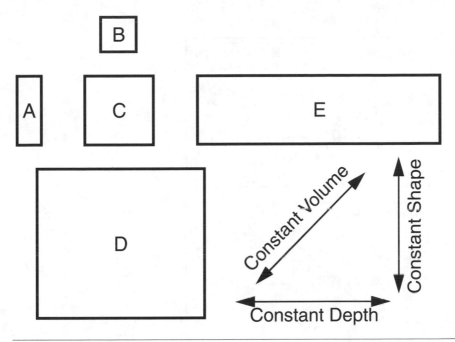

FIGURE 9•4 *Experimental Ecosystem Size and Shape Experiment*
Scale diagram of relative sizes and shapes of cylindrical experimental ecosystems viewed from the side. Constant-depth series (A, C, E; depth = 1 m) and constant-shape series (B, C, D; radius/depth = 0.56) intersect at the intermediate 1–m^3 C mesocosms. A and B both have a volume of 0.1 m^3; D and E both have a volume of 10 m^3. Three replicates of each mesocosm type were constructed. (From Petersen et al. 1997)

In a modeling exercise, Englund (1997) examined how enclosure size affects the interpretation of predator-prey interaction patterns, and the mechanisms generating these patterns. His models suggested that the size of enclosures in predation experiments that allow prey to move in and out of the experimental unit affects the perceived importance of particular mechanisms such as prey behavior or direct predation compared to underlying predator-prey interactions. Prey movements had less effect on prey densities in larger enclosures, whereas predation mortality was independent of experimental-unit size.

The effects of an herbivorous zooplankter, *Daphnia*, on a natural community of microzooplankton were compared across two different-sized mesocosms and in a whole-lake setting where a fish kill provided a natural experiment (Sarnelle 1997). *Daphnia* effects varied substantially with the type of experiment. The magnitude and direction of *Daphnia* effects on microzooplankton depended on the experimental setting. Not surprisingly, and perhaps reassuringly, results in the larger mesocosms

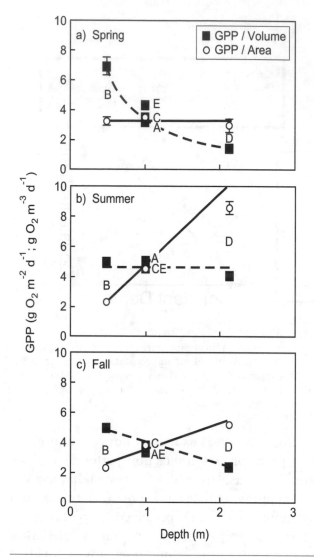

FIGURE 9•5 *Primary Productivity in Size and Shape Experiment*
Gross primary productivity averaged over prenutrient pulse period for each dimension of mesocosm in spring, summer, and fall experiments (GPP). GPP is expressed both per unit volume (closed squares) and per unit area (open circles). In (a), the dashed line through GPP per unit volume is from equation 6 in Petersen et al. (1997). Note that this describes a hyperbolic relationship between GPP per unit volume and depth. In (b), the continuous line through GPP per unit area is from equation 9 in Petersen et al. (1997). Continuous and dashed lines in (c) are lines of best fit. Error bars are ±1 SE of the mean. Bars are excluded in cases where the error is smaller than the diameter of the symbols. (From Petersen et al. 1997)

were most similar to those that occurred in the whole lake. Understanding the reasons for such scale-induced differences can reveal fundamental aspects of the nature of interactions among different ecosystem components.

Finally, considering a different sort of gradient, Gonzalez and Frost (1994) tested the extent to which zooplankton responses during a whole-lake acidification experiment could be predicted by standard laboratory bioassays. Two rotifer species had responded markedly to acid additions in Little Rock Lake, Wisconsin. *Keratella cochlearis* populations decreased with acidification (figure 9.1) but *K. taurocephala* abundance increased dramatically (figure 9.2), becoming the dominant rotifer species at the lowest pH stage of the experiment. Laboratory assays were not at all effective at predicting in-lake responses, however. Assays were able to predict that *K. cochlearis* would decline, but only when increased acidity was combined with reduced food availability. Laboratory experiments with feeding *ad libidum* demonstrated no sensitivity by *K. cochlearis* to acid levels. For *K. taurocephala*, responses differed even more sharply between the lake and the laboratory. Laboratory assays suggested that *K. taurocephala* population growth rates would be diminished under acid conditions. They gave no indication of its eventual tremendous proliferation in the lake's treatment basin during experimental acidification. Certainly predictions based upon these laboratory experiments were of limited utility in predicting responses when a whole lake was acidified.

The extreme differences reported between laboratory and lake conditions emphasize how difficult extrapolations from artificial to natural conditions can be. At the same time, systematic consideration of scale differences in experimental systems can reveal much about ecosystem conditions. Even the work that demonstrated the difficulties of predicting in-lake effects from laboratory bioassays (Gonzalez and Frost 1994) was also able to confirm that the shifts occurring with acidification in the treatment basin were driven by indirect, food-web related mechanisms rather than the direct effects of acidification (Webster et al. 1992). Scaling issues have a bearing on the numerous aquatic experiments that make use of enclosures but they are not considered explicitly in much of this work (Petersen et al. 1999), even to the point that many papers do not report the dimensions of the enclosures used in the experiments. Enclosure experiments are essential in increasing the basic understanding of ecosystem function but they can be much more useful if they consider scale issues directly as the examples we have provided demonstrate.

LESSONS TO BE LEARNED FROM UNREALISTIC EXPERIMENTS

Experimentalists frequently attempt to simulate natural ecosystems in order to evaluate fundamental processes, but realistic experiments do not provide the only opportunity to develop an understanding of factors controlling ecosystems. Allen (this volume) illustrates some very different ways that realistic and unrealistic experiments can be used to understand ecosystem processes. In a simple dichotomy, the assumptions underlying experiments can be classed as realistic or unrealistic. Likewise, the results of experiments can be categorized as reasonably representative of natural conditions or not. When realistic assumptions generate reasonable results, experiments can be used to evaluate the effects of factors that have been manipulated as reflecting close-to-natural conditions. Such situations can be replicated in a systematic fashion to evaluate the effects of a factor being manipulated. This is the standard situation that is usually sought in experimental work.

A less obvious source of insights can also derive from unrealistic assumptions in experiments. If results clearly do not represent natural situations, an experimenter has results that are highly predictable but not very useful. For example, if the oxygen is exhausted from a mesocosm, it is not surprising if fish do not survive. Unexpected results, however, can provide very important insights into fundamental system-organizing processes. The mud minnow, *Umbra limi*, does not die in chambers from which oxygen has been exhausted if there are bubbles present because it is capable of gathering oxygen from bubbles that are typical under ice in many mud-minnow habitats (Magnuson et al. 1983). This result has helped to explain the overall distribution pattern of fish communities in lakes across northern Wisconsin (Tonn and Magnuson 1982). It also illustrates critical tradeoffs that occur between adaptations to extreme environments and survival capabilities in interactions with other species. The mud minnow's unusual ability to survive oxygen stress allows it to prosper where other fishes can not. Mud minnows are, however, extremely vulnerable to the effects of oxygen-stress-intolerant fishes when oxygen is present. Mud minnows dominate only in habitats with oxygen stress. A fish experiment without oxygen may have seemed unrealistic but it revealed the mud minnow's unusual adaptations to a unexpectedly wide range of habitats.

Experiments that are too long but that continue to produce realistic results can serve as other examples of surprising situations that have the potential to provide insights into the nature of ecosystem processes. In such cases, ecosystem processes are much more resilient to the time-related effects of an experiment than would originally have been anticipated.

One major justification for the use of simple experimental systems is their ability to eliminate variables to determine the degree to which unnatural conditions can result in natural behavior. This situation parallels what occurs in successful system modeling. For example, axenic cultures can be used to assure that nutrient uptake or formation of a particular compound is caused by a particular alga alone and not by associated microbes or by microbe-algal interactions. Information on single and paired algal cultures has been useful in explaining dominance shifts in more complex systems that contain multiple species of algae, bacteria, and grazers (Rose et al. 1988).

Properties controlled by a dominant process can be easier to predict, repeat, and model than properties that result from several interacting processes. For example, in interlaboratory testing of Standardized Aquatic Microcosms (SAMs) (Taub 1993; American Society for Testing and Materials 1996), nitrate depletion in controls was highly reproducible, and fairly easy to model, whereas the time for recovery from $CuSO_4$ toxicity was more variable between experiments, and more difficult to model (Swartzman et al. 1990; Taub 1993). The $CuSO_4$ recovery process depended on the balance of many variables (e.g., growth of the least sensitive algal species, production of dissolved organic matter, and development of alkaline conditions) and, whereas the sequences of recovery were similar in different experiments (Conquest and Taub 1989), the time of recovery was highly variable (Meador et al. 1993). Experimental work can be used to develop an understanding of ecosystem processes by evaluating simple phenomena rather than mimicking a natural system.

In natural temperate ecosystems, the intensity of light and its spectral distribution undergo highly predictable diurnal and seasonal changes. In many indoor microcosm facilities such as the SAMs, however, light is supplied on an on/off basis for the same number of hours per day and at the same intensity throughout an experiment. In spite of the simplification of the light regime, SAM communities underwent sequences similar to spring algal blooms, zooplankton blooms, and summer "clear-water" phases in a 63-day experiment (Taub 1993). The mechanism driving these phenomena, as demonstrated in a simulation model, involved nutrient dynamics and toxicants that delayed nutrient uptake by the algae and shifted the overall alga cycle (Swartzman et al. 1990; Conquest and Taub 1989). These results suggest that the natural light cycle (varying intensity during the day and increasing duration from spring to summer) was not required to generate typical seasonal behavior. Indeed, many aquatic microcosm facilities have not attempted to supply light in a realistic manner. In other situations where the effects of different light regimes have been tested experimentally, however, they have been linked with the overall structure of freshwater phytoplankton communities (Litchman 1998). Different ecological processes appear to vary in their sensitivity to light regimes but light conditions must be considered as a factor in many enclosure-experiment designs.

If a model system is developed so that specific simple properties or purposes are met, artifacts do not have the same effects that they would if a system were designed to simulate natural situations. This is analogous to a physics experiment that explores the properties of an object moving on a frictionless plane. As long as the properties of the plane are reasonably frictionless, the experimental results can be informative, although they do not completely simulate a natural system, and ideal conditions cannot be met. Taub (1969a, 1969b) examined the behavior of three gnotobiotic ecosystems that contained (1) a single species of alga, (2) that species with a single protozoan grazer species, and (3) the alga and grazer with three species of bacteria. Some responses of the systems were reasonably similar. Of course, the outcomes could have been different if container size, light intensity, nutrient levels, and so on had been different. These experiments were not intended to replicate natural systems but to use fundamental principles to explore the behavior of ecological systems. Experimental systems can be heuristic even when they do not mimic natural ecosystems.

FRESHWATER VERSUS ESTUARINE EXPERIMENTS AND ECOSYSTEMS

This chapter's arising from a workshop on scale considerations in experiments sponsored by the Multiscale Experimental Ecosystem Research Center (MEERC) warrants some direct comparisons between estuarine and freshwater systems. Contrasts between these habitat types reveal how different controlling variables are emphasized in manipulative work in estuaries and fresh water but also illustrate some features common to experiments in both habitats.

Critical chemical factors provide some of the sharpest contrasts between estuarine and freshwater experiments. Ionic strength, and its variability, is one of the defining differences between estuarine and freshwater ecosystems. Fresh water typically contains only limited amounts of dissolved salts, whereas estuaries can have much larger amounts of salts, and they can be quite variable. Estuarine experiments must consider salinity as a controlling variable for a wide range of processes, whereas it is a minor factor in most freshwater manipulations. Salinity is characterized by mass, which does not lend itself to the same types of scale considerations as extensive variables such as time or size.

Another chemical factor, the concentration of hydrogen ions, generates major differences both among and within freshwater ecosystems but only exhibits minor differences within estuaries. The pH of freshwater habitats has figured extensively in investigations of critical processes in lake ecosystems (e.g., Schindler et al. 1991), but pH is not typically given special consideration in estuarine studies. Of course, other intensive chemical variables such as the concentrations of N or P are critical in both freshwater and estuarine ecosystems, but these intensive variables too are not subject to the same scale considerations as extensive variables.

Nutrient levels have certainly figured prominently in MEERC programs and other estuarine research, but considerations for estuaries do not typically extend to oligotrophic conditions. Are there any modes of behavior in oligotrophic aquatic ecosystems that are not found in relatively more productive estuaries? One fundamental difference may be an absence of mutualistic or positive feedback interactions among organisms. Such relationships can be fairly common in nutrient-poor environments (e.g., tropical oceans, open-ocean gyres, and oligotrophic lakes and wetlands) where algal-invertebrate symbioses occur, in sponges (Wilkinson 1987; Frost 1991) and corals (Reisser 1992), and where periphyton-carnivorous plant interactions (Knight and Frost 1991; Ulanowicz 1995) exist; they are relatively rare in estuaries.

Most important perhaps, freshwater and estuarine ecosystems differ in their degree of openness or their connection with surrounding ecosystems. Many freshwater ecosystems occupy a concave topography, that is, they are situated at a low point of the surrounding landscape. Gravity tends to bring not only water into such systems but also nutrients, sediments, and detritus that have been washed off the surrounding watershed. As a result, lakes tend to become progressively more eutrophic and experience directional succession over time. Estuaries also serve as traps of nutrients, sediments, and toxins (Odum 1970), but they are fundamentally different physical habitats than lakes. Long-term trends in conditions in estuaries may exhibit different or less directional trends over time than in lakes and other freshwater ecosystems because estuaries are open to the sea. Differences in the degree of openness of estuaries and freshwater habitats may lead to fundamental differences in the ways that microcosms relate to natural freshwater and estuarine ecosystems.

These differences do not suggest, however, that there are not common processes and problems in freshwater and estuarine ecosystems. Controls over primary production by nutrients or light (Petersen et al. 1997) and the combined potential influence of bottom-up and top-down processes on food webs (Carpenter and Kitchell 1993; Micheli 1999) are two obvious examples of phenomena that characterize both habitat types. Similarly, human-generated eutrophication is a major problem in both habitats (Paerl 1988). Eutrophication is driven primarily by P inputs in freshwater ecosystems but by N in estuaries (Paerl 1988), of course, again reiterating fundamental differences between estuaries and fresh water. Most important, however, some techniques for extrapolating from experiments to understanding the functioning of natural ecosystems are applicable in fresh water and in estuaries.

COMMENTS

The points that we have raised have shown that a number of cautions must be borne in mind when evaluating freshwater ecosystems by experiments. Care in interpretation is necessary even when experiments are conducted in whole ecosystems and certainly when they are carried out in enclosures. At the same time, we have presented ample evidence for the diverse ways that the fundamental understanding of ecosystems depends upon experimental work that is typically conducted in enclosures. Some experiments can be designed to reasonably simulate natural ecological processes. Other experiments that have been conducted under conditions that are far from realistic can still be useful when they systematically evaluate gradients of experimental effects or when they produce surprisingly realistic results. A wide range of differently scaled experiments that are interpreted with proper care is essential for increasing the general understanding of ecosystem patterns and processes.

ACKNOWLEDGMENTS

Thomas Frost's contribution to this chapter was supported by several grants from the National Science Foundation. In addition, the authors thank Vic Kennedy, Michael Pace, John Petersen, and two anonymous reviewers for helpful suggestions on this chapter.

LITERATURE CITED

American Society for Testing and Materials. 1996. *Standard Practice E1366-91: Standardized Aquatic Microcosms: Fresh Water*. Vol. 11.05, *Water and Environmental Technology: Biological Effects and Environmental Fate; Biotechnology, Pesticides*, pp. 734–768. Annual Book of ASTM Standards. West Conshohocken, PA: ASTM.

Bloesch, J., P. Bossard, H. Buhrer, H. R. Burgi, and U. Uehlinger. 1988. Can results from limnocorral experiments be transferred to in situ conditions? *Hydrobiologia* 159:297–308.

Blumenshine, S. C., D. M. Lodge, and J. R. Hodgson. 2000. Gradient of fish predation alters body size distributions of lake benthos. *Ecology* 81:374-386.

Blumenshine S. C., Y. Vadeboncoeur, D. M. Lodge, K. L. Cottingham, and S. E. Knight. 1997. Benthic-pelagic links: Responses of benthos to water-column nutrient enrichment. *Journal of the North American Benthological Society* 16:466–479.

Brett, M. T., and C. R. Goldman. 1997. Consumer versus resource control in freshwater pelagic food webs. *Science* 275:384–386.

Brezonik, P. L., J. G. Eaton, T. M. Frost, P. J. Garrison, T. K. Kratz, C. E. Mach, J. H. McCormick, J. A. Perry, W. A. Rose, C. J. Sampson, B. C. L. Shelley, W. A. Swenson, and K. E. Webster. 1993. Experimental acidification of Little Rock Lake, Wisconsin: Chemical and biological changes over the pH range 6.1 to 4.7. *Canadian Journal of Fisheries and Aquatic Sciences* 50:1101–1121.

Carlton, R. G., and R. G. Wetzel. 1988. Phosphorus flux from lake sediments: Effect of epipelic algal oxygen production. *Limnology and Oceanography* 33:562–570.

Carpenter, S. R., S. W. Chisholm. C. J. Krebs, D. W. Schindler, and R. F. Wright. 1995. Ecosystem experiments. *Science* 269:324–327.

Carpenter, S. R., T. M. Frost, D. Heisey, and T. K. Kratz. 1989. Randomized intervention analysis and the interpretation of whole-ecosystem experiments. *Ecology* 70:1142–1152.

Carpenter, S. R., and J. F. Kitchell, eds. 1993. *The Trophic Cascade in Lakes*. New York: Cambridge University Press.

Chen C. -C., J. E. Petersen, and W. M. Kemp. 1997. Spatial and temporal scaling of periphyton growth on walls of estuarine mesocosms. *Marine Ecology Progress Series* 155:1–15.

Cole, J. J., N. F. Caraco, and G. E. Likens. 1990. Short-range atmospheric transport: A significant source of phosphorus to an oligotrophic lake. *Limnology and Oceanography* 35:1230–1237.

Conquest, L. L., and F. B. Taub. 1989. Repeatability and reproducibility of the Standardized Aquatic Microcosm: Statistical properties. In U. Cowgill, ed., *Aquatic Toxicology and Hazard Assessment*, vol. 12, pp. 159–177. ASTM STP 1027. Philadelphia: American Society for Testing and Materials.

Cottingham, K. L., and S. R. Carpenter. 1998. Population, community, and ecosystem variates as ecological indicators: Phytoplankton responses to whole-lake enrichment. *Ecological Applications* 8:508–530.

Cottingham, K. L., S. E. Knight, S. R. Carpenter, J. J. Cole, M. L. Pace, and A. E. Wagner. 1997. Response of phytoplankton and bacteria to nutrients and zooplankton: A mesocosm experiment. *Journal of Plankton Research* 19:995–1010.

Daehler, C. C., and D. R. Strong. 1996. The role of microcosms in ecological research [special feature]. *Ecology* 77:663–705.

DeAngelis, D. L., W. M. Post, and C. C. Travis. 1986. *Positive Feedback in Natural Systems*. New York: Springer-Verlag.

Denbigh, K. 1964. *The Principles of Chemical Equilibrium*. Cambridge: Cambridge University Press.

Diehl S., and R. Kornijow. 1997. The influence of submerged macrophytes on trophic interactions between fish and macroinvertebates. In E. Jeppesen, M. Sondergaard, M. Sondergaard, and K. Christoffersen, eds., *The Structuring Role of Submerged Macrophytes in Lakes*, pp. 24–46. New York: Springer-Verlag.

Eklov, P. 1995. Effects of behavioural flexibility and habitat complexity on predator-prey interactions in fish communities. Ph.D. diss., Umeå University, Umeå, Sweden.

Englund, G. 1997. Importance of spatial scale and prey movements in predator caging experiments. *Ecology* 78:2316–2325.

Fisher, S. G. 1994. Pattern, process, and scale in freshwater systems: Some unifying thoughts. In P. S. Giller, A. G. Hildrew, and D. Rafaelli, eds., *Aquatic Ecology: Scale, Pattern and Process*, pp. 575–591. Oxford: Blackwell Science.

Forrester, J. W. 1987. Nonlinearity in high-order models of social systems. *European Journal of Operational Research* 30:104–109.

Frost, T. M. 1991. Porifera. In J. H. Thorp and A. P. Covich, eds., *Ecology and Classification of North American Freshwater Invertebrates*, pp. 95–124. New York: Academic Press.

Frost, T. M., S. R. Carpenter, A. R. Ives, and T. K. Kratz. 1995. Species compensation and complementarity in ecosystem function. In C. G. Jones and J. H. Lawton. eds., *Linking Species and Ecosystems*, pp. 224–239. New York: Chapman and Hall.

Frost, T. M., S. R. Carpenter, and T. K. Kratz. 1992. Choosing ecological indicators: Effects of taxonomic aggregation on sensitivity to stress and natural variability. In D. H. McKenzie, D. E. Hyatt, and V. J. McDonald, eds., *Ecological Indicators*, vol. 1, pp. 215–227. Kidlington, UK: Elsevier Science.

Frost, T. M., D. L. DeAngelis, S. M. Bartell, D. J. Hall, and S. H. Hurlbert. 1988. Scale in the design and interpretation of aquatic community research. In S. R. Carpenter, ed., *Complex Interactions in Lake Communities*, pp. 229–258. New York: Springer-Verlag.

Frost, T. M., P. K. Montz, T. K. Kratz, T. Badillo, P. L. Brezonik, M. J. Gonzalez, R. G. Rada, C. J. Watras, K. E. Webster, J. G. Wiener, C. E. Williamson, and D. P. Morris. 1999. Multiple stresses from a single agent: Diverse responses to the experimental acidification of Little Rock Lake, Wisconsin. *Limnology and Oceanography* 44:784–794.

Gonzalez, M. J., and T. M. Frost. 1994. Comparisons of laboratory bioassays and a whole-lake experiment: Rotifer responses to experimental acidification. *Ecological Applications* 4:69–80.

Hairston, N. G., Sr. 1989. *Ecological Experiments: Purpose, Design, and Execution.* Cambridge: Cambridge University Press.

Hansson, L. -A. 1988. Effects of competitive interactions on the biomass development of planktonic and periphytic algae in lakes. *Limnology and Oceanography* 33:128–135.

Hansson, L. -A., and S. R. Carpenter. 1993. Relative importance of nutrient availability and food chain for size and community composition in phytoplankton. *Oikos* 67:257–263.

Inouye, R. S., and D. Tilman. 1988. Convergence and divergence of old-field plant communities along experimental nitrogen gradients. *Ecology* 69:995–1004.

Ives, A. R. 1995. Predicting the response of populations to environmental change. *Ecology* 76:926–941.

Ives, A. R., J. Foufopoulos, E. D. Klopfer, J. L. Klug, and T. M. Palmer. 1996. Bottle or big scale studies: How do we do ecology? *Ecology* 77:681–685.

Knight, S. E., and T. M. Frost. 1991. Bladder control in *Utricularia macrorhiza*: Lake-specific variation in plant investment in carnivory. *Ecology* 72:728–734.

Leibold M. A., J. M. Chase, J. B. Shurin, and A. L. Downing. 1997. Species turnover and the regulation of trophic structure. *Annual Review of Ecology and Systematics* 28:467–494.

Litchman, E. 1998. Population and community responses of phytoplankton to fluctuating light. *Oecologia* 117:247–257.

Lodge, D. M., S. C. Blumenshine, and Y. Vadeboncouer. 1998. Insights and applications of large-scale, long-term ecological observations and experiments. In W. J. Resetarits and J. Bernardo, eds., *Experimental Ecology: Issues and Perspectives*, pp. 202–235. Oxford: Oxford University Press.

Lodge, D. M., R. A. Stein, K. M. Brown, A. P. Covich, C. Bronmark, J. E. Garvey, and S. P. Klosiewski. 1998. Predicting impact of freshwater exotic species on native biodiversity: Challenges in spatial scaling. *Australian Journal of Ecology* 23:53–67.

Lyche, A., T. Andersen, K. Christoffersen, D. O. Hessen, P. H. B. Hansen, and A. Klysner. 1996. Mesocosm tracer studies. 2. The fate of primary production and the role of consumers in the pelagic carbon cycle of a mesotrophic lake. *Limnology and Oceanography* 41:475–487.

Magnuson, J. J., J. W. Keller, A. L. Beckel, and G. W. Gallepp. 1983. Breathing gas mixtures different from air: an adaptation for survival under the ice of a facultative air-breathing fish. *Science* 220:312–314.

Mazumder, A., W. D. Taylor, D. J. McQueen, and D. R. S. Lean. 1989. Effects of nutrients and grazers on periphyton phosphorous in lake enclosures. *Freshwater Biology* 22:405–415.

Meador, J. P., F. B. Taub, and T. H. Sibley. 1993. Copper dynamics and the mechanism of ecosystem level recovery in a standardized aquatic microcosm. *Ecological Applications* 3:139–155.

Micheli, F. 1999. Eutrophication, fisheries, and consumer-resource dynamics in marine pelagic ecosystems. *Science* 285:1396–1398.

Odum, W. E. 1970. Insidious alteration of the estuarine environment. *Transactions of the American Fisheries Society* 99:836–847.

Paerl, H. 1988. Nuisance phytoplankton blooms in coastal, estuarine, and inland waters. *Limnology and Oceanography* 33:823–847.

Persson, A. 1997. Effects of fish predation and excretion on the configuration of aquatic food webs. *Oikos* 79:137–146.

Petersen, J. E., C. -C. Chen, and W. M. Kemp. 1997. Scaling aquatic primary productivity experiments under nutrient- and light-limited conditions. *Ecology* 78:2326–2338.

Petersen, J. E., J. C. Cornwell, and W. M. Kemp. 1999. Implicit scaling in the design of experimental aquatic ecosystems. *Oikos* 85:3–18.

Peterson, B. J., J. E. Hobbie, A. E. Hershey, M. A. Lock, T. E. Ford, J. R. Vestall, V. L.McKinley, M. A. J. Hular, M. C. Miller, R. M. Ventulo, and G. S. Volk. 1985. Transformation of a tundra river from hetertrophy to autotrophy by addition of phosphorus. *Science* 229:1383–1386.

Polis G. A., R. D. Holt, B. A. Menge, and K. O. Winemiller. 1996. Time, space, and life history: Influences on food webs. In G. A. Polis and K. O. Winemiller, eds., *Food Webs: Integration of Patterns and Dynamics*, pp. 435–460. New York: Chapman and Hall.

Reisser, W. 1992. *Algae and Symbioses: Plants, Animals, Fungi, Viruses, Interactions Explored.* Bristol, U.K.: Biopress.

Resetarits, W. J., and J. Bernardo. 1998. *Experimental Ecology: Issues and Perspectives.* Oxford: Oxford University Press.

Reynolds, C. S. 1997. *Vegetation Processes in the Pelagic: A Model for Ecosystem Theory.* Excellence in Ecology No. 9. Oxford: Ecology Institute.

Rose, K. A., G. L. Swartzman, A. C. Kindig, and F. B. Taub. 1988. Stepwise iterative calibration of a multispecies phytoplankton-zooplankton simulation model using laboratory data. *Ecological Modelling* 42:1–32.

Sand-Jensen, K., and J. Borum. 1991. Interactions among phytoplankton, periphyton, and macrophytes in temperate freshwaters and estuaries. *Aquatic Botany* 41:137–175.

Sarnelle, O. 1997. *Daphnia* effects on microzooplankton: Comparisons of enclosure and whole-lake responses. *Ecology* 78:913–928.

Schindler, D. W. 1990. Experimental perturbations of whole lakes as tests of hypotheses concerning ecosystem structure and function. *Oikos* 57:25–41.

Schindler, D. W., T. M. Frost, K. H. Mills, P. S. S. Chang, I. J. Davies, L. Findlay, D. F. Malley, J. A. Shearer, M. A. Turner, P. J. Garrison, C. J. Watras, K. E. Webster, J. M. Gunn, P. L. Brezonik, and W. A. Swenson. 1991. Comparisons between experimentally and atmospherically acidified lakes during stress and recovery. *Proceedings of the Royal Society of Edinborough* 97B:193–226.

Schindler, D. W., K. H. Mills, D. F. Malley, D. L. Findlay, J. A. Shearer, I. J. Davies, M. A. Turner, G. A. Linsey, and D. R. Cruikshank. 1985. Long-term ecosystem stress: The effects of years of experimental acidification on a small lake. *Science* 228:1395–1401.

Stewart-Oaten, A., J. R. Bence, and C. W. Osenberg. 1992. Assessing effects of unreplicated perturbations: No simple solutions. *Ecology* 73:1396–1404.

Stewart-Oaten, A., W. Murdoch, and K. Parker. 1986. Environmental impact assessment: "Pseudoreplication" in time? *Ecology* 67:929–940.

Swartzman, G. L., F. B. Taub, J. Meador, C. S. Huang, and A. Kindig. 1990. Modeling the effects of algal biomass on multispecies aquatic microcosms: Response to copper toxicity. *Aquatic Toxicology* 17:93–118.

Taub, F. B. 1969a. Gnotobiotic models of freshwater communities. *Internationale Vereinigung für theoretische und angewandte Limnologie, Verhandlungen* 17:485–496.

Taub, F. B. 1969b. A biological model of a freshwater community: A gnotobiotic ecosystem. *Limnology and Oceanography* 14:136–142.

Taub, F. B. 1993. Standardizing an aquatic microcosm test. In A. Soares and P. Calow, eds., *Progress in Standardization of Aquatic Toxicity Tests*, pp. 159–188. Oxford: Pergamon Press.

Tonn, W. M., and J. J. Magnuson. 1982. Patterns in the species composition and richness of fish assemblages in northern Wisconsin lakes. *Ecology* 63:1149–1166.

Ulanowicz, R. E. 1991. Formal agency in ecosystem development. In M. Higashi and T. D. Burns, eds., *Theoretical Studies of Ecosystems: The Network Perspective*, pp. 58–70. Cambridge: Cambridge University Press.

Ulanowicz, R. E. 1995. *Utricularia*'s secret: The advantage of positive feedback in oligotrophic environments. *Ecological Modelling* 79:49–57.

Vanni, M. J., C. D. Layne, and S. E. Arnott. 1997. "Top-down" trophic interactions in lakes: Effects of fish on nutrient dynamics. *Ecology* 78:1–20.

Webster, K. E., T. M. Frost, C. J. Watras, W. A. Swenson, M. Gonzalez, and P. J. Garrison. 1992. Complex biological responses to the experimental acidification of Little Rock Lake, Wisconsin, USA. *Environmental Pollution* 78:73–78.

Wilbur, H. M. 1997. Experimental ecology of food webs: Complex systems in temporary ponds—The Robert H. MacArthur Award Lecture. *Ecology* 78:2279–2302.

Wilkinson, C. R. 1987. Interocean differences in size and nutrition of coral-reef sponge populations. *Science* 236:1654–1657.

CHAPTER 10

Terrestrial Perspectives on Issues of Scale in Experimental Ecology

Anthony W. King, Robert H. Gardner, Colleen A. Hatfield,
Shahid Naeem, John E. Petersen, and John A. Wiens

THE 1980S WERE A WATERSHED FOR THE CONSIDERATION OF SCALE in terrestrial ecology—a period when developments in diverse disciplines crystallized into a new and coherent perspective on the importance of the role of spatial and temporal dimensions of ecological experiments and observations. Allen and Starr (1982) heightened the formal discussion of scale in their seminal treatise on hierarchy theory as a perspective for ecological systems. From their treatment of scale, couched largely in terms of information flow and filters, emerged a strong argument for the fundamental importance of the scale of observation in system description. Later work reinforced that argument (Allen et al. 1984, 1987). A second factor providing new perspectives on scale was the continuing debate concerning equilibrium versus nonequilibrium dynamics in ecological communities that prompted Wiens (1984) to recognize that observations made at different scales are likely to produce conflicting results and, therefore, differing views of equilibrium. Giller and Gee (1987) made similar arguments.

Another important factor was the consideration of scale commonplace in the marine and aquatic sciences (see Monin et al. 1977; Steele 1989a) that began to penetrate into discussions of terrestrial ecology (Dayton and Tegner 1984; Steele 1985). Time-space diagrams similar to the Stommel diagrams of marine science (Stommel 1963) appeared frequently in papers on terrestrial ecosystems (e.g., Delcourt et al. 1983; Urban et al. 1987; King et al. 1990). Fostered by the growth of landscape ecology as a discipline, geography's perspective on scale was also introduced into discussions in terrestrial ecology (Wiens 1981; Meentemyer and Box 1987; Meentemyer 1989). Indeed, during the 1980s the science of landscape ecology became almost synonymous with the science of scale in terrestrial ecology (e.g., Dale et al. 1989). Workshops and proceedings on scale were common by the middle to late 1980s (e.g., Risser 1986; Roswall et al. 1988; Dale et al. 1989). By the turn of the decade, Wiens (1989) was calling for the consideration of scaling issues as a primary focus of ecological research, and Levin (1992) had identified the problem of pattern and scale as the central problem in ecology. Although the expanded discussion of scale in terrestrial ecology (also see Schneider 1994) might be attributed to the emergence of scale as a buzzword (Wiens 1989) or part of the phenomenon of "trendy" ecology (Abrahamson et al. 1989), the increased attention to scale in terrestrial ecology has been substantive—representing a real recognition of the critical importance of scale in describing and understanding ecological systems.

Coincident with the expanded discussion of scale, the 1980s also saw a growth in experimental ecology, particularly the use of terrestrial experiments as tests of competition (Connell 1983; Schoener 1983; Underwood 1986). Hairston (1989) documented a general trend of increasing numbers of papers concerning field experiments during the 1980s, and Ives et al. (1996) described an increase in ecological field manipulations since 1975. It does not appear, however, that the coincident awareness of the importance of scale substantially affected methods of experimental field ecology, at least as it applies to terrestrial ecosystems. Kareiva and Andersen (1988), for example, expressed concern over a bias toward small plot sizes in experimental community ecology. One-half of 97 studies published in *Ecology* from 1980 to 1986 used plots no larger than 1 m in diameter; one-quarter used plots no larger than 0.25 m in diameter. Tilman (1989) describes how ecological experiments have been mostly conducted at small spatial and short temporal scales. It appears that plot size (scale) is determined more by convenience and convention than by an explicit consideration of scale as part of the experimental design (see Hoekstra et al. 1991). The scale of these experiments may have been appropriate to the hypotheses being tested, but the bias toward smaller plot sizes suggests an inattention to how issues associated with scale might affect experimental results.

This chapter reviews the importance of a reciprocal dialogue between the theory and implementation of scale with a special focus on terrestrial ecology (see chapters 9, 11, and 12 of this volume for discussions of freshwater, land-margin, and marine systems, respectively). Our chapter attempts to answer a series of questions, including: Has the dialogue between theory and experimental science been fruitful and productive?

Have operational guidelines been developed that are being empirically tested and, conversely, are empirical results affecting the refinement of theory? Are there standard practices and procedures to avoid the pitfalls and failures to consider scale in experiments? And, finally, what are the challenges for the future?

THE EXPERIMENTALISTS' VIEW OF SCALE

The lack of attention to scale (e.g., size of experimental units or duration of experiment) in the design of terrestrial field experiments is due in part to the neglect of scale considerations in prescriptions and critiques of experimental design (but see Manley 1992; Resetartis and Bernardo 1998). We examined a variety of textbooks used to train graduate students in the design of ecological studies. Although indices are imperfect reflections of the contents of a book, they still provide an objective measure of the importance the authors assign to particular topics. Sir Ronald Fisher (1960) indexed factorial design, null hypothesis, replication, and shape of blocks and plots, but did not index scale or size of plots. Likewise, Lewis and Taylor's (1967) *Introduction to Experimental Ecology* did not index these terms. The authors did, however, illustrate how changing sample unit size (area) can change the estimate of sample mean and variance (pp. 93–97). Green (1979) indexed sample unit size and shape, and actually included the need to verify the appropriateness of sample unit size as one (no. 8) of ten principles of sampling design in environmental studies. In *Ecological Experiments: Purpose, Design, and Execution*, Hairston (1989) did mention "large-scale ecological experiments" (p. 29), but in the context of "conditions in which replication cannot be achieved" (p. 29). Hairston (1989) also discussed (pp. 58–59) the problem of the necessary size of experimental units in field experimentation, but limited the discussion essentially to the problem of movements by mobile animals. He almost trivialized the issue for sessile or sedentary organisms, stating that: "Except for trees, sessile or sedentary organisms pose few problems, the size of experimental units being determined by density and the number of individuals needed for adequate population estimates" (p. 58).

Scale was not indexed in Underwood's (1997) *Experiments in Ecology*, but scale was indexed in *Design and Analysis of Ecological Experiments* by Scheiner and Gurevitch (1993) with reference to large-scale experiments where replication is not practical. Dutilleul (1998a) discussed adjustments in experimental design to accommodate heterogeneity at different scales. The "size and shape of" sampling units was indexed in Ludwig and Reynolds's (1988) *Statistical Ecology*, but this book is not about the design of experimental manipulations (pp. 3–4). Rather, it is a book about "pattern detection methods" (p. 4) rooted in the observational ecology and pattern analysis of I. Noy-Meir, E. C. Pielou, and P. Greig-Smith (p. 3). Interestingly, a meta-analysis of competition in field experiments did not address the effect of spatial scale on experimental outcomes (Gurevitch et al. 1992), although the effect of the duration (temporal scale) of the experiments was analyzed. Greater competitive effects were seen in experiments using cages rather than those using free-roaming organisms, implying an effect of experimental scale, but spatial scale per se (e.g., plot size) was not addressed in the meta-analysis.

The near absence of prescriptions for incorporating scale in experimental design may partially explain why explicit consideration of scale is not more prevalent in the design of terrestrial field experiments. To assess current trends of treating scale-dependent relationships in ecological experiments, we analyzed in detail all experimental studies in the most recent volumes of four (*Ecology, Ecological Applications, Journal of Animal Ecology*, and *Landscape Ecology*) ecological journals (table 10.1). In addition, we surveyed the ten most recent volumes of six other ecological journals (*Canadian Journal of Forest Research, Forest Science, Journal of Ecology, Journal of Vegetation Science, Oikos*, and *Oecologia*) for evidence of explicit consideration of scale in the design of published terrestrial field experiments. Our criteria for judging whether scale was explicitly considered included (1) treatment plots or experimental units were systematically arranged at more than one temporal or spatial scale, and/ or (2) there was an explicit discussion of how consideration of the natural scales of the system affected the choice of scale of experimental units (e.g., area of treatment plots). An example of the former is found in Collins and Wein (1998) where fertilized plots of three different sizes were established to "create three heterogeneity scales" (p. 239). Another example is Banks's (1998) manipulation of the scale of fragmentation in a study of herbivore response to vegetation heterogeneity. In these studies, scale is effectively or explicitly an experimental treatment. An example of the second type of evidence for explicit consideration of scale is the results of Aars et al. (1995, 1999) showing that "[t]he distance between

TABLE 10•1 *The Number of Articles in the Ten Most Recent Volumes of Four Journals that Involved Experimental Manipulations and Either Did Not Consider Scale Issues (No Reference to Scale) or Reported Results in Terms of Scale (Scale Sensitive Articles)*

Journal	Articles Reviewed	Experimental Manipulation	No Reference to Scale (%)	Scale Sensitive Articles (%)
Ecology	2,032	300	232 (11%)	67 (3%)
Ecological Applications	561	67	51 (9%)	29 (5%)
Journal of Animal Ecology	601	80	58 (10%)	25 (4%)
Landscape Ecology	226	4	1 (0%)	3 (1%)
Totals	3,420	451	342 (10%)	124 (4%)

habitat patches [within an enclosed plot] was approximately twice the length of a male root vole home range diameter."

We limited our survey to terrestrial field experiments involving experimental manipulations (e.g., fertilizer applications, herbivore exclosures). This limitation is not a comment on the contribution of greenhouse or growth chamber experiments, but merely a device to control the scope of our effort and to define our universe of inference. We identified 513 terrestrial field manipulation experiments in our survey. Only 106 of these (21 percent) exhibited evidence of an explicit consideration of scale in their design (see details for four leading journals in table 10.1). We believe these results show a slow but steady effect of scale-related issues on the design of experiments (see figure 1.1 for secular trends in reference to scale concepts). However, issues of scale have yet to make a significant effect on the design of most terrestrial field experiments.

THEORETICAL PERSPECTIVE OF SCALING IN EXPERIMENTS

If experimental terrestrial ecology has been less than aggressive in its consideration of scale in experimental design and analysis, it is also true that the practical limitations of experimental design have not been a high priority in theoretical considerations of scale. Again, an index is an imperfect reflection of the contents of a book, but Allen and Starr's book (1982) does not index experiments in their important contribution to scale in ecology. They index "monkey typists" but not "experiment" or "experimental design." Although Allen and Starr (1982:135) do note that problems of scale "are of critical importance in experimental design," they do not specifically discuss experimental design. Jeffers (1988) discusses the need for a rigorous approach to the design of multifactorial experiments, but provides no guidance on how attention to scale might be part of that approach. O'Neill (1989) alludes to the scale-dependence of experimental results, but does not provide specific guidance for experimental design. How choice of scale and scale of observation influence choice of organism or concept to be studied is often discussed (e.g., Allen et al. 1984, 1987; Hoekstra et al. 1991), but these discussions rarely, if ever, address how choice of scale might influence experimental design and interpretation. Allen and Hoekstra (1992) have much to say about scale and how careful consideration of scale might influence the management of ecological systems, but little to say about how the same consideration of scale might influence the design and interpretation of ecological experiments. O'Neill and King (1998) talk about a theory of scale in ecology and how it interacts with hierarchy theory, but they do not discuss scale and ecological experiments. In general, theoretical treatments of scale are essentially mute on how scale should or might be incorporated into experimental design.

Considering the lack of guidance from the theoreticians (who rarely comment on the details of experimental methods), or other prescriptive sources, it is not surprising that explicit and rigorous consideration of scale is often lacking in the design of terrestrial field experiments. In light of the attention given to scale in ecology from the 1980s onward, experimentalists are almost certainly aware that choices of scale are likely to affect experimental results. The experimenter probably even makes a conscious decision that the scale of the experiment is appropriate to the question being asked or hypothesis being tested (Kimberly A. With, personal communication, 1998, Bowling Green, Ohio). However, in the absence of specific guidelines from either a theory of scale in ecology or discussions of scale in experimental design, the decision is likely to be subjective rather than objective. Even if the intuition of the designer proves to be correct, the logic of the consideration of scale may be difficult to communicate, and the logic and correct decision may be difficult to replicate elsewhere. There are more objective criteria for other aspects of good design (e.g., replication). The same should be true for considerations of scale.

In summary we find that issues of scale (e.g., size of experimental units or plots or duration of experiment) are rarely discussed as part of experimental design, and then only superficially. We also find that discussions of principles of scale or a theory of scale in ecology are virtually silent on how scale should influence experimental design and interpretation. We perceive a two-part reciprocal challenge: on the one hand, a challenge for the more theoretical treatments of scale to explicitly address scale as it affects the design and interpretation of terrestrial experiments. On the other hand, there is a challenge for the design of terrestrial experiments to be more explicit in their considerations of scale and to design more scale-sensitive experiments. This challenge must be met by improved dialogue between theoreticians and experimentalists as, obviously, neither see these matters as a central part of their activities.

THE NEED TO INTEGRATE THEORY AND DESIGN

Choices of scale (e.g., plot size, sampling extent, duration of experiment) are dominated by pragmatic considerations of feasibility and practicality. Considerations of available space, accessibility, measurement technology, available personnel, costs, and duration of funding play a much larger role in the design of terrestrial experiments than do theoretical considerations of scale guided by theory. Although there is, perhaps, an increasing awareness of the desirability and need for "large-scale" and "long-term" experiments, ascribing this awareness to a "theory of scale" is unwarranted. It might be tempting to credit instances of measurement at multiple scales to theory, but these instances are more characteristic of observational studies (surveys) than experiments. It is fair to ascribe a heightened desire for measurement at multiple scales to the general discussion of scale in ecology over the past ten to fifteen years, but once again we find this to be such a minimal use of scale theory as to be almost trivial. Recognition that "scale matters" is not synonymous with a theory of scale. That recognition motivates the development of a theory of scale and scaling (or at least it should), but it does not define that theory.

There are, of course, important exceptions to this general characterization—exceptions in which an explicit consideration of scale was an integral part of the experimental design. The whole watershed experiments at Hubbard Brook (Bormann and Likens 1979) and Coweeta (Swank and Douglas 1974) involved the recognition that the ecosystem processes under investigation could be observed only at the scale of the watershed (i.e., the "whole" ecosystem). We were able to locate other examples of consideration of scale in experiments scattered throughout the recent literature. The examples are diverse. For instance, Aars et al. (1999) used an experimental design that examined the effect of different fragmentation patterns on two strains of root voles over the course of nine years. Fragmentation patterns were varied by size and inter-fragment connectivity. The results showed altered sex ratios associated with habitat fragmentation. The authors argue that adaptive sex ratio variation was likely due to spatial variability of "fitness returns from daughters than from sons" (females voles compete over smaller areas than males).

Other examples include Ash and McIvor (1998) who varied the size of their experimental grazing plots to achieve observed levels of herbage consumption. Banks et al. (1999:248) "assumed that . . . distances [of 7–72 km between valleys with experimental fox removal] provided some barrier to fox movement and hence independence between sites." Pither and Taylor (1998) did a preliminary experiment to verify if the experimental displacement of damselflies up to 700 m from their stream habitat would provide reasonable rates of reobservation in their study of movement and landscape connectivity. Brewer et al. (1998) chose an experimental wrack (detritus) mat size (0.25 m^2) that was similar to commonly observed wrack mats, but smaller than average. Herbers and Banschbach (1998:146) chose an experimental plot size for manipulation of food supply that "typically encompasses 20–40 nests" of the ant species under study. Imhoof and Schmid-Hempel (1998) selected bumblebee queens from locations separated by at least 20 km for their experimental infection study of local adaptation by a parasite. The 10 m x 10 m experimental gap size in Kneeshaw et al. (1998) was chosen to be consistent with naturally occurring forest gaps. Moloney and Chiariello (1998:751) chose experimental pots deployed in the field with a surface area "large enough to support an average of seven to nine species and over 200 individual plants . . . in undisturbed serpentine grassland."

Other experiments that consider multiple space and/or time scales have not used theoretical concepts to determine the appropriate scale of experiment or manipulation, but nevertheless exhibit a sensitivity to scale-dependencies not common to most terrestrial field experiments. For example, it is not uncommon for studies to show shifts in treatment response as a function of the duration of an experiment (e.g., Milchunas and Lauenroth 1995; Mal et al. 1997; Herbers and Banschbach 1998). Recognizing that the scale of forest patches and gaps has a significant impact on nutrient dynamics (Didham 1998), Denslow et al. (1998) felled trees to create experimental canopy openings ranging in size from 65 to 611 m^2. Similarly, Golden and Crist (1999) used subplot patch sizes of 1 × 1 m, 2 × 2 m, and 3 × 3 m in their experimental study of habitat fragmentation effects. Collins and Wien (1998) created three scales of resource heterogeneity with fertilizer manipulations in three plot sizes. Cresswell (1997) conducted experiments on bumblebee movement at two scales, with experimental plantings in short (10 m) rows and long (20 m) rows. Herben et al. (1997) observed responses to experimental species removal at fine spatial scale in 3.3 × 3.3 cm subplots and at a coarser scale in 25 × 25 cm plots. Riba (1997) used experimental clear-cutting events at three different intervals: 30, 5, and 2 yr. Turkington et al. (1998) applied fertilizer treatments in large 1 km^2 grids and in smaller 5 × 5 m plots. Gibson et al. (1993) burned and mowed in large-scale plots (10,000 m^2) and in small-scale plots (100 m^2). Gusewell et al. (1998) assessed treatment effects on species richness in experimental mowings at 1 m^2 and 25 m^2.

Phillips and Shure (1990) studied the effects of disturbance size on early successional community structure and function. Four different-sized forest clearings were established ranging from 0.016 to 2.0 ha and physical variables and vegetation composition response was monitored. Above ground net primary productivity was higher in response to higher light availability in the larger clearings resulting in differences in species composition and richness. Milchunas and Lauenroth (1995) used a factorial application of water, nitrogen, and water-plug nitrogen to a shortgrass steppe community and noted that the plant population changed with time. Results showed that short- and long-term responses were significantly different. Thus, conclusions made early in the study did not "scale" with time. The authors suggested that biotic regulation accounted for temporal shifts and that "time lags in initial response means that an ecosystem can pass a threshold leading to transitions to altered states before it is evident in structural characteristics such as species composition."

Methods of ecosystem management have recognizable scale-dependent effects, and these have been experimentally investigated by

Gibson et al. (1993), who contrasted management practices (i.e., mowing, fertilization, burning at large and small spatial scales) on the species composition of tallgrass prairie, and Riba (1997), who examined the effect of three different intervals between consecutive clear-cutting events on forest regeneration patterns. Other examples of experiments sensitive to scale include growth responses of *Eucalyptus* clones (Bouvet 1997) and shifts in plant community dynamics due to the duration and timing of fertilizer addition (Turkington et al. 1998).

These multiscale experiments provide evidence of how experimental response may vary with the scale of treatments—evidence that provides feedback to a theory of scale in terrestrial ecology. However, the empirical evidence that has had the greatest influence on theoretical developments has come primarily from observational studies rather than experimental field manipulations of scale (e.g., Krummel et al. 1987; Wiens 1989; O'Neill et al. 1991; Turner et al. 1991). One exception is the results from aquatic microcosms of different sizes (see Petersen et al. 1999), which have greatly influenced theoretical developments (R. V. O'Neill, personal communication, 1998, Oak Ridge, Tennessee). This is not surprising because many of these studies (including those reported in this volume) have been motivated directly from theoretical concepts of scale-dependencies. Although concepts of scale are widely regarded as important, the evidence shows that they are only now beginning to affect the design of terrestrial field experiments

Challenges for the Future

The challenge to the theory of scale in ecology is clear. Theoretical investigations must be expanded to explicitly consider how concepts of scale are useful in the design and interpretation of experiments. It would be helpful if individuals promoting and developing theoretical insights developed specific and testable hypotheses that can be experimentally verified. Until this occurs, the criticism that theoretical ecology has given insufficient attention to how issues of scale might affect experimental design and interpretation will remain valid. The current situation may exist because individuals developing theoretical insights may not have had strong links or experience with experimental ecology, or that the theory has yet to mature to the point where operational guidelines can be expressed.

Regardless of the causes for the current state, as experimental ecology grows as an approach to understanding ecological systems, proponents and developers of an ecological theory of scale should accept and respond to the challenge of explicitly addressing issues of scale in experimental terrestrial systems. Only by doing so can we expect empirical insights to affect theoretical advances, and vice versa.

Individuals with a more empirical and experimental bent are usually willing, indeed anxious, to test theoretical expectations, if in fact those expectations are formulated as specific and testable hypotheses. This is as true for a theory of ecological scale and scaling as for other ecological theory (Kareiva 1989). O'Neill (1989:147) hypothesized that "the probability of success in competition experiments should be inversely related to the scale at which a species operates and to life history characteristics such as dispersal ability," and suggested field experimentation to test the idea. Scale and the associated hierarchy theory in ecology has been criticized for being more philosophy than theory, more a conceptual framework than a generator of testable (e.g., quantitative) hypotheses with links to experimental studies (e.g., Steele 1989b). O'Neill's (1989) hypothesis is one example of a move toward testable hypotheses about scale. Others who promote a theory of scale in ecology should also respond to the challenge to develop testable hypotheses. As Kareiva (1989) has noted, however, theoreticians should not be the primary testers of their own hypotheses. Good experimental tests require a collaborative dialogue between theoreticians and experimentalists. Part of that dialogue is refining theoretical consequences and expectations into testable hypotheses. Another part is providing guidelines for how scale should be incorporated into experimental design and analysis (Dutilleul 1998a, 1998b; Hobbs 1998:482).

Terrestrial experiments that explicitly include scale (e.g., experimental plot size or duration of experiment) as a treatment are highly desirable. There are several examples from aquatic ecology (e.g., Gascon and Travis 1992; Petersen, Chen et al. 1997; Petersen, Cornwell et al. 1999), but far fewer from terrestrial ecology. These multiscale experiments provide the most direct evidence of how experimental response varies with scale, and invaluable empirical evidence for the development and testing of ecological scaling theory, but how does one know which scale to use?

Terrestrial experiments need not include experimental units across a range of scales or sizes to be "scale-sensitive" and to contribute to understanding scale and scaling of terrestrial systems. By "scale-sensitive" we mean an *a priori* consideration of scale that recognizes that scale may affect the system under consideration and, therefore, results in a corresonding experimental design (see Pither and Taylor 1998). For example, if patches of vegetation occur at a characteristic scale on the landscape, the patches may well reflect processes and interactions operating at particular scales. A scale-sensitive experiment would match the size of experimental units or plots with the size of these patches so that treatments could interact with whatever coherent processes are responsible for the scale structure (e.g., patterns of soil homogeneity and heterogeneity or patterns of herbivory). Treatments that are not coincident with these characteristic scales are likely to generate inaccurate or confounded results. For instance, an enclosure smaller than the home range of the organism is likely to alter patterns of behavior and lead to anomalous results. Less obviously, applying treatments (e.g., fertilizer) across natural, albeit inconspicuous, patches (e.g., unknowingly applying fertilizer to half of one patch type and half to another) could likewise produce anomalous and confounding results. Designing terrestrial experiments with knowledge of the spatial autocorrelation of the system should improve the interpretation of the experimental results and enhance the reliability of extrapolation of those results to other systems or other scales.

Designing scale-sensitive experiments requires knowledge of the system under study that may, unfortunately, be difficult to acquire *a priori* for most terrestrial experiments. The need for pretreatment understanding of the system is not, however, unique to the design of scale-sensitive experiments. Good experimental design generally requires pretreatment knowledge of initial conditions, spatial variability, and so on and may often require prior sampling to estimate how much replication is required. Characterizing the system as part of a scale-sensitive design should be combined with the pretreatment characterization that is part of good experimental design.

A variety of methods exists to characterize the scale structure of terrestrial systems, including geostatistics, spectral analysis, and other statistical approaches (e.g., Carlile et al. 1989; Isaaks and Srivastava 1989; Turner et al. 1991; Underwood and Chapman 1998). The analysis of scale for both the design of ecological experiments and extrapolation or interpretation of results is not so much limited by the availability of quantitative tools as by (1) the absence of explicit guidance from theory, (2) the lack of a true conviction that scale is critical (not just important, but critical), (3) the absence of a true commitment to multiscale analysis, (4) a failure to think carefully about the issues of scale involved, and (5) an unwillingness perhaps to tackle unfamiliar quantitative analyses. If practitioners are looking for a "PROC SCALE" or a "click here," we fear they will be disappointed. The field is not yet that mature. If, however, they are willing to dig in, think about the problem, and rummage through an admittedly somewhat disorganized set of concepts and tools, we believe that what will be found will be sufficient to dramatically alter both their approach and interpretation of results.

But even then, if the experimentalist is truly convinced that "scale *really* matters," and all of the above limitations are overcome, the reality of limited resources, financial and otherwise, is still present. Considerations of scale ("Should plots of different size be included?") will be weighed in the balance with more conventional considerations of experimental design ("How many replicates?"), which are already compromises with available resources. How do considerations of scale fare in the balance with these other considerations? Which are more important? To which is a reviewer going to be more sensitive?

Having now seen that scale matters, we cannot ignore scale any more than we could ignore replication and control. What is needed is an enhanced awareness that scale matters, an increase in multiscale experiments to establish critical relationships, and empirical verification of our ability to extrapolate across scale. This is a daunting task requiring active participation of scientists in both the development of theory and experimental design.

ACKNOWLEDGMENTS

Support for C. A. Hatfield was provided by the COASTES program funded by Coastal Ocean Program Office of NOAA.

LITERATURE CITED

Aars, J., E. Johannesen, and R. A. Ims. 1995. Root voles: Litter sex ratio variation in fragmented habitat. *Journal of Animal Ecology* 64:459–472.

Aars, J., E. Johannesen, and R. A. Ims. 1999. Demographic consequences of movements in subdivided root vole populations. *Oikos* 85: 204–216.

Abrahamson, W. G., T. G. Whitham, and P. W. Price. 1989. Fads in ecology. *BioScience* 39:321–325.

Allen, T. F. H., and T. W. Hoekstra. 1992. *Toward a Unified Ecology.* New York: Columbia University Press.

Allen, T. F. H., R. V. O'Neill, and T. W. Hoekstra. 1984. *Interlevel Relations in Ecological Research and Management: Some Working Principles from Hierarchy Theory.* USDA Forest Service General Technical Report RM-110. Fort Collins, Colo.: Rocky Mountain Forest and Range Experiment Station.

Allen, T. F. H., R. V. O'Neill, and T. W. Hoekstra. 1987. Interlevel relations in ecological research and management: Some working principles from hierarchy theory. *Journal of Applied Systems Analysis* 14:63–79.

Allen, T. F. H., and T. B. Starr. 1982. *Hierarchy: Perspectives for Ecological Complexity.* Chicago: University of Chicago Press.

Ash, A. J., and J. G. McIvor. 1998. How season of grazing and herbivore selectivity influence monsoon tall grass communities of northern Australia. *Journal of Vegetation Science* 9:123–132.

Banks, J. E. 1998. The scale of landscape fragmentation affects herbivore response to vegetation heterogeneity. *Oecologia* 117:239–246.

Banks, P. B., I. D. Hume, and O. Crowe. 1999. Behavioural, morphological, and dietary response of rabbits to predation risk from foxes. *Oikos* 85:247–256.

Bormann, F. H., and G. E. Likens. 1979. *Pattern and Process in a Forested Ecosystem.* New York: Springer-Verlag.

Bouvet, J. M. 1997. Effect of spacing on juvenile growth and variability of *Eucalyptus* clones. *Canadian Journal of Forest Research* 27:174–179.

Brewer, J. S., J. M. Levine, and M. D. Bertness. 1998. Interactive effects of elevation and burial with wrack on plant community structure in some Rhode Island salt marshes. *Journal of Ecology* 86:125–136.

Carlile, D. W., J. R. Skalski, J. E. Batker, J. M. Thomas, and V. I. Cullinan. 1989. Determination of ecological scale. *Landscape Ecology* 2:203–213.

Collins, B., and G. Wien. 1998. Soil resource heterogeneity effects on early succession. *Oikos* 82:238–245.

Connell, J. H. 1983. On the prevalence and relative importance of interspecific competition: Evidence from field experiments. *American Naturalist* 122:661–696.

Cresswell, J. E. 1997. Spatial heterogeneity, pollinator behaviour and pollinator-mediated gene flow: Bumblebee movements in variously aggregated rows of oil-seed rape. *Oikos* 78:546-556.

Dale, V. H., R. H. Gardner, and M. G. Turner. 1989. Predicting across scales—
Comments of the guest editors of *Landscape Ecology*. *Landscape Ecology* 3:147–
151.

Dayton, P. K., and M. J. Tegner. 1984. The importance of scale in community
ecology: A kelp forest example with terrestrial analogs. In P. W. Price, C. N.
Slobodchikoff, and W. S. Gaud, eds., *A New Ecology: Novel Approaches to
Interactive Systems*, pp. 457–481. New York: Wiley.

Delcourt, H. R., P. A. Delcourt, and T. Webb III. 1983. Dynamic plant ecology: The
spectrum of vegetational change in space and time. *Quaternary Science Reviews*
1:153–175.

Denslow, J. S., A. M. Ellison, and R. E. Sanford. 1998. Treefall gap size effects on
above- and below-ground processes in a tropical wet forest. *Journal of Ecology*
86:597–609.

Didham, R. K. 1998. Altered leaf-litter decomposition rates in tropical forest
fragments. *Oecologia* 116:397–406.

Dutilleul, P. 1998a. Incorporating scale in ecological experiments: Study design. In
D. L. Peterson and V. T. Parker, eds., *Scale Issues in Ecology*, pp. 369–386. New
York: Columbia University Press.

Dutilleul, P. 1998b. Incorporating scale in ecological experiments: Data analysis.
In D. L. Peterson and V. T. Parker, eds., *Scale Issues in Ecology*, pp. 387–425. New
York: Columbia University Press.

Fisher, R. A. 1960. *The Design of Experiments*. 7th ed. New York: Hafner.

Gascon, C., and J. Travis. 1992. Does the spatial scale of experimentation matter?
A test with tadpoles and dragonflies. *Ecology* 73:2237–2243.

Gibson, D. J., T. R. Seastedt, and J. M. Briggs. 1993. Management practices in
tallgrass prairie: Large- and small-scale experimental effects on species
composition. *Journal of Applied Ecology* 30:247–255.

Giller, P. S., and J. H. R. Gee. 1987. The analysis of community organization: The
influence of equilibrium, scale, and terminology. In J. H. R. Gee and P. S.
Giller, eds., *Organization of Communities: Past and Present*, pp. 519–541. Oxford:
Blackwell Science.

Golden, D. M., and T. O. Crist. 1999. Experimental effects of habitat
fragmentation on old-field canopy insects: Community, guild and species
responses. *Oecologia* 118:371–380.

Green, R. H. 1979. *Sampling Design and Statistical Methods for Environmental
Biologists*. New York: Wiley.

Gurevitch, J., L. L. Morrow, A. Wallace, and J. S. Walsh. 1992. A meta-analysis of
competition in field experiments. *American Naturalist* 140:539–572.

Gusewell, S., A. Buttler, and F. Klotzli. 1998. Short-term and long-term effects of
mowing on the vegetation of two calcareous fens. *Journal of Vegetation Science*
9:861–872.

Hairston, N. G. 1989. *Ecological Experiments: Purpose, Design, and Execution.*
Cambridge: Cambridge University Press.

Herben, T, F. Krahulec, V. Hadincova, and S. Pechackova. 1997. Fine-scale species
interactions of clonal plants in a mountain grassland: A removal experiment.
Oikos 78:299–310.

Herbers, J. M., and V. S. Banschbach. 1998. Food supply and reproductive allocation in forest ants: Repeated experiments give different results. *Oikos* 83:145–151.

Hobbs, R. J. 1998. Managing ecological systems and processes. In D. L. Peterson and V. T. Parker, eds., *Scale Issues in Ecology*, pp. 459–497. New York: Columbia University Press.

Hoekstra, T. W., T. F. H. Allen, and C. H. Flather. 1991. Implicit scaling in ecological research. *BioScience* 41:148–154.

Imhoof, B., and P. Schmid-Hempel. 1998. Patterns of local adaptation of a protozoan parasite to its bumblebee host. *Oikos* 82:59–65.

Isaaks, E. H., and R. M. Srivastava. 1989. *An Introduction to Applied Geostatistics*. New York: Oxford University Press.

Ives, A. R., J. Foufopouls, E. D. Klopfer, J. L. Klug, and T. M. Palmer. 1996. Bottle or big-scale studies: How do we do ecology? *Ecology* 77:681–685.

Jeffers, J. N. R. 1988. Statistical and mathematical approaches to issues of scales in ecology. In T. Rosswall, R. G. Woodmansee, and P. G. Risser, eds., *Scales and Global Change: Spatial and Temporal Variability in Biospheric and Geospheric Processes*, pp. 47–56. SCOPE 35. Chichester, U.K.: Wiley.

Kareiva, P. 1989. Renewing the dialogue between theory and experiments in population ecology. In J. Roughgarden, R. M. May, and S. A. Levin, eds., *Perspectives in Ecological Theory*, pp. 68–88. Princeton: Princeton University Press.

Kareiva, P., and M. Andersen. 1988. Spatial aspects of species interactions: The wedding of models and experiments. In A. Hastings, ed., *Community Ecology*, pp. 35–50. Lecture Notes in Biomathematics 77. Berlin: Springer-Verlag.

King, A. W., W. R. Emanuel, and R. V. O'Neill. 1990. Linking mechanistic models of tree physiology with models of forest dynamics: Problems of temporal scale. In R. K. Dixon, R. S. Meldahl, G. A. Ruark, and W. G. Warren, eds., *Process Modeling of Forest Growth Responses to Environmental Stress*, pp. 241–248. Portland, Ore.: Timber Press.

Kneeshaw, D., Y. Bergeron, and L. De Grandpre. 1998. Early response of *Abies balsamea* seedlings to artificially created openings. *Journal of Vegetation Science* 9:543–550.

Krummel, J. R., R. H. Gardner, G. Sugihara, and R. V. O'Neill. 1987. Landscape patterns in a disturbed environment. *Oikos* 48:321–324.

Levin, S. 1992. The problem of pattern and scale in ecology. *Ecology* 73:1943–1967.

Lewis, T., and L. R. Taylor. 1967. *Introduction to Experimental Ecology*. London: Academic Press.

Ludwig, J. A., and J. F. Reynolds. 1988. *Statistical Ecology: A Primer on Methods and Computing*. New York: Wiley.

Mal, T. K., J. Lovett-Doust, and L. Lovett-Doust. 1997. Time-dependent competitive displacement of *Typha angustifolia* by *Lythrum salicaria*. *Oikos* 79:26–33.

Manley, B. F. J. 1992. *The Design and Analysis of Research Studies*. Cambridge: Cambridge University Press.

Meentemyer, V. 1989. Geographical perspectives of space, time, and scale. *Landscape Ecology* 3:163–173.

Meentemeyer, V., and E. O. Box. 1987. Scale effects in landscape studies. In M. G. Turner, ed., *Landscape Heterogeneity and Disturbance*, pp. 15–34. New York: Springer-Verlag.

Milchunas, D. G., and W. K. Lauenroth. 1995. Inertia in plant community structure: State changes after cessation of nutrient enrichment stress. *Ecological Applications* 5:452–458.

Moloney, K. A., and N. Chiariello. 1998. Yield-density functions as predictors of community structure in a serpentine annual grassland. *Journal of Ecology* 86:749–764.

Monin, A. S., V. M. Kamenkovich, and V. G. Kort. 1977. *Variability of the Oceans.* New York: Wiley.

O'Neill, R. V. 1989. Perspectives in hierarchy and scale. In J. Roughgarden, R. M. May, and S. A. Levin, eds., *Perspectives in Ecological Theory*, pp. 140–156. Princeton: Princeton University Press.

O'Neill, R. V., and A. W. King. 1998. Homage to St. Michael; or, why are there so many books on scale? In D. L. Peterson and V. T. Parker, eds., *Scale Issues in Ecology*, pp. 3–15. New York: Columbia University Press.

O'Neill, R. V., S. J. Turner, V. I. Cullinan, D. P. Coffin, T. Cook, W. Conley, J. Brunt, J. M. Thomas, M.R. Conley, and J. Gosz. 1991. Multiple landscape scales: An intersite comparison. *Landscape Ecology* 137–144.

Petersen, J. E., C. -C. Chen, and W. M. Kemp. 1997. Scaling aquatic primary productivity: Experiments under nutrient- and light-limited conditions. *Ecology* 78:2326–2338.

Petersen, J. E., J. C. Cornwell, and W. M. Kemp. 1999. Implicit scaling in the design of experimental aquatic ecosystems. *Oikos* 85:3–18.

Phillips, D. L., and D. J. Shure. 1990. Patch size effects on early succession in southern Appalachian forests. *Ecology* 71:204–212.

Pither, J., and P. D. Taylor. 1998. An experimental assessment of landscape connectivity. *Oikos* 83:166–174.

Resetartis, W. J., Jr., and J. Bernardo, eds. 1998. *Experimental Ecology: Issues and Perspectives*. New York: Oxford University Press.

Riba, M. 1997. Effects of cutting and rainfall pattern on resprouting vigour and growth of *Erica arborea* L. *Journal of Vegetation Science* 8:401–404.

Risser, P. G. 1986. *Spatial and Temporal Variability of Biospheric and Geospheric Processes: Research Needed to Determine Interactions with Global Environmental Change*. Paris: ICSU Press.

Roswall, T., R. G. Woodmansee, and P. G. Risser, eds. 1988. *Scales and Global Change: Spatial and Temporal Variability in Biospheric and Geospheric Processes*. SCOPE 35. Chichester, U.K.: Wiley.

Scheiner, S. M., and J. Gurevitch, eds. 1993. *Design and Analysis of Ecological Experiments*. London: Chapman and Hall.

Schneider, D. C. 1994. *Quantitative Ecology*. San Diego: Academic Press.

Schoener, T. W. 1983. Field experiments on interspecific competition. *American Naturalist* 122:240–285.

Stommel, H. 1963. Varieties of oceanographic experience. *Science* 139:572–576.

Steele, J. H. 1985. Comparison of marine and terrestrial ecological systems. *Nature* 313:355–358.

Steele, J. H. 1989a. The ocean "landscape." *Landscape Ecology* 3:185–192.

Steele, J. H. 1989b. Discussion: Scale and coupling in ecological systems. In J. Roughgarden, R. M. May, and S. A. Levin, eds., *Perspectives in Ecological Theory*, pp. 177–180. Princeton: Princeton University Press.

Swank, W. T., and J. E. Douglass. 1974. Streamflow greatly reduced by converting deciduous hardwood stands to pine. *Science* 185:857–859.

Tilman, D. 1989. Ecological experimentation: Strengths and conceptual problems. In G. E. Likens, ed., *Long-term Studies in Ecology: Approaches and Alternatives*, pp. 136–157. New York: Springer-Verlag.

Turkington, R. E. John, C. J. Krebs, M. R. T. Dale, V. O. Nams, R. Boonstra, S. Boutin, K. Martin, A.R.E. Sinclair, and J. N. M. Smith. 1998. The effects of NPK fertilization for nine years on boreal forest vegetation in northwestern Canada. *Journal of Vegetation Science* 9:333–346.

Turner, S. J., R. V. O'Neill, W. Conley, M. R. Conley, and H. C. Humphries. 1991. Pattern and scale: Statistics for landscape ecology. In M. G. Turner and R. H. Gardner, eds., *Quantitative Methods in Landscape Ecology*, pp.17–49. New York: Springer-Verlag.

Underwood, A. J. 1986. The analysis of competition by field experiments. In J. Kikkawa and D. J. Anderson, eds., *Community Ecology: Pattern and Process*, pp. 240–268. Victoria, Australia: Blackwell Science.

Underwood, A. J. 1997. *Experiments in Ecology: Their Logical Design and Interpretation Using Analysis of Variance*. Cambridge: Cambridge University Press.

Underwood, A. J., and M. G. Chapman. 1998. A method for analysing spatial scales of variation in composition of assemblages. *Oecologia* 117:570–578.

Urban, D. L., R. V. O'Neill, and H. H. Shugart Jr. 1987. Landscape ecology. *BioScience* 37:119–127.

Wiens, J. A. 1981. Single sample surveys of communities: Are the revealed patterns real? *American Naturalist* 117: 90–98.

Wiens, J. A. 1984. On understanding a non-equilibrium world: Myth and reality in community patterns and processes. In D. L. Strong Jr., D. Simberloff, L. G. Abele, and A. B. Thistle, eds., *Ecological Communities: Conceptual Issues and the Evidence*, pp. 3–13. Princeton: Princeton University Press.

Wiens, J. A. 1989. Spatial scaling in ecology. *Functional Ecology* 3:385–397.

CHAPTER 11

Issues of Scale in Land-Margin Ecosystems

Walter R. Boynton, James D. Hagy,
and Denise L. Breitburg

If we study a system at an inappropriate scale, we may not detect
its actual dynamics and patterns but may instead identify patterns that are artifacts of scale. Because we are clever at devising
explanations of what we see, we may think we understand the
system when we have not even observed it correctly.

—J. A. Wiens, *Spatial Scaling in Ecology* (1989)

It always seemed a fine idea to me to build a showboat with just
one big flat open deck on it, and to keep a play going continuously. The boat wouldn't be moored, but would drift up and
down the river on the tide, and the audience would sit along
both banks. They could catch whatever part of the plot happened to unfold as the boat floated past, and then they'd have to
wait until the tide ran back again to catch another snatch of it, if
they still happened to be sitting there. To fill in the gaps they'd
have to use their imaginations, or ask more attentive neighbors,
or hear the word passed along from upriver or downriver. Most
times they wouldn't understand what was going on at all, or
they'd think they knew, when actually they didn't. Lots of times
they'd be able to see the actors, but not hear them. I needn't
explain that that's how much of life works.

—John Barth, *The Floating Opera* (1967)

WE DECIDED TO BEGIN THIS CHAPTER WITH THE ABOVE QUOTES, which come from very different perspectives, to entice the reader into reading on and to remind ourselves that scale is so very central to our understanding of most things, be they arts or ecology. This chapter discusses issues of scale primarily in those ecosystems that form the interface between the terrestrial and marine realms, the estuaries, coastal embayments, lagoons, and salt marshes we refer to as land-margin ecosystems.

On a global basis, land-margin ecosystems constitute a small percentage (~0.5%) of the world's oceanic areas. However, the high fisheries production, proximity to major urban areas and associated transportation networks, and the use of these areas for recreational purposes makes them far more important than indicated by spatial extent alone (Houde and Rutherford 1993). Because of the location of these systems at the margin between land and ocean, serious degradation has become widespread during the last few decades. Current demographic projections indicate that human activities in the coastal zone will continue to intensify. For example, the average population density in coastal counties in the northeast region of the United States (Maine to Virginia) was about 340 people per square mile in 1988 and is expected to increase by an additional 30 percent by 2010 (Culliton et al. 1990). Of the 140 land-margin ecosystems recently reviewed by NOAA (1998), about 80 percent were classified as moderately to seriously impacted by excessive nutrient inputs and other materials.

Sediments, nutrients, and an array of toxic materials will continue to find their way into these aquatic systems, leading to further declines in water quality, habitat conditions, and living resources, especially if these areas do not have effective management programs. In addition, increased human activities will intensify pressures on the habitats and living resources characteristic of these systems (Matson et al. 1997; Vitousek et al. 1997). In many such systems, seagrass communities and other habitats have already been lost or degraded, tidal wetlands filled, and fish and shellfish stocks overfished or contaminated (e.g., Duarte 1995; Rabalais et al. 1996). Rapid and poorly designed development and other activities within adjacent drainage basins have destroyed or negatively impacted the very resources that were the prime reasons stimulating development in the first place.

In the last few decades the rapid growth in human activities in watersheds surrounding land-margin ecosystems has resulted in numerous resource use conflicts. One of the pressing needs in many of these areas concerns establishment of cause-effect linkages. Basic questions and debate often focus on cause(s) of some ecosystem change. For example, what was the factor or factors causing the great decline in seagrass communities in many land-margin systems? On the other hand, questions are asked about the consequences of increasing or decreasing some input (e.g., nitrogen) or extraction (e.g., fish catch) from land-margin ecosystems. Answers to these types of questions are often developed through some form of controlled experimental study, and mesocosms of some shape and size are often used as primary tools. The appeal of mesocosms is based on several characteristics, including the facts that (1) experiments are often impossible, too costly, or too risky to perform on natural land-margin systems, (2) experiments testing a variety of cause-effect linkages can be done relatively quickly in mesocosms compared to waiting for a time-series of measurements to be collected from natural land-margin systems, (3) mesocosm approaches often, but not always, allow for some replication and hence added confidence in results, and (4) there is a reasonable amount of control inherent in this approach that is generally lacking in natural land-margin systems. In short, mesocosm approaches offer an experimental tool to answer pressing management questions of what is causing what and will this happen if we do such and such (Crossland and La Point 1992). The problem is that mesocosm approaches, almost without exception, operate at small spatial and short temporal scales but basic questions about how ecosystems operate and more applied questions about which of many management options will provide desired results mainly occur at larger and longer scales (Hoekstra et al. 1991). So, how do we scale-up when using mesocosm approaches in land-margin systems and how do we conduct better comparative ecological analyses of these systems in search of general principles?

This chapter resulted from discussions, presentations, and literature reviews during and following a workshop on scaling relations in experimental ecology sponsored by the Multiscale Experimental Ecosystem Research Center (MEERC). Its focus is on scaling issues in land-margin systems. Specifically, we considered the following: (1) definitions of scaling and scale–dependent behavior; (2) characteristics of land-margin systems as these relate both to the need for scaling mesocosm-based studies of these systems and the possibilities of using the systems in a comparative mode to develop better understanding of scale-dependent behavior in these systems; (3) examples of scaling in land-margin systems and mesocosms to assess the current state of the art and the degree to which scaling is an explicit part of land-margin science; and (4) recommendations about what might be done in the short- and long-term to improve our understanding of scale-dependent behavior in these systems and use of mesocosms in land-margin studies.

DEFINITIONS OF SCALE, SCALING, AND SCALE-DEPENDENT BEHAVIOR

Although it might seem unnecessary to provide definitions of the central topic of this workshop, we found within our working group a diversity of opinion concerning what was meant by scale, scaling, scale–dependent behavior, and other associated terms. To clarify this, and to make the topic more accessible to those who have not had dealings with ecological scaling, we have assembled some definitions from the recent literature.

Definitions of Scale

- A change in pattern as determined by the spatial and/or temporal extent of measurements necessary to detect significant differences in the variability of the quantity of interest (Gardner 1998).

- Scale refers to resolution (spatial grain, time step, or degree of complication) and to extent (in time, space, and number of components included) (Costanza et al. 1993).

- Scale denotes the resolution within the range of a measured quantity (Schneider 1994).

- Scale refers to the resolution (grain) and to the range (extent) of research activities and ecological rates (Schneider et al. 1997).

- Scale is defined by the temporal and spatial characteristics of energy and matter within and among ecological systems. The scale of a study is determined by the size and extent of the observations in time and space, as well as by the resolving power of the individual measurements (Hoekstra et al. 1991).

Definitions of Scaling

- Scaling refers to the application of information or models developed at one scale to problems at other scales (Costanza et al. 1993).

- Scaling is that which is needed when the "ballgame rules" change; when relationships break down; when size (for example) changes and some other feature does not change in a linear fashion (from workshop notes).

- Scaling pertains to the use of rules to explain the manner in which processes change when extrapolating from short to long and small to large time and space scales; all measurements are related to scale (from workshop notes).

Definitions of Scale-Dependence

- Scale-dependent pattern can be defined as a change in some measure of pattern with a change in either the resolution or range of measurement (Schneider 1994).

- Scale-dependent pattern refers to the way in which a statistical summarization of a quantity changes with spatial or temporal scale.

- Scale-dependent processes are those where the ratio of one rate to another varies with either resolution or range of measurement (Schneider 1994).

Other Definitions

- *Extent* is the overall area encompassed by a study.

- *Grain* is the size of the individual units of observation. Extent and grain define the upper and lower limits of resolution of a study (O'Neill et al. 1986).

- *Allometric* relationships are those wherein a part of an organism or a process is related to or scaled to another and the scaling is not a direct proportion (i.e., the scaling factor is not unity); *isometric* relationships are those related by direct proportion (i.e., the scaling exponent is unity) (Calder 1983).

- *Normalization* is scaling that is done with simple direct proportions; identical with isometric scaling.

- The scope of a natural phenomenon is defined as the ratio of the upper to lower limit. Scope is thus the ratio of the extent to the grain or the ratio of the largest to the smallest size scale of measurement. Scope also provides an indication of the degree of extrapolation that is needed or used (Schneider 1998).

- Multiscale analysis is simply the recognition that there is no single "right" scale for analysis at levels of organization above an individual organism (Schneider 1994).

The workgroup distinguished between scaling to properly understand scale-dependent features in land-margin ecosystems from scaling so as to avoid including experimental artifacts, particularly those associated with mesocosm studies, in results and interpretations. Initially, the issue seemed clear; the type of scaling we are talking about involves primarily the former and the latter is simply to be avoided. However, if as suggested at the beginning of this chapter, all measurements are matters of scale, both need to be considered although the exact methodological approaches for dealing with these problems are not clear.

The workgroup also identified several categories of scaling issues, including (1) isometric scaling or normalization (e.g., converting total nutrient loads to an estuary to load m^{-2} based simply on the size of the estuary); (2) homogeneous spatial scaling (i.e., the scaling exponent is not unity, as in normalization, but the systems to be scaled one to the other are considered to be relatively homogeneous); (3) heterogeneous system scaling (i.e., deals with nonuniform processes in time and space in the systems being scaled; scaling with measures of variance are included here); and (4) scaling for sharp gradients or discontinuous properties (i.e., new organisms are added to an experiment, water transparency changes so sediments have the potential to have autotrophic components, water flow characteristics change from laminar to turbulent).

CHARACTERISTICS OF LAND-MARGIN ECOSYSTEMS

A complete list of all characteristics of land-margin ecosystems is not the goal here. Rather, a subset of those systems that occur at the boundary of land and water were considered and these mainly included estuaries, coastal embayments, and lagoons. We recognize that there are other land-margin systems, including tidal and nontidal marshes and forested wetlands and bogs, but have chosen to not explicitly consider those systems in order to limit the task at hand.

The following characteristics are particularly relevant descriptors of land-margin systems, although they are not necessarily unique to such systems; however, all have some bearing on the issue of scaling and mesocosm design.

- Land-margin systems tend to be small. For example, the median size of coastal lagoons was found to be about 7,800 ha with a range of size from 3 to 800,000 ha (Nixon 1982). On a global basis, such systems comprise less than 0.5 percent of the world's oceanic areas.

- These systems tend to exhibit relatively high rates of primary and secondary production and contribute about 20 percent of the world fisheries catch (Houde and Rutherford 1993). At broad temporal and spatial scales (annual or multiyear and whole system), there is a strong correlation between primary production and fishery yields despite generally poor understanding of food web structure and dynamics. Fishery yields per unit primary production also appear to be considerably higher than for lakes, but the exact reasons are not clear (Nixon et al. 1986).

- The suite of organic matter sources in land-margin systems is often diverse, including phytoplankton, seagrasses, macroalgae, benthic microalgae, and inputs from terrestrial sources. Typically, not all sources are important at the same time of the year or in all years.

- Many land-margin systems exhibit strong physical, chemical, and biological gradients. The relatively small size of the systems coupled with the enclosed nature of estuaries contributes to the formation of gradients. In stratified systems, such as the Chesapeake Bay, some of the strongest gradients occur in the vertical direction over spatial scales of meters.

- In part because of close association with the land, these systems are often exposed to high temporal variability in inputs, particularly from rivers. Costanza et al. (1995) have argued that this variability limits the development of complexity in land-margin systems but favors high net production. Input variability is not limited to land linkages; ocean inputs of water, salt, nutrients, and biological populations also exhibit considerable variability in land-margin systems (Smith et al. 1991).

- Land-margin systems often have complex and variable rates of physical transport and mixing, including those associated with seasonal scale weather, local and nonlocal storm effects, and short scale tidal and local wind forcing.

- Benthic-pelagic coupling is characteristic of most land-margin systems and the strength of the coupling appears to be directly related to water depth (Kemp and Boynton 1992). The generally shallow nature of the systems makes depth a particularly important feature. The sediment-water interface is one of several physical interfaces characteristic of these systems and not so characteristic of many others.

- These systems are often characterized by seasonally migrating sub-systems of organisms. Larval and juvenile stages of many fish species move from the coastal ocean to estuaries for at least part of their life cycle (blue crab [*Callinectes sapidus*], Atlantic croaker [*Micropogonias undulatus*], menhaden [*Brevoortia tyrannus*]) and anadromous fish spawn in estuarine systems and then participate in seasonal coastal migrations (American shad [*Alosa sapidissima*], striped bass [*Morone saxatili*]).

Steele (1985, 1991) and Steele et al. (1993) have compared and contrasted characteristics of terrestrial and aquatic systems and physical forcing on these systems with the goal of better understanding the utility of theory and measurement across system types. Marine and terrestrial systems both exhibit a variety of temporal patterns, but in marine systems large-scale factors (e.g., El Niño) have been typically invoked as prime factors causing variability and dramatic change whereas predator-prey and other community-level explanations are invoked for terrestrial systems. Steele and Henderson (1994) indicate that these different explanations are based on the overlap of time and space scales of physics and biology in the ocean and the wider separation of these in atmospheric and terrestrial systems.

The above authors and others (e.g., Stommel 1963; Haury et al. 1978; Denman 1992; Hildrew and Giller 1992) have developed versions of time-space diagrams to examine characteristics of terrestrial and marine system physics and biology and to look for overlaps of time and space scales that may suggest patterns of control. However, we are not aware of such a synthesis for land-margin systems and have included a preliminary version of such a diagram which includes time-space characteristics of physics, weather, biology, and commonly used research approaches in these systems (figure 11.1). As expected, there are some similarities with previously developed diagrams, including weather events (thunderstorms and hurricanes) and large time and space scale atmospheric perturbations (droughts). Some important ocean physics (ocean basin gyres and deep ocean circulation) do not register on the spatial scales of land-margin systems. However, several physical processes are unique and prominent features of many of these systems, including turbidity maxima, river floods, and sediment resuspension events. Most representations of marine communities show a progression from autotrophs at short and small time and space scales to large and long-lived heterotrophs at large time and space scales.

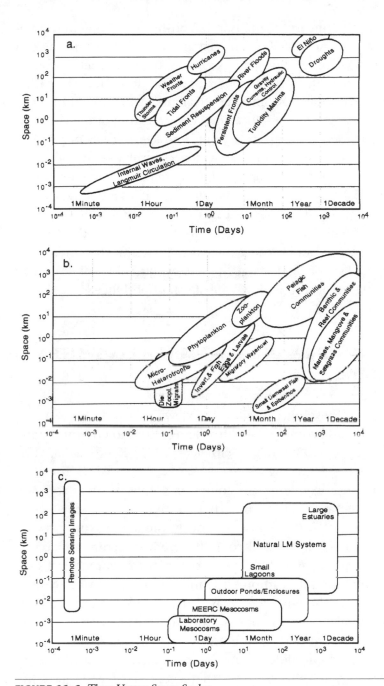

FIGURE 11•1 *Time Versus Space Scales*
Approximate time and space scales relevant to land margin (LM) ecosystems for (a) physical circulation, weather, and climate, (b) biotic components, and (c) commonly applied research approaches.

This simple continuum from small and short-lived to large and long-lived is certainly not the case for land-margin systems; autotrophs can occur at extremes of time and space scales as do some heterotrophic components; land-margin system representations look like a partial combination of terrestrial and marine diagrams, which is intuitively pleasing because these systems are located at the transition of the land and sea. There is considerable overlap in the time and space scales of physical processes and biological components, which is probably involved in creating the qualitative idea that land-margin systems are complex, messy places of substantial variability both because of changing physics and interacting biological communities. Because only a portion of the time-space domain of land-margin processes can directly be examined with even the largest of mesocosms, and because there is the potential for interaction among the rich milieu of land-margin processes, questions of scale and problems of scaling-up or scaling-down are of real importance.

SCALED RELATIONSHIPS IN LAND-MARGIN ECOSYSTEMS AND MESOCOSMS

Several authors have developed literature reviews that indicate a strong increase in interest and activity related to scaling issues in environmental research since the mid-1980s (e.g., Schneider 1994; Kemp et al., this volume). Schneider (1992) suggested a number of reasons for this, including great improvements in computing power and software capability, instrumentation of all sorts that has allowed investigators to measure environmental features not previously measured, and growing recognition that biological-physical coupling, environmental variability, and propagation of physical and biological effects through food webs are all scale-related and central to improving understanding of how natural systems operate. Although we have not conducted another comprehensive literature review, our distinct impression is that scaling issues have received considerably more attention in terrestrial ecology (landscape ecology), ecophysiology, and oceanography than in studies of land-margin systems, despite the fact that there have been pressing practical needs for such activities for quite some time. We examined results from several electronic searches (*Aquatic Sciences and Fisheries Abstracts, 1978–1998; Ecology Abstracts, 1981–1998*) and found one to two papers per year concerned primarily with issues of scale in land-margin systems. We also reviewed publications in several journals that emphasize studies of these systems (*Estuaries, Estuarine Coastal and Shelf Science,* and *Marine Ecology Progress Series*) and found relatively few citations (1 to 3 papers per year).

However, some examples of scaling of land-margin organisms, ecosystem processes, and mesocosms are available, although not in the numbers that would indicate scaling to be a prominent component of land-margin research thinking. We have assembled representative results from some studies in Table 11.1. Our goal is to provide a flavor of some scaling activities in land-margin (and closely related) systems rather than organize a comprehensive synthesis, although this needs to be done. Several of the ecosystem-level examples can be thought of as comparative ecology studies where the scaling turned out to be isometric, or at least was treated as isometric. The relationships observed by Monbet (1992) between nutrient concentrations and algal stocks were weak among estuarine study sites until observations were categorized into micro- and macro-tidal range sites, a nice example where issues of scale explain variability in the fashion suggested by Carpenter and Kitchell (1987). We did not find many examples of scaled relationships involving data collected from both small-scale (e.g., mesocosm) and natural ecosystems, but of those we did, system depth was an important scaling variable. We found several examples of artifacts in mesocosm systems and the effect these have on relating results to larger systems at lower (e.g., Chen et al. 1997; Petersen et al. 1997) and higher (de Lafontaine and Leggett 1987a) trophic levels. The research area with perhaps the most information is that of ecophysiology of land-margin (and related) organisms. We readily found allometric relationships relating a variety of processes to functions of phytoplankton cell size (e.g., Chisholm 1992), zooplankton size (e.g., Kiorboe and Sabatini 1995), and seagrass morphology (e.g., Duarte 1991). These latter results are not surprising because ecophysiology has used allometric scaling as a tool for some time and this class of measurement is relatively easy to make.

We conclude this section with several examples that emphasize the use of scale concepts in design and analysis of environmental data sets. These studies did not produce general scaled relationships (e.g., allometric relationships of phytoplankton cell size to growth rate) but issues of scale and scale-dependent pattern were central to design and analysis.

TABLE 11·1 *Examples of Scaled Relationships Based on Data Collected from Land Margin (LM), Marine, and Lake Ecosystems and Land Margin (LM) Mesocosms; and Some Allometric Relationships for Various Organism Groups*

Scaled Feature	System Size or Type	Scaling Variable	Comments	Reference
Nutrient-loading rates	LM ecosystems	Size (area or volume)	Simple, linear normalization	Boynton et al. (1982, 1995)
Phytoplankton production and biomass	LM ecosystems	Nitrogen load (areal or volumetric)	Log-log relationship of load vs pri prod rate or biomass	Nixon (1992)
Phytoplankton biomass	LM ecosystems	Nitrogen concentration and tidal range	Log-log relationships	Monbet (1992)
Plankton C:N ratios	LM mesocosms and Narragansett Bay	Phytoplk production rate	Phytoplk C:N Ratio=23.8 + 6.0 Log (pri prod rate)	Nixon (1992)
Autotroph/heterotroph ratios	LM, marine and lake ecosystems	Autotrophic biomass	Negative log-normal relationships for ocean, LMs, and lakes	Gasol et al. (1997)
Nitrogen export	LM, marine and lake ecosystems	Water residence time	%TN exported= 64.8 - 27.0*log (water residence time)	Nixon et al. (1996)
Fisheries yield	LM, marine and lake ecosystems	Phytoplk production rate	Log-log relationship between pri production and fisheries yield	Nixon et al. (1986)
Sediment-water	LM mesocosms	Depth	Sediment NH4 Flux/TN Load = $0.85 \cdot \text{Depth}^{-2.00}$	Kemp and Boynton (MEERC data)
Nutrient exchanges	Chesapeake Bay subestuaries	Depth	Sediment NH4 Flux/TN Load = $26.9 \cdot \text{Depth}^{1.71}$	(MEERC data)
Water column DIN loss rate	LM mesocosms and Chesapeake Bay	Depth	DIN loss rate(uM day^{-1}) = $0.084 + 0.99 * \text{Depth}^{-1}$	Kemp (MEERC data)
Spring zooplk abundance	LM mesocosms and Chesapeake Bay	Depth	Zooplk abundance (g C m^{-3}) = $0.0005 + 1.07 * \text{Depth}^{-1}$	Kemp (MEERC data)
Phytoplankton prod rates	LM mesocosms (0.1-10 m^3)			
Spring, light-limited		Depth	Volumetric prod (P_v) = $P_a * z^{-1}$; P_a = production per area	Peterson et al. (1997)
Summer, nutrient-limited		Depth	Areal prod (P_a) = $P_v * z$	
Periphyton biomass	LM mesocosms (0.1-10 m^3)	Size (mesocosm volume and wall area)	Quadratic relationship between periphyton biomass and wall area (Aw) to water volume (V) ratio (Aw:V)	Chen et al. (1997)
Larval fish predation/mortality	Mesocosms (0.3 - 6.4 m^3)	Mesocosm volume	lnY = 4.3 - 0.76 ln x : Y = mortality rate; x = mesocosm volume	de Lafontaine and Leggett (1987a)
Marine phytoplankton	Individual size classes and species	Size	Multiple relationships in the form of Y = aX^b	Chisholm (1992)
Marine phytoplankton	Individual size classes and species	Cell carbon	Cell growth = a(cell carbon)^b, b~ 0.75	Banse (1976)
Phytoplankton nutrient uptake	Different species	Body size	Multiple relationships in the form of Y = aX^b	Moloney and Field (1989)
Marine copepods	Female zooplankton	Body weight and egg number	Multiple relationships in the form of Y = aX^b	Kiorboe and Sabatini (1995)
Seagrass form and productivity	Different seagrass species	Rhizome diameter and shoot weight	Multiple relationships in the form of Y = aX^b	Duarte (1991)

Notes: phytoplk = phytoplankton; pri prod = primary production; TN = total nitrogen; 300plk = 300 plankton

The first is a well-known and instructive analysis of scale-dependent pattern involving sea bird distributions (common murres [*Uria aalge*] and Atlantic puffins [*Fratercula arctica*]) relative to prey (capelin [*Mallotus villosus*]) in the northwest Atlantic (Schneider and Piatt 1986). In this study, counts of birds and fish were made along a 15 km transect with a minimum resolution of 0.25 km. Counts were grouped into mean counts at a variety of spatial scales (0.25, 1.5, 2.5, and 3.0 km). Spatial association between birds and prey was strongest at the larger spatial scales, which are the scales at which birds were adjacent to the fish schools between feeding periods. The multiple scales of observation used were critical in understanding these predator-prey interactions.

A number of coordinated studies were conducted on an intertidal sandflat in Manukau Harbour, New Zealand, by Thrush and his colleagues. The theme of these studies was to investigate issues of scale using field measurements and experiments. Specific elements of these studies considered factors controlling spatial distributions of dominant bivalves and adult-juvenile interactions (Legendre et al. 1997), factors controlling macrofaunal recolonization rates (Thrush et al. 1995, 1996), evaluations of methods for scaling up small-scale manipulation experiments to large-scale areas (Thrush, Schneider et al. 1997), scale-dependent patterns of sediment reworking and transport (Grant et al. 1997), and others (Thrush, Cummings et al. 1997; Thrush, Pridmore et al. 1997). An important aspect of the study design was that routine sampling and experiments were conducted at several spatial and temporal scales and a variety of modeling and statistical techniques were used for analysis. Investigators found negative correlations between juvenile and adult bivalves at spatial lags of 1 m but positive correlations at spatial lags of 5 m; sediment reworking and transport were variable over spatial scales of 10 to 100 m and temporal scales of several days; at larger spatial scales, physical factors (elevation, wave disturbance) were important in determining bivalve densities but at smaller spatial scales intrinsic factors were more important. The authors observed that had these studies been conducted at but one scale, much of the interesting information concerning factors apparently controlling biology and physics of the sandflat would have not emerged or would have appeared to be contradictory. They also observed that despite increasing interest in issues of ecological scale, practical advice for field ecologists relative to study design and analysis is limited.

A final example is from Rose and Leggett (1990), who used simulation models and acoustic data to examine signs and strengths of spatial correlations between Atlantic cod (*Gadus morhua*) and its prey, capelin. They found that when the prey species was not in a thermal refuge, capelin and cod densities were coherent across spatial scales of 2 to 30 km. At spatial scales > 4 to 10 km, densities of both species were positively correlated; at scales less than school sizes (3 to 5 km), densities were negatively correlated; and at the smallest scales (2 to 3 km) negative correlations were strongest. When capelin were in thermal refuges, predator-prey relations were in phase only at scales larger than the thermal refuges.

It comes as no surprise, indeed it is intuitively satisfying, that there are scale-dependent patterns that emerged in all of these studies. However, as Rose and Leggett (1990) note "the spatial dynamics of predator-prey interactions are clearly complex, and quantification is particularly difficult in the wild." Tools are more available now than in the recent past for making similar measurements in land-margin systems at scales of grain and extent that are appropriate for a variety of organisms. It is probable that some of the conflicting or uncertain interpretations of cause and effect in these systems will be resolved when issues of scale are considered more explicitly.

ISSUES OF SPECIAL CONCERN IN DESIGN AND USE OF LAND-MARGIN MESOCOSMS

There are many publications on mesocosms, including experimental ecosystem design (Kemp et al. 1980; Ives 1996; Lawton 1996; Petersen et al. 1999; Kemp et al., this volume), turbulence and mixing in experimental systems (Oviatt 1981; Sanford 1990), fish and other predators (Gamble 1985; Houde 1985), benthic studies (Bakke 1990), benthic-pelagic interactions and land-margin system responses to nutrient additions (Pilson 1990), toxic contaminant effects (Morgan et al. 1990), and others. It is not the intent of this chapter to review all or even most of these. Rather, we summarize some of the pressing issues of scale in the design of mesocosm research concerned with land-margin ecosystems; the above sources have been indicated as a point of departure for those interested in specific design issues.

As Petersen et al. (1999) have pointed out, mesocosms are smaller than natural systems, have reduced spatial and biological complexity, are surrounded by walls for organisms to grow on, have modified physics compared to natural systems, and are often operated for relatively short periods of time. Encapsulated in this summary statement are many of the design issues of interest to those working in land-margin ecosystems.

Distinguishing fundamental effects of scale from artifacts of the mesocosm itself is clearly a prominent and difficult problem. In experiments reported by Chen et al. (1997), periphyton biomass and metabolism increased as a quadratic function of increasing mesocosm wall-area-to-volume ratio. In addition, in these experiments there were significant negative correlations between periphyton biomass and measures of phytoplankton and zooplankton abundance. In mesocosms of this size (0.1 to 10 m^3), periphyton exerted major artifactual effects on important components in only several weeks. In similar studies focusing on plankton dynamics, Petersen et al. (1997) found fundamental relationships of production to scaling factors as well as artifactual effects on plankton metabolism associated with wall effects of periphyton and light availability.

To examine the artifactual effect of walls and periphyton at a larger scale, we combined several comparative analyses based on field data and included similar data collected from MEERC mesocosms (figure 11.2). In the top panel, when areal nutrient loading rates for a variety of land-margin systems are compared to rates in MEERC mesocosms, there is little disparity, as expected in these well-controlled systems. However, the percentage of total nitrogen exported (middle panel) as a function of water-residence time for a variety of land-margin systems is similar to that estimated in MEERC mesocosms in spring (period of low periphyton growth) but very different in summer when wall growth is actively sequestering added nitrogen. In the lower panel, annual average water column chlorophyll a biomass in plankton in the mesocosms had a strong relationship relative to scaled nutrient loads for a number of sites in the Chesapeake Bay. However, chlorophyll a associated with periphyton on the mesocosm walls departed strongly from other observations, probably because periphyton were not subject to export from the mesocosms as were phytoplankton. Clearly, wall effects can be important artifacts in mesocosms. The degree of interference can be minimized by using large-diameter systems of mesocosms and by keeping experimental periods relatively short.

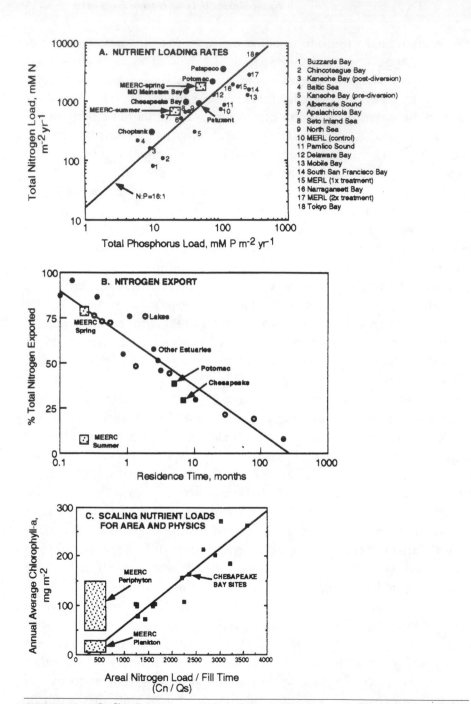

FIGURE 11•2 *Scaling Patterns*

Examples of preliminary scaling exercises using data from coastal systems, Chesapeake Bay subsystems and MEERC mesocosms. (A) adapted from Boynton et al. (1995); (B) from Nixon et al. (1996); (C) from Boynton (unpublished data). Cn and Qs in the lower panel refer to areal total nitrogen loading rate (g N m^{-2} y^{-1}) and freshwater fill time (years).

Often land-margin research and management questions involve higher trophic level components. Questions such as why shad are not reproducing well, what controls bay anchovy (*Anchoa mitchilli*) growth rate, how important is jellyfish predation on larval fish survival, and others are common. Inclusion of upper trophic levels in mesocosm design has been, and probably will continue to be, a difficult issue for several reasons. The more obvious concern is the larger size of higher trophic level organisms relative to mesocosm space and the behavioral patterns of these organisms. Petersen et al. (1999) reported that in a large sampling of mesocosm experiments there was a significant increase in volume between mesocosms that contain only one as opposed to two or three trophic levels. As mesocosm size is decreased it is clear that artifacts come into play; survival of fish larvae, for example, increases sharply as mesocosm size increases (Theilacker 1980; de Lafontaine and Leggett 1987a). It seems that there is emerging consensus that larval fish studies need mesocosms on the order of 3 m^3 or larger and that depth characteristics of the mesocosms also come into play (de Lafontaine and Leggett 1987b). Also, the use of "pulse" experiments (fish are added, then removed) with larger fish may be useful (Frost et al. 1988) and smaller fish predators can be used in a "press" mode (fish are added and remain in the mesocosms) for some types of experiments. The careful consideration of species-specific (and life-stage specific) neighborhood size evaluations may also be useful in designing mesocosm experiments with larger animals (Addicott et al. 1987).

One of the benefits of new technology in the last decade has been the much enhanced ability to measure variability on large and small spatial scales. For example, Harding et al. (1992) used a radiometer (ODAS) on a fixed-wing plane and conducted weekly or bi-weekly surveys of chlorophyll, temperature, and salinity throughout Chesapeake Bay (300 km × 20 km) in a few hours. Patches of high chlorophyll water were almost always evident, with patch size varying from fractions of a km^2 to hundreds of km^2. The point that spatial and temporal heterogeneity is important in structuring ecosystems has been made repeatedly in recent years (Wiens 1976; Barry and Dayton 1991; O'Neill et al. 1991; Levin 1992). However, creating and maintaining or realistically evolving heterogeneity in mesocosms is difficult at best, although at small enough scales it may persist even if not wanted (Lehman and Scavia 1982; Atkinson et al. 1987; Duarte and Vaqué 1992). Short of building very large mesocosms or using semi-enclosed natural systems (Davies and Gamble 1979; Oiestad 1990), some degree of variability can be achieved by connecting mesocosms one to another and allowing different regimes to develop while still maintaining some degree of connectiveness. Alternatively, some experiments could be done in separate mesocosms (with different treatments) and the results used in spatially explicit simulation models as suggested by Denman (1992). We suggest careful consideration of what can be left out of mesocosm studies for the sake of tractability, but also recognize that suppression of variability in and among mesocosms may be a mistake if ecosystem depends on some feature of variance.

One of the approaches that has been successfully used elsewhere in dealing with ecological questions that appear too large to be studied directly in mesocosms has been called the "scale-down approach" (Vitousek 1993). In this approach some "whole system" measurements are made and then, where appropriate, mesocosm experiments are conducted to estimate rates and examine mechanisms. For example, in Hubbard Brook Experimental Forest, nutrient and other types of budgets were developed under undisturbed and perturbed circumstances (Vitousek 1993; Carpenter 1996). Smaller and more controlled experiments were then done to investigate rates; these experimentally developed rates could then be compared to whole system rates as a check for realism. In some land-margin systems similar "whole system" estimates of rates are also possible. For example, Hagy (1996) developed an estuarine box model to generate transport estimates and used these, in conjunction with nutrient and carbon concentration measurements, to compute net fluxes of nitrogen, phosphorus, and carbon in different sections of the estuary. These large-scale measurements were then compared to small-scale and short-term mesocosm rate estimates. In several cases mesocosm estimates appeared to be low, possibly because of variability in sediment properties that were not captured in the small-scale studies.

Finally, the time might be right for a manual of mesocosm design for land-margin systems. More specifically, this would be an engineering-like handbook that contains scaled relationships and could be used to develop issue-specific mesocosm systems. Our sense is that enough is now known to produce a first edition similar to the handbook developed by Jorgensen et al. (1991) for ecological and ecotoxicological parameters. This manual could also contain evaluations of various scaling variables such as depth, mixed layer depth, photic depth, important system component neighborhood sizes, surface area to volume effects, residence time and mixing schemes, and the like. The manual would be an effort to move mesocosm design from the artful or intuitive stage it is now in to a more quantitative stage. It probably would go a long way toward limiting re-invention of mesocosm wheels and avoiding results that are dominated by experimental artifacts.

TOOLS FOR ANALYSIS OF SCALE AND EXTRAPOLATING AMONG SCALES IN LAND-MARGIN SYSTEMS

This section begins with a few examples of scope to help examine the adequacy of typical environmental measurements and the degree of extrapolation required in land-margin studies. In other words, how big is the problem of going from measurement scales to land-margin ecosystems or how big is the jump from mesocosm scales to natural land-margin systems?

In Table 11.2, we assembled size data (volume) for a range of land-margin ecosystems and compared these to the sizes of three mesocosm systems. The scopes range from large to huge and make the point that guidance for extrapolation is very important. Even if we take the point of view that just one mesocosm would not be used as representative of Chesapeake Bay, or even a small lagoon, we still could not possibly use enough mesocosms to reduce the scope to more comfortable levels. In Table 11.3 we made some assumptions about the volume of water swept (i.e., volume represented by a measurement) and compared these to volumes of water in several natural systems. These estimates involve approximations as to effective swept volume and reasonable effort levels and thus the focus should be on the order of magnitude obtained rather than the actual estimate of swept volume. Moreover, a comparison of swept volume does not encompass a complete comparison of the effectiveness of a technology for any particular purpose. For example, for chlorophyll-*a*, ODAS offers synopticity over a large area at the expense of very fine-scale resolution that can be obtained by SAIL or DATAFLOW. Metabolism is simply harder to measure than basic water quality variables. Nonetheless, for water quality variables, the significance of the orders of magnitude increase in swept volume per boat day with newer technologies should not be lost as scientists and environmental managers struggle to obtain better information at lower cost.

The above concerns aside, the results indicate that unless much more generous assumptions are made about the degree of spatial auto-

TABLE 11•2 *A Comparison of Several Size Classes of Land Margin Mesocosm and Land Margin Ecosystem Volumes*

Land Margin Ecosystem Examples	Ecosystem Volume (10^6 m^3)	Ecosystem: Mesocosm Ratio		
		MEERC "C" Tank (1 m^3)	URI MERL Tank (13 m^3)	EXP Ponds (400 m^3)
Chesapeake Bay, MD/VA	74,000	7.4E+10	5.7E+09	1.9E+08
Patuxent River, MD	650	6.5E+08	5.0E+07	1.6E+06
West River, MD	12	1.2E+07	9.2E+05	3.0E+04
Small Coastal Lagoon, MD	4	4.0E+06	3.1E+05	1.0E+04
Global lagoons				
Smallest	0.03	3.0E+04	2.3E+03	7.5E+01
Mode	35	3.5E+07	2.7E+06	8.8E+04
Median	78	7.8E+07	6.0E+06	2.0E+05
Largest	8,000	8.0E+09	6.2E+08	2.0E+07

NOTE The MEERC "C" mesocosms are a part of an experimental mesocosm facility at the University of Maryland, Center for Environmental Science (Petersen et al. 1999), the MERL mesocosms are a facility at the University of Rhode Island and the Experimental Ponds (EXP Ponds) are at the University of Maryland, Center for Environmental Science.

SOURCE Lagoon data are from Nixon (1982), and other site data are from Boynton et al. (1995).

correlation, all of these methods are inadequate to sample all of Chesapeake Bay with a high degree of synopticity. For smaller systems, the more advanced technologies approach the ability to obtain more complete information in a short period of time, even without more generous assumptions about spatial autocorrelation. Older technologies for water quality measurements and the more challenging measurements (e.g., metabolism, nutrient flux) face more severe limitations regarding undersampling. In large land-margin systems, sampling spatial and temporal variability is clearly difficult at the whole system level, even with the most advanced technologies. So, we are often left with difficult issues of scaling up from local measurements and experiments to the scale of natural systems.

A number of analytical and statistical approaches have been developed in recent years for issues relating to scale, particularly in terrestrial and marine systems, and several reviews of these methods and approaches have been published. For example, Turner (1989) and Garcia-Moliner et

TABLE 11•3 *A Summary of Volumes of Water Swept with a Variety of Sampling Techniques Often Used in Land Margin Ecosystems and Estimates of the Time Required to Completely Sample Land Margin Ecosystems of Several Classes Using These Techniques*

Sampling Method	Volume Swept[a] $(10^4 m^3 d^{-1})$	Time Required to Sample System Volume, d			
		Chesapeake Bay	Patuxent River, MD	West River, MD	Small Coastal Bay, MD
In situ nutrient concentrations	79	94,752	831	15.3	5.10
Plankton metabolism (bottle incubations)	39	189,262	1,659	30.5	10.18
Sediment nutrient fluxes (in situ incubations)	8	947,516	8,306	152.9	50.96
Traditional CTD casts	157	47,376	415	7.6	2.55
Continuous surface water sampling (large ship)	4,260	1,746	15	0.3	0.09
Continuous surface water sampling (fast small boat)	2,220	3,350	29	0.5	0.18
Undulating CTD (e.g., GMI-scanfish)	16,700	445	4	0.1	0.02
Aerial remote sensing (e.g., ODAS radiometer system)	22,224	2,010	18	0.3	0.11
Ichthyplankton tows (e.g., tucker trawl)	247	30,113	264	4.9	1.62
Fish tows (e.g., mid-water trawls)	370	20,103	176	3.2	1.08
Chesapeake Bay Monitoring Program (physical)	785	9,475	83	1.5	0.51
Chesapeake Bay Monitoring Program (chemical)	314	23,688	208	3.8	1.27
Fish acoustics (sensor deployed from large ship)	11,112	669	6	0.1	0.04

[a]Calculating reasonable estimates of these scaling comparisons requires some assumptions that qualify the value. For example, the interpretation of a scaling comparison for chlorophyll a measurements made in a laboratory from filters depends on the number of samples that is reasonable to run and any underlying model of spatial correlation that is assumed. At the extremes, it isn't reasonable to assume that a single measurement represents the whole system, nor is it reasonable to assume that the sample represents nothing more than the water that was actually filtered. The range of reasonable values may span a large range, but probably does not include these extremes. In these comparisons, we assume homogeneity over a horizontal distance of 100 m and a vertical distance of 1 m. Under this assumption, the volume "swept" by a sampling method allowed a minimum sampled volume of 7.85×10^3 m^3 per sample. Sampling at high resolution was considered continuous sampling, where the unit of sampling was not the sample, but a tow or cast. These types of sampling techniques result in volumes swept which have either cylindrical or rectangular parallelepiped geometries. The amount of effort applied to the method is standardized to one team of scientists on one boat for one day. The volumes of Chesapeake Bay, Patuxent River, West River and a small coastal bay used to compute days needed for complete sampling were 7.44×10^{10} m^3, 6.52×10^8 m^3, 1.2×10^7 m^3, 4×10^6 m^3, respectively.

al. (1993) provided useful examples of techniques used for description and analysis of spatial patterns including autocorrelation, spectral density functions, wavelet analysis, and fractal geometry. More recently, Gardner (1998) has reviewed reasons for investigation of spatial scales and patterns and also suggested analytical approaches including dimensional analysis, allometric relationships, and use of variance estimates such as semivariograms and correlograms. Some interesting characteristics of normalized biomass and allometric relationships have been used to estimate trophic biomass and production in the Great Lakes (Sprules and Stockwell 1995), and Plotnick et al. (1993) applied a technique called lacunarity to determine characteristics of spatial dispersion. It is clear that an impressive array of techniques is available for investigation of issues related to scale. However, it also appears that these techniques have not been used to any large extent in land-margin systems. The reasons for this are not completely clear to us but the lack of time-series data and the relatively rare fine-scale spatial data sets needed for some of these analyses are certainly part of the reason. Application of newer technologies capable of making routine measurements at appropriate time and space scales is needed in land-margin system studies.

A number of authors have considered practical approaches to the issue of relating results obtained at one scale to issues at other scales in terrestrial and marine systems and some may have direct application to land-margin ecosystem scaling issues. At one end of the spectrum, Carpenter (1996) argued that mesocosm experiments (bottom-up approach) are not very useful and probably misleading when not done in conjunction with large-scale field studies. He urged extreme caution in scaling when using data obtained from such studies. The size and duration of mesocosm studies either exclude or distort important features and some relevant components or important/interesting processes simply will not fit in any mesocosm. Carpenter (1996) suggested that results of large-scale studies (e.g., whole lake experiments, forest manipulations) be used to guide the selection of mesocosm experiments and that these be conducted to estimate rates and to evaluate alternative hypotheses. Although there is certainly merit in this argument, this approach still begs the question of how to scale rate measurements conducted in mesocosms and, as Petersen et al. (1999) have noted, the results of whole ecosystem experiments (usually conducted in small whole ecosystems) often need to be scaled to even larger whole ecosystems (i.e., ponds to lakes). Given the large number of lagoonal systems in the United States and elsewhere, it might be worth considering using some selection of these system types for whole system experiments analogous to the way the Experimental Lakes are used as a research tool. Earlier, Vitousek (1993) suggested that "interaction between the top-down (whole-system experiments) and bottom-up approaches can yield better results more rapidly than can bottom-up analyses alone; the interaction can yield an understanding (of mechanisms) that is wholly inaccessible to top-down analyses."

Schneider (1994) has suggested a series of practical strategies (strengths and weaknesses) for scaling up environmental data. These included (1) multiplication of small-scale measurement to the size of the natural system; (2) use of linear scaling only for those quantities that have limited scope values (ratio of largest to smallest expected values); (3) use of large extent, fine-grain data (e.g., ocean color images) coupled with small time and space scale measurements (e.g., direct algal biomass estimates) to obtain large-scale estimates via summation rather than multiplication; (4) use of statistical models to scale limited measurements to larger areas; (5) use of hierarchy theory to identify nested systems (with internal similarities) within larger systems; and (6) use of dimensionless ratios to compute large-scale estimates from small-scale measurements. With respect to this last strategy, Miller et al. (1984) and Horne and Schneider (1994) used dimensionless ratios to evaluate some aspects of benthic deposit feeding and temporal and spatial dynamics of a coastal fish stock, respectively.

Other investigators have recommended practical approaches for dealing with scale-up issues and these are also worth noting for possible use in land-margin system research. Root and Schneider (1995) summarized problems associated with projecting results of studies to the scale of the natural world. For example, they note that use of the "scale-up" paradigm (building the whole from detailed, mechanistic, small-scale results) suffers from the fact that conspicuous features at small scales may not reveal dominant processes that generate large-scale pattern. However, the "scale-down" paradigm suffers from an inability to reveal cause-effect relationships. Root and Schneider (1995) suggest use of "strategic cyclical scaling" wherein both scale-down and scale-up approaches are cyclically used and strategically applied to practical problems. In short, they recommend using the larger scale to guide studies at the smaller scale. In fact, this theme is repeated in the literature, often prefaced by some statement to the effect that this is the best that can be done at present because of a lack of solid and practical theory (e.g., Turner 1989).

Vitousek (1993) also argued for use of detailed satellite data to establish pattern coupled with plot-level measurements and experiments for examination of mechanisms and the like. Other authors have called for use of mathematical models in the extrapolation process. Rastetter et al. (1992) and Caldwell et al. (1993) reported on using both statistical tests and models to better use fine-scale data in models of regional-level dynamics. Reynolds et al. (1993) suggested identifying hierarchical levels of the system of interest and then developing nested sets of models.

RECOMMENDATIONS CONCERNING FUTURE WORK ON SCALING OF LAND-MARGIN SYSTEM PROPERTIES

During the past fifteen years or so, substantial progress has been made relative to scaling issues in terrestrial, freshwater, and marine ecosystems. Scaling activities have ranged from microscopic to global and from physiological to ecological. However, it appears to us that scaling has not yet entered the mainstream of thought in land-margin research, despite motivation that might have come from successful applications in other fields and despite some of the characteristics of land-margin systems and land-margin mesocosms that indicate that issues of scale are particularly important. There is a great deal of room for increased activity in this area. Toward this end we have assembled a series of suggestions as to what might be done to initiate and augment future work concerning scaling of land-margin system properties and mesocosms.

- Start regular graduate-level teaching of scaling concepts and method-ologies as an important tool for ecologists and others involved in field and mesocosm studies. It appears to us that other disciplines (e.g., landscape ecology, elements of oceanography, physiological ecology) are more advanced in this area. Fisheries scientists have what amounts to a "methodological toolbox" from which to select appropriate methods for particular problems, some of which involve scaling. While land-margin scientists have a variety of methods at their disposal, an active appreciation of scale does not seem to be one of them. In the past few years, books have been published on various scaling issues (Ehleringer and Field 1993; Schneider 1994; Peterson and Parker 1998) that could serve as a basis for advanced courses. Scaling, both theory and practice, needs to find its way into the land-margin research toolbox.

- Syntheses of known scaling relationships are available in a few disciplines. Possibly the most diverse concerns allometric scaling relationships for mammals and birds (e.g., Calder 1981, 1983) but also for marine plankton and benthic invertebrates (e.g., Banse 1976; Chisholm 1992). Some particularly instructive examples of successful use of scaling ideas are available, particularly from terrestrial ecology (e.g., Wiens 1989). However, a synthesis of scaling relationships (allometric or otherwise) is not available for land-margin systems. Such a synthesis, possibly beginning with plant and animal allometric relationships, would be immediately useful for a number of investigations (bioenergetics, food web analyses) and would stimulate further interest in scaling applications.

- In a similar fashion, continued syntheses of land-margin mesocosm design is needed. In this area, MEERC has a good start in examining effects of size and shape and small-scale physics in experimental mesocosms. Petersen et al. (1999) reviewed design characteristics of 360 experimental mesocosm studies and, based on this review, urged mesocosm users to more explicitly assess the effects of scale with "scale-sensitive" experiments. For example, the depth, volume, and area of mesocosms need to be considered in light of ecological characteristics such as home range, organism size, and life history stage. Crossland and La Point (1992) also urged continued investigation of scale issues (e.g., scaling up mesocosm results to natural ecosystems) in mesocosms used for toxicological studies, particularly for pesticide registration.

- There is a pressing need for the use of new measurement technologies in land-margin research. The use of technologies that allow for greater temporal and spatial resolution will be of prime importance in determining what characteristic spatial and temporal scales are important for a variety of organisms and processes in these systems. In many cases, available data sets have increased in quality and size in the last several decades. However, the temporal and spatial sampling regimes have not been determined with scale in mind. Sampling regimes for dissolved nutrients, phytoplankton, zooplankton, benthos, or fish tend to be the same as each other. This being the case, examination of these data sets for scaling relations is of limited potential. We suggest use of high frequency moored and towed devises that use optical, electronic, and acoustic sensors. Aerial remote sensing in land-margin systems is also rapidly improving. Certainly, components such as zooplankton, fish larvae, and adult fish have been highly undersampled in the past; however, new technologies could go a long way toward providing data sets that would be rich in scaling issues and answers to pressing ecosystem questions.

- Likewise, there are a number of statistical tools that have been applied to landscape and some oceanographic issues. These need to be tested for application in land-margin systems; such applications may be more attractive when more high resolution data are available.

• Uncertainties about system function span large scales of time, space, and organism/community characteristics. Commonly used mesocosms, as described by Petersen et al. (1999), can be used to good effect for some problems, have limited use for others, and cannot be used at all for still others. The scope calculations we presented earlier suggest the size of the problem. Use of spatial models and linked mesocosms are two ways to scale-up from smaller to larger sizes and from shorter to longer time periods, but there are obvious limits to this approach. Limnologists have had good success using whole lake ecosystem approaches (Schindler 1987; Carpenter et al. 1995), but these also have problems and limitations. It seems to us that land-margin scientists might consider establishing a whole ecosystem approach with lagoons as a primary focus. These systems are common, often somewhat protected from weather extremes, small enough to be readily sampled, and large enough to be relevant for experiments that can not be done in traditional mesocosms. An experimental lagoon facility would extend land-margin mesocosms by the same large-scope factor that experimental lakes extended limno-corrals and other lake mesocosms.

ACKNOWLEDGMENTS

R. Bartleson, J. Dewar, W. M. Kemp, I. Mendelssohn, W. J. Mitsch, L. Murray, H. Neckles, S. Nixon, C. Reid, and B. Sturgis were active participants in the workshop discussions that provided the foundation for this chapter.

LITERATURE CITED

Addicott, J. F., J. M. Aho, M. F. Antolin, D. K. Padilla, J. S. Richardson, and D. A. Soluk. 1987. Ecological neighborhoods: Scaling environmental patterns. *Oikos* 49:340–346.

Atkinson, M. J., T. Berman, B. B. Allanson, and J. Imberger. 1987. Fine-scale oxygen variability in a stratified estuary: Patchiness in aquatic environments. *Marine Ecology Progress Series* 36:1–10.

Bakke, T. 1990. Benthic mesocosms: II, Basic research in hard-bottom benthic mesocosms. In C. M. Lalli, ed., *Enclosed Experimental Marine Ecosystems: A Review and Recommendations*, pp. 122–135. New York: Springer-Verlag.

Banse, K. 1976. Rates of growth, respiration and photosynthesis of unicellular algae as related to cell size: A review. *Journal of Phycology* 12:135–140.

Barry, J. P., and P. K. Dayton. 1991. Physical heterogeneity and the organization of marine communities. In J. Kolasa and S. T. A. Pickett, eds., *Ecological Heterogeneity*, pp. 270–320. New York: Springer-Verlag.

Barth, J. 1967. *The Floating Opera*. New York: Doubleday.

Boynton, W. R., J. H. Garber, R. Summers, and W. M. Kemp. 1995. Inputs, transformations, and transport of nitrogen and phosphorus in Chesapeake Bay and selected tributaries. *Estuaries* 18(1B):285–314.

Boynton, W. R., W. M. Kemp, and C. W. Keefe. 1982. A comparative analysis of nutrients and other factors influencing estuarine phytoplankton production. In V. S. Kennedy, ed., *Estuarine Comparisons*, pp. 69–90. New York: Academic Press.

Calder, W. A. III. 1981. Scaling of physiological processes in homeothermic animals. *Annual Review of Physiology* 43:301–322.

Calder, W. A., III. 1983. Ecological scaling: Mammals and birds. *Annual Review of Ecology and Systematics* 14:213–230.

Caldwell, M. M., P. A. Matson, C. Wessman, and J. Gamon. 1993. Prospects for scaling. In J. R. Ehleringer and C. B. Field, eds., *Scaling Physiological Processes: Leaf to Globe*, pp. 223–230. San Diego: Academic Press.

Carpenter, S. R. 1996. Microcosm experiments have limited relevance for community and ecosystem ecology. *Ecology* 77:677–680.

Carpenter, S. R., S. W. Chisholm, C. J. Krebs, D. W. Schindler, and R. F. Wright. 1995. Ecosystem experiments. *Science* 269:324–327.

Carpenter, S. R., and J. F. Kitchell. 1987. The temporal scale of variance in limnetic primary production. *American Naturalist* 129:417–133.

Chen, C.-C., J. E. Petersen, and W. M. Kemp. 1997. Spatial and temporal scaling of periphyton growth on walls of estuarine mesocosms. *Marine Ecology Progress Series* 155:1–15.

Chisholm, S. W. 1992. Phytoplankton size. In P. G. Falkowski and A. D. Woodhead, eds., *Primary Productivity and Biogeochemical Cycles in the Sea*, pp. 213–237. New York: Plenum.

Costanza, R., M. Mageau, R. E. Ulanowicz, and L. Wainger. 1993. Scaling aspects of modeling complex systems. MEERC Synthesis Paper. University of Maryland Center for Environmental Science, Cambridge.

Costanza, R., W. M. Kemp, and W. R. Boynton. 1995. Scale and biodiversity in coastal and estuarine ecosytems. In C. Perrings, K. Maler, C. Folke, C. S. Holling, and B. Jansson, eds., *Biodiversity Loss: Economic and Ecological Issues*, pp. 84–125. New York: Cambridge University Press.

Crossland, N. O., and T. W. La Point. 1992. The design of mesocosm experiments. *Environonmental Toxicology and Chemistry* 11:1–4.

Culliton, T. J., M. A. Warren, T. R. Goodspeed, D. G. Remer, C. M. Blackwell, and J. J. McDonough, III. 1990. *The Second Report of a Coastal Trends Series: 50 Years of Population Change Along the Nation's Coasts, 1960–2010*. Rockville, Md.: Strategic Assessment Branch, Ocean Assessments Division, Office of Oceanography and Marine Assessment, National Ocean Service, National Oceanic and Atmospheric Administration (NOAA).

Davies, J. M., and J. C. Gamble. 1979. Experiments with large enclosed ecosystems. *Philosophical Transactions of the Royal Society of London, B. Biological Sciences* 286:523–544.

de Lafontaine, Y., and W. C. Leggett. 1987a. Effect of container size on estimates of mortality and predation rates in experiments with macrozooplankton and larval fish. *Canadian Journal of Fisheries and Aquatic Sciences* 44:1534–1543.

de Lafontaine, Y., and W. C. Leggett. 1987b. Evaluation of in-situ enclosures for larval fish studies. *Canadian Journal of Fisheries and Aquatic Sciences* 44:54–65.

Denman, K. L. 1992. Scale-determining biological-physical interactions in oceanic food webs. In P. S. Giller, A. G. Hildrew, and D. G. Raffaelli, eds., *Aquatic Ecology: Scale, Pattern and Process*, pp. 377–402. London: Blackwell Science.

Duarte, C. M. 1991. Allometric scaling of seagrass form and productivity. *Marine Ecology Progress Series* 77:289–300.

Duarte, C. M. 1995. Submerged aquatic vegetation in relation to different nutrient regimes. *Ophelia* 41:87–112.

Duarte, C. M., and D. Vaque. 1992. Scale dependence of bacterioplankton patchiness. *Marine Ecology Progress Series* 84:95–100.

Ehleringer, J. R., and C. B. Field, eds. 1993. *Scaling Physiological Processes: Leaf to Globe*. New York: Academic Press.

Frost, T. M., D. L. DeAngelis, S. M. Bartell, D. J. Hall, and S. H. Hurlbert. 1988. Scale in the design and interpretation of aquatic community research. In S. Carpenter, ed., *Complex Interactions in Lake Communities*, pp. 229–257. New York: Springer-Verlag.

Gamble, J. C. 1985. More space for the sparsely distributed: An evaluation of the use of large enclosed experimental systems in larval fish research. International Council for the Exploration of the Sea, C. M. 1985/Mini-Symposium No. 2.

Garcia-Moliner, G., D. M. Mason, C. H. Greene, A. Lobo, B. Li, J. Wu, and G. A. Bradshaw. 1993. Description and analysis of spatial patterns. In S. Levin, T. M. Powell, and J. H. Steele, eds., *Patch Dynamics*, pp. 70–89. New York: Springer-Verlag.

Gardner, R. H. 1998. Pattern, process, and the analysis of spatial scales. In D. L. Peterson and V. T. Parker, eds., *Ecological Scale: Theory and Applications*. New York: Columbia University Press.

Gasol, J. M., P. A. del Giorgio, and C. M. Duarte. 1997. Biomass distribution in marine planktonic communities. *Limnology and Oceanography* 42:1353–1363.

Grant, J., S. J. Turner, P. Legendre, T. M. Hume, and R. G. Bell. 1997. Patterns of sediment reworking and transport over small spatial scales on an intertidal sandflat, Manukau Harbour, New Zealand. *Journal of Experimental Marine Biology and Ecology* 216:33–50.

Hagy, J. D. III. 1996. Residence times and net ecosystem processes in Patuxent River estuary. Master's thesis, University of Maryland.

Harding, Jr., L. W., E. C. Itsweire, and W. E. Esaias. 1992. Determination of phytoplankton chlorophyll concentrations in the Chesapeake Bay with aircraft remote sensing. *Remote Sensing of the Environment* 40:79–100.

Haury, L. R., J. A. McGowan, and P. H. Weibe. 1978. Patterns and processes in the time-space scales of plankton distributions. In J. H. Steele, ed., *Spatial Patterns in Plankton Communities*, pp. 277–327. New York: Plenum.

Hildrew, A. G., and P. S. Giller. 1992. Patchiness, species interactions, and disturbance in the stream benthos. In P. S. Giller, A. G. Hildrew, and D. G. Raffaelli, eds., *Aquatic Ecology: Scale, Pattern and Process*, pp. 21–62. London: Blackwell Science.

Hoekstra, T. W., T. F. H. Allen, and C. H. Flather. 1991. Implicit scaling in ecological research. *BioScience* 41:148–154.

Horne, J. K., and D. C. Schneider. 1994. Analysis of scale-dependent processes with dimensionless ratios. *Oikos* 70:201–211.

Houde, E. D. 1985. Mesocosms and recruitment mechanisms. International Council for the Exploration of the Sea, C. M. 1985/Mini-Symposium No. 2.

Houde, E. D., and E. S. Rutherford. 1993. Recent trends in estuarine fisheries: Predictions of fish production and yield. *Estuaries* 16:161–176.

Ives, A. R., J. Foufopoulos, E. D. Klopfer, J. L. Klug, and T. M. Palmer. 1996. Bottle or big-scale studies: How do we do ecology? *Ecology* 77:681–685.

Jorgensen, S. E., S. N. Nielsen, and L. A. Jorgensen. 1991. *Handbook of Ecological Parameters and Ecotoxicology.* New York: Elsevier.

Kemp, W. M., and W. R. Boynton. 1992. Benthic-pelagic interactions: Nutrient and oxygen dynamics. In D. E. Smith, M. Leffler, and G. Mackiernan, eds., *Oxygen Dynamics in the Chesapeake Bay: A Synthesis of Recent Research*, pp. 149–221. Maryland Sea Grant Book. College Park: University of Maryland.

Kemp, W. M., M. R. Lewis, J. F. Cunningham, J. C. Stevenson, and W. R. Boynton. 1980. Microcosms, macrophytes, and hierarchies: Environmental research in the Chesapeake Bay. In J. P. Giesy Jr, ed., *Microcosms in Ecological Research*, pp 911–936. Springfield, Va.: National Technical Information Service.

Kiorboe, T., and M. Sabatini. 1995. Scaling of fecundity, growth and development in marine planktonic copepods. *Marine Ecology Progress Series* 120:285-298.

Lawton, J. H. 1996. The Ecotron facility at Silwood Park: The value of "big bottle" experiments. *Ecology* 77:665–669.

Legendre, P., S. F. Thrush, V. J. Cummings, P. K. Dayton, J. Grant, J. E. Hewitt, A. H. Hines, B. H. McArdle, R. D. Pridmore, D. C. Schneider, S. J. Turner, R. B. Whitlach, and M. R. Wilkinson. 1997. Spatial structure of bivalves in a sandflat: Scale and generating processes. *Journal of Experimental Marine Biology and Ecology* 216:99–128.

Lehman, J. T., and D. Scavia.1982. Microscale patchiness of nutrients in plankton communities. *Science* 216:729–730.

Levin, S. A. 1992. The problem of pattern and scale in ecology. *Ecology* 73:1943–1967.

Matson, P. A.,W.J. Parton, A. G. Power, and M. J. Swift. 1997. Agricultural intensification and ecosystem properties. *Science* 277:504–509.

Miller, D. C., P. A. Jumars, and A. R. M. Nowell. 1984. Effects of sediment transport on deposit feeding: Scaling arguments. *Limnology and Oceanography* 29:1202–1217.

Moloney, C. L., and J. G. Field. 1989. General allometric equations for rates of nutrient uptake, ingestion, and respiration in plankton organisms. *Limnology and Oceanography* 34:1290–1299.

Monbet, Y. 1992. Control of phytoplankton biomass in estuaries: A comparative analysis of microtidal and macrotidal estuaries. *Estuaries* 15:563–571.

Morgan, R. P., J. E. Baker, J. W. Gooch, and R. B. Brinsfield. 1990. Comparative analysis of scales of contaminant inputs. MEERC Synthesis Report. Horn Point Laboratories, University of Maryland Center for Environmental Science, Cambridge.

Nixon, S. W. 1982. Nutrient dynamics, primary production and fisheries yields of lagoons. *Oceanologica Acta* 4(suppl.):357–371.

Nixon, S. W. 1992. Quantifying the relationship between nitrogen input and the productivity of marine ecosystems. *Proceedings of the Advanced Marine Technology Conference* 5:57–83.

Nixon, S. W., J. W. Ammerman, L. P. Atkinson, V. M. Berounsky, G. Billen, W. C. Boicourt, W. R. Boynton, T. M. Church, D. M. Ditoro, R. Elmgren, J. H. Garber, A. E. Giblin, R. A. Jahnke, N. J. P. Owens, M. E. Q. Pilson, and S. P. Seitzinger. 1996. The fate of nitrogen and phosphorus at the land-sea margin of the North Atlantic Ocean. *Biogeochemistry* 35:141–180.

Nixon, S. W., C. A. Oviatt, J. Frithsen, and B. Sullivan. 1986. Nutrients and the productivity of estuarine and coastal marine ecosystems. *Journal of the Limnology Society of South Africa* 12(1/2):43–71.

NOAA (National Oceanic and Atmospheric Administration). 1998. *NOAA's Estuarine Eutrophication Survey*, vols. 1–5. Silver Spring, MD: Office of Ocean Resources Conservation and Assessment, National Ocean Service, NOAA.

Oiestad, V. 1990. Specific application of meso- and macrocosms for solving problems in fisheries research. In C. M. Lalli, ed., *Enclosed Experimental Marine Ecosystems: A Review and Recommendations*, pp. 136–154. New York: Springer-Verlag.

O'Neill, R. V., D. L. DeAngelis, J. B. Wade, and T. F. H. Allen. 1986. *A Hierarchical Concept of Ecosystems*. Princeton: Princeton University Press.

O'Neill, R. V., R. H. Gardner, B. T. Milne, M. G. Turner, and B. Jackson. 1991. Heterogeneity and spatial hierarchies. In J. Kolasa and S. T. A. Pickett, eds., *Ecological Heterogeneity*, pp. 86–96. New York: Springer-Verlag.

Oviatt, C. A. 1981. Effects of different mixing schedules on phytoplankton, zooplankton and nutrients in marine mesocosms. *Marine Ecology Progress Series* 4:57–67.

Petersen, J. E., C.-C. Chen, and W. M. Kemp. 1997. Scaling aquatic primary productivity: Experiments under nutrient- and light-limited conditions. *Ecology* 78:2326–2338.

Petersen, J. E., J. Cornwell, and W. M. Kemp. 1999. Implicit scaling in the design of experimental aquatic ecosystems. *Oikos* 85:3–15.

Peterson, D. L., and V. T. Parker, eds. 1998. *Ecological Scale: Theory and Applications*. New York: Columbia University Press.

Pilson, M. E. 1990. Application of mesocosms for solving problems in pollution research. In C. M. Lalli, ed., *Enclosed Experimental Marine Ecosystems: A Review and Recommendations*, pp. 122–135. New York: Springer-Verlag.

Plotnick, R. E., R. H. Gardner, and R. V. O'Neill. 1993. Lacunarity indices as measures of landscape texture. *Landscape Ecology* 8:201–211.

Rabalais, N. N., W. J. Wiseman, Jr., R. E. Turner, B. K. S. Gupta, and Q. Dortch. 1996. Nutrient changes in the Mississippi River and system responses on the adjacent continental shelf. *Estuaries* 19:386–407.

Rastetter, E. B., A. W. King, B. J. Cosby, G. M. Hornberger, R.V. O'Neill, and J. E. Hobbie. 1992. Aggregating fine-scale ecological knowledge to model coarser-scale attributes of ecosystems. *Ecological Applications* 2:55–70.

Reynolds, J. F., D. W. Hilbert, and P. R. Kemp. 1993. Scaling ecophysiology from the plant to the ecosystem: A conceptual framework. In J. R. Ehleringer and C. B. Field, eds., *Scaling Physiological Processes: Leaf to Globe*, pp. 127–140. San Diego: Academic Press.

Root, T. L., and S. H. Schneider. 1995. Ecology and climate: Research strategies and implications. *Science* 269:334–341.

Rose, G. A., and W. C. Leggett. 1990. The importance of scale to predator-prey spatial correlations: An example of Atlantic fishes. *Ecology* 71:33–43.

Sanford, L. 1990. Controlled turbulence generation and mixing in laboratory microcosms. MEERC Synthesis Report. Horn Point Laboratories, University of Maryland Center for Environmental Science, Cambridge.

Schindler, D. W. 1987. Detecting ecosystem responses to anthropogenic stress. *Canadian Journal of Fisheries and Aquatic Sciences* 44(suppl. 1):6–25.

Schneider, D. C. 1992. Scale-dependent patterns and species interactions in marine nekton. In P. S. Giller, A. G. Hildrew, and D. G. Raffaelli, eds., *Aquatic Ecology: Scale, Pattern and Process*, pp. 441–467. Oxford: Blackwell Science.

Schneider, D. C. 1994. *Quantitative Ecology: Spatial and Temporal Scaling*. New York: Academic Press.

Schneider, D. C. 1998. Applied scaling theory. In D. L. Petersen and V. T. Parker, eds., *Ecological Scale: Theory and Applications*, pp. 253–269. New York: Columbia University Press.

Schneider, D. C., and J. F. Piatt. 1986. Scale-dependent correlation of seabirds with schooling fish in a coastal ecosystem. *Marine Ecology Progress Series* 32:237–246.

Schneider, D. C., R. Walters, S. F. Thrush, and P. Dayton. 1997. Scale-up ecological experiments: Density variation in the mobile bivalve *Macomona liliana*. *Journal of Experimental Marine Biology and Ecology* 216:129–152.

Smith, S. V., J. T. Hollibaugh, S. J. Dollar, and S. Vink. 1991. Tomales Bay metabolism: C-N-P stoichimetry and ecosystem heterotrophy at the land-sea interface. *Estuarine and Coastal Shelf Science* 33:223–257.

Sprules, W. G., and J. D. Stockwell. 1995. Size-based biomass and production models in the Saint Lawrence Great Lakes. *ICES Journal of Marine Science* 52:705–710.

Steele, J. H. 1985. A comparison of terrestrial and marine ecological systems. *Nature* 313:355–358.

Steele, J. H. 1991. Can ecological theory cross the land-sea boundary? *Journal of Theoretical Biology* 153:425–436.

Steele, J. H., S. R. Carpenter, J. E. Cohen, P. K. Dayton, and R. E. Ricklefs. 1993. Comparing terrestrial and marine ecological systems. In S. Levin, T. M. Powell, and J. H. Steele, eds., *Patch Dynamics*, pp. 1–12. New York: Springer-Verlag.

Steele, J. H, and E. W. Henderson. 1994. Coupling between physical and biological scales. *Philosophical Transactions of the Royal Society of London, B. Biological Sciences* 343:5–9.

Stommel, H. 1963. Varieties of oceanographic experience. *Science* 139:572–576.

Theilacker, G. H. 1980. Rearing container size affects morphology and nutritional condition of larval jack mackerel, *Trachurus symmetricus*. *Fishery Bulletin* 78:789–791.

Thrush, S. F., V. J. Cummings, P. K. Dayton, R. Ford, J. Grant, J. E. Hewitt, A. H. Hines, S. M. Lawrie, R. D. Pridmore, P. Legendre, B. H. McArdle, D. C. Schneider, S. J. Turner, R. B. Whitlach, and M. R. Wilkinson. 1997. Matching the outcome of small-scale density manipulation experiments with larger scale patterns: An example of bivavle adult/juvenile interactions. *Journal of Experimental Marine Biology and Ecology* 216:153–169.

Thrush, S. F., J. E. Hewitt, V. J. Cummings, and P. K. Dayton. 1995. The impact of habitat disturbance by scallop dredging on marine benthic communities: What can be predicted from the results of experiments? *Marine Ecology Progress Series* 129:141–150.

Thrush, S. F., R. D. Pridmore, R. G. Bell, V. J. Cummings, P. K. Dayton, R. Ford, J. Grant, M. O. Green, J. E. Hewitt, A. H. Hines, T. M. Hume, S. M. Lawrie, P. Legendre, B. H. McArdle, D. Morrisey, D. C. Schneider, S. J. Turner, R. A. Walters, R. B. Whitlach, and M. R. Wilkinson. 1997. The sandflat habitat: Scaling from experiments to conclusions. *Journal of Experimental Marine Biology and Ecology* 216:1–9.

Thrush, S. F., D. C. Schneider, P. Legendre, R. B. Whitlach, P. K. Dayton, , J. E. Hewitt, A. H. Hines, V. J. Cummings, S. M. Lawrie, J. Grant, R. D. Pridmore, S. J. Turner, and B. H. McArdle. 1997. Scaling-up from experiments to complex ecological systems: Where to next? *Journal of Experimental Marine Biology and Ecology* 216:243–254.

Thrush, S. F., R. B. Whitlatch, R. D. Pridmore, J. E. Hewitt, V. J. Cummings, and M. R. Wilkinson. 1996. Scale dependent recolonization: The role of sediment stability in a dynamic sandflat habitat. *Ecology* 77:2472–2487.

Turner, M. G. 1989. Landscape ecology: The effect of pattern on process. *Annual Review of Ecology and Systematics* 20:171–197.

Vitousek, P. M. 1993. Global dynamics and ecosystem processes: Scaling up or scaling down? In J. R. Ehleringer and C. B. Field, eds., *Scaling Physiological Processes: Leaf to Globe*, pp. 169–177. San Diego: Academic Press.

Vitousek, P. M., H. A. Mooney, J. Lubchenco, and J. M. Melillo. 1997. Human domination of earth's ecosystems. *Science* 277:494–499.

Wiens, J. A. 1976. Population responses to patchy environments. *Annual Review of Ecology and Systematics* 7:81–120.

Wiens, J. A. 1989. Spatial scaling in ecology. *Functional Ecology* 3:385–397.

CHAPTER 12

Scaling Issues in Marine Experimental Ecosystems
The Role of Patchiness

David L. Scheurer, David C. Schneider,
and Lawrence P. Sanford

IF ONE EXPANDS THE DEFINITION OF A MARINE EXPERIMENTAL ECOSYSTEM to include any artificial system that isolates a parcel of water from the natural environment, then most in situ aquatic studies (e.g., bottle experiments, floating pens, etc.), as well as typical mesocosm experiments, can be included under this heading. Any type of containment also entails a physical and biological isolation, dependent on the size of the enclosure (Frost et al. 1988). Problems can arise, however, when the results from these simplified and truncated systems are extrapolated to longer and broader time-space scales (Carpenter 1996). The implicit assumption for extrapolation is that the experimental ecosystems are analogs of larger natural systems. But within an ecosystem the relative importance of controlling biological and physical processes can change with scale, creating scale-dependence and difficulty in extrapolation (Wiens 1989; Allen and Hoekstra 1992; Levin 1992; Peterson and Parker 1998).

"Patchiness" is one source of scale-dependent effects in pelagic environments. It is also a source of error when results from fine-scale studies are extrapolated to broader scales in natural ecosystems (Stommel 1963; Allen 1977; Frost et al. 1988; Giller et al. 1992). Many different descriptions for patchiness exist in the marine literature. In this chapter, we use two of the more common definitions for patchiness. The first is when patchiness refers to the presence of identifiable structures (e.g., clusters and mosaics) that exhibit elevated levels against a contrasting background environment (figure 12.1a, b). Clusters are groups of individuals that form a discrete unit (e.g., school of fish, swarm of krill), whereas mosaics are diffuse regions containing a concentration gradient (e.g., bloom of phytoplankton, nutrient plume). The second usage refers to measures of heterogeneity (e.g., variance, autocorrelation, semivariance, spectral analysis) describing the spatial distribution of some quantity of interest. The spectral slope of temperature or phytoplankton (i.e., how the variance changes with spatial scale over a given distance) is one example (figure 12.1c).

In marine systems, the gradients at the "edges" of patches vary a great deal in their intensity, making it difficult to identify patches in the ocean. Figure 12.2 illustrates two idealized gradients. Often the steepness of the gradient will change depending on the spatial scale (e.g., resolution of the sampling technique, area over which measurements are taken) and on the threshold used to determine what constitutes elevation above background levels.

Patchiness, both physical and biological, exists at all scales within the pelagic marine environment (Steele 1978). The physical and biological scales of patchiness, along with their interactions, have a long history in oceanographic research (Platt 1972; Haury et al. 1978; Harris 1980; Okubo 1980; Mackas et al. 1985; Powell 1989; Steele 1991). Patchiness in the pelagic ocean is important for many reasons. Patches concentrate resources and are, therefore, critical for organisms living in areas were the average distribution of a resource is so dilute that growth and survival are reduced (LeBrasseur et al. 1969; Lasker 1975). Patchiness is also important because it provides protection from predation (Turner and Pitcher 1986; Inman and Krebs 1987). In a statistical sense, the chances of being consumed decreases as the school (i.e., patch) gets larger and saturation of predators due to handling time constraints and satiation becomes more probable. Patches can create advantages for reproduction (Brown 1975). Mate acquisition is facilitated in aggregations of individuals, and synchronized spawning adaptations can increase fertilization success. From a life-history standpoint, a large reproductive output released at one time will aid survival chances, thereby augmenting recruitment and future cohorts. In addition, the magnitude or degree of patchiness will influence processes such as the exchange of material and organisms across boundaries, predator-prey interactions, effects of disturbance, life-history traits, and genetic transfer between populations (Marquet et al. 1993).

In this chapter we focus on the role that patchiness plays in limiting extrapolation from experimental ecosystems to marine ecosystems. We review the characteristics and causes of patchiness in the natural oceanic environment, address some of the problems associated with the altered patchiness that are inherent in mesocosm experiments, and discuss the linkage and extrapolation of results obtained from small-scale ecosystem studies to broad-scale environmental issues.

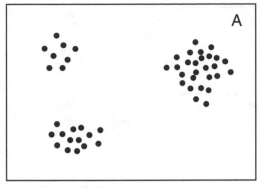

Clusters

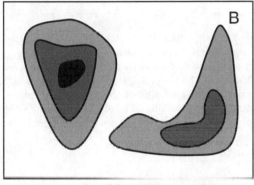

Mosaics

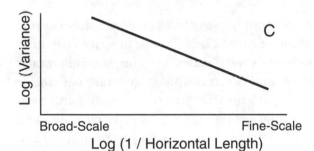

FIGURE 12•1 *Types of Patchiness*
Diagram of the various forms of patchiness used in the chapter. (A) Identifiable clusters of individuals (e.g., schools, swarms). (B) Mosaic regions containing a gradient in concentration. Commonly found with continuously variable attributes (e.g., temperature, nutrients, chlorophyll). (C) A graphical measure (e.g., spectral analysis) of spatial variability (i.e., patchiness). For the graph shown the most variability occurs at the broadest space scales. This relationship need not be linear with distance as illustrated but may be nonlinear or irregular. See text for additional details.

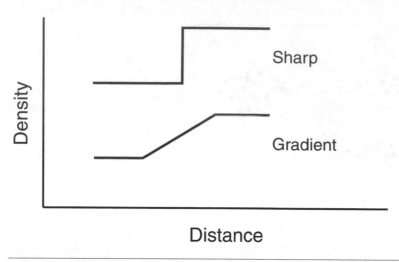

FIGURE 12•2 *Edges Between Patches*

Conceptual illustration of two types of patch edges found in aquatic systems. The upper line (i.e., sharp) shows a sudden change in concentration or magnitude such as might be encountered between the separation of warm and cold water by a thermocline or a school of fish against a background of lower fish density. The lower line (i.e., gradient) shows a gradual increase in density that might occur due to the diffusion of a substance from a concentrated source (e.g., a nutrient plume from upwelling). The resolution of the sampling technique will often determine the type of edge observed for a given quantity (e.g., biomass, temperature).

PATCHINESS IN THE PELAGIC OCEAN

There are many variance-generating structures within the oceanic environment that create patchiness over a broad range of space (10^{-3} m to 10^4 km) and time (seconds to years) scales. At the broadest scales (> 10,000 km), the atmosphere and ocean couple to generate basin-wide circulation (gyres) and current patterns that give rise to eddies and rings at the 10 to 1000 km scale. Other processes, such as internal waves and fronts, occur in the 100 m to 10 km range and give way to turbulence and viscous dissipation at the finest scales (Kolmogorov 1941; Mackas et al. 1985; figure 12.3).

Even though there is a continuous transfer of variance from broad to fine scales, with the largest variability occurring at the longest time and broadest space scales, certain processes dominate others at particular scales (Saunders 1992). For example, at broad scales the oceanic environment resembles a two-dimensional "flat" system. Geostrophic motions (i.e., oceanic current systems) drive the dynamics in the horizontal direction while vertical processes play a minor role. At finer scales, the variability is more three-dimensional because turbulence becomes the main driving force rather than geostrophic motion (Denman 1992).

Recognition that marine organisms are not randomly distributed but are patchy in time and space dates back to the early part of this century (Hardy and Gunther 1935). Since then, many investigators have searched for mechanisms that may produce biological patchiness within an environment dominated by physical processes (e.g., Steele 1978; Okubo 1980; Mackas et al. 1985). The early work of Skellam, Kierstead, and Slobodkin (KISS model) showed that biological patchiness in plankton could be maintained through reproductive means despite the highly dispersive physical environment found in aquatic systems (Skellam 1951; Kierstead and Slobodkin 1953). In the natural environment under optimal conditions (high nutrients, stable water column, convergence zones), phytoplankton patch maintenance through reproductive means can be very large, such as when an algal bloom occurs. But in general, growth-induced

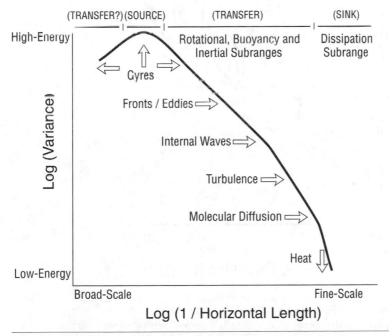

FIGURE 12•3 *Transfer of Variability Across Scales*
Schematic showing the transfer of variability (i.e., energy) from broad to fine scales for oceanic systems and some of the associated physical processes involved. (Modified from Mackas et al. 1985; Denman 1992)

patches are ephemeral because nutrients become exhausted, predators respond to higher densities, or large mixing events break down and disperse the patch. Subsequently, for passively dispersed plankton their distribution or patchiness is largely governed by physical processes (Denman 1976; Denman and Platt 1976; Gower et al. 1980).

As organism size increases, generation time also lengthens in the oceanic environment (figure 12.4), reducing the ability of organisms to create growth-induced patchiness (Sheldon et al. 1972). These larger, often mobile, organisms (e.g., zooplankton, fish, etc.) have the ability to create biological patchiness through directed movement in the form of aggregations, schools, swarms, and as a result of taxis behavior (Hamner et al. 1983; Okubo 1986; Norris and Schilt 1987; figure 12.5).

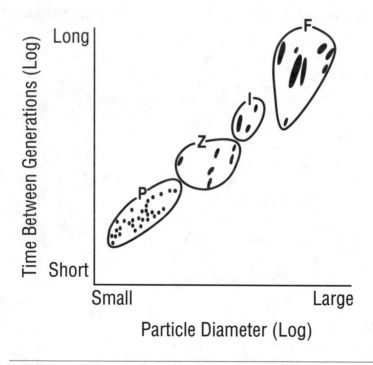

FIGURE 12•4 *Generation Time Versus Particle Diameter*
Schematic of the relationship between generation time (i.e., reproductive rate) and particle diameter (i.e., organismal size). Groupings are as follows: phytoplankton (P), zooplankton (Z), invertebrates (I), and fish (F). (After Sheldon et al. 1972; Steele 1978)

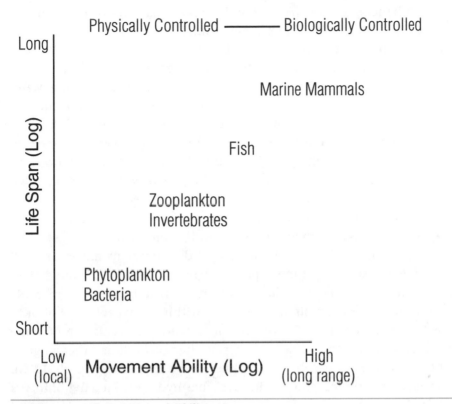

FIGURE 12•5 *Life Span Versus Mobility*
Relationship between life span and movement ability. As organism life span increases, size also increases allowing for greater movement ability and a shift from physically controlled patchiness to biological mechanisms. (From Steele 1978; Raffaelli et al. 1992)

Behavioral responses produce patchiness that can differ fundamentally from growth-induced patchiness. Organisms overcome or modify the dispersive forces around them and, with sufficient mobility, become decoupled from the physical environment that plays such a dominant role in structuring groups of strictly planktonic organisms. This decoupling often results in patchiness at smaller scales than would be predicted if the patchiness followed that of the physical environment (Weber et al. 1986; Levin et al. 1989; Schneider 1992; Horne and Schneider 1997; see figure 1.11).

Patch-generating processes that interact with one another lead to complex dynamics that cannot be explained solely by one process. This is especially true when the time and space scales of ecosystem processes are similar to those of physical processes. When the processes (physical and biological) are in synchrony, there is a potential for the dynamics to become amplified, thereby enhancing the interactions and observed patchiness (Denman and Powell 1984; Abbot 1993). For example, frontal zones (e.g., Gulf Stream) are often areas of enhanced biological productivity. Organisms, especially planktonic species, can become trapped in convergence zones where nutrient concentrations may be higher than in the surrounding water column (Franks 1992). If the phytoplankton are retained long enough, they can take advantage of the elevated nutrients and reduced physical dispersal to greatly increase biomass levels (Strass 1992). This effect is not limited to just frontal boundaries but can apply to any physical discontinuity from the scale of a thermocline up to mesoscale oceanic eddies (Gower et al. 1980; Venrick 1982; Oschlies and Garcon 1998).

Many organisms benefit from physical and biological patch-generating processes to increase their survival and reproductive success. The congregation of sea birds, pelagic fish, and marine mammals on krill patches are all examples of this type of behavior (Gaskin 1976; Sund et al. 1981; Schneider 1991). Vertical migration is used by some planktonic organisms to either maintain their position within a favorable habitat or to move to a new location. This is seen in estuaries when larvae of some organisms migrate into the surface water on the flood tide and then migrate to the bottom on the ebb tide (Staples 1979; Rothlisberg et al. 1983). Occurrence of breeding grounds in favorable hydrographic environments provides free transport mechanisms that can aid in the survival of larval offspring (Sinclair 1987).

PATCHINESS ISSUES ASSOCIATED WITH EXPERIMENTAL ECOSYSTEMS

Typical mesocosm experiments have limited temporal (days to months) and spatial (1 to 10's m) scales that create corresponding truncated physical and biological dynamics (Petersen et al. 1999). Given these constraints, the full suite of patch-generating mechanisms operating in the marine environment cannot be reproduced in experimental ecosystems, which can lead to scale-dependent results and a corresponding limit on extrapolation. We will address four such patch-generating mechanisms operating in the marine environment (space, physical processes, biological processes, and large-scale long-term dynamics) and how they are modified, distorted, or absent in experimental ecosystems.

Space

Marine mesocosms are small, typically isolated, and mixed differently from marine systems. This can affect the outcome of experiments significantly. Phenomena such as clumping, species coexistence, and stabilization of predator-prey cycles are examples of spatial dynamics altered in theoretical or experimental studies where space is explicitly included (Tilman and Kareiva 1997). The isolation caused by mesocosm boundaries may also result in species extinctions either from competitive exclusion or unstable population cycles, with subsequent changes in ecosystem function.

The walls of mesocosms will also affect the spatial dynamics and environment that an organism experiences. Some effects include the increased attenuation of light through the water column (Petersen et al. 1997), altered benthic-pelagic coupling due to enhanced exchanges with wall organisms (Chen et al. 1997), and artificial gradients (e.g., dead zones) resulting from uneven mixing (Sanford 1997). While such effects occur in the natural environment from interactions with the bottom and attached structures, within mesocosms the effect is much greater because of the higher surface area to volume ratio found within these enclosed systems. For pelagic systems, bottom or wall effects are usually nonexistent.

In the natural environment, the exchange of material and individuals (fluxes) is a key process (Marquet et al. 1993). At the mesocosm scale, these processes (emigration, immigration, transport) are difficult to include and are an often-cited reason why mesocosms diverge from natural systems over time (Bloesch et al. 1988). The biological development of a mesocosm can be highly sensitive to the assemblage of species present at the start of an experiment and to mesocosm size. Consequently, concepts such as extinction, competitive exclusion, and equilibrium are likely to have limited applicability in "open" marine systems where local populations are replenished from other locations and where recruitment is highly variable from year to year (NRC 1995). Moving from the well-mixed reactor model for experimental ecosystems to a spatial and heterogeneous (i.e., patchy) environment can cause serious problems for extrapolation.

Physical Processes

The type of mixing (rate, frequency, etc.) applied to mesocosms affects a number of important physically mediated ecosystem processes (Sanford 1997). Some of these include diffusion across boundaries, uptake kinetics, predator-prey contact rates, time scales for mixing and flushing, and shear effects on organisms. Often, balancing for one process distorts another. For example, a mesocosm that mimics realistic water column turbulence may result in unrealistic fluxes across the sediment-water interface. Alternatively, using a mixing rate to achieve a particular light regime may give altered shear velocities or predator-prey contact rates. Fine-scale studies in aquatic experimental ecosystems tend to be dominated by diffusive processes, whereas advection plays a more prominent role in larger systems.

Because the strength of turbulence depends on spatial scale (Denman 1992), the extrapolation of its effects is challenging in mesocosms. Approximating the effects of small-scale turbulence is feasible, but mimicking the effects of large-scale turbulence-generating processes is usually impossible. Turbulent effects may include those due to shear flows across boundaries (pycnoclines, thermoclines), patchiness generated through advection and eddies, and stratification/destratification events. Mesocosms are generally too limited in their size to include these important larger-scale effects that can drive community- (succession, bloom events) and ecosystem-level processes (nutrient recycling).

Biological Processes

A potentially serious problem of mesocosm studies is the lack of trophic complexity that arises because of truncated space and time scales. The effects of larger organisms or organisms with broad home ranges are particularly problematic (Carpenter et al. 1995). These organisms can have important feedbacks with ecosystem processes or can have a direct "top-down" effect on dynamics. The effects created by the organisms at the top of the food chain can be dramatic but very transient or "patchy" in nature. The extrapolation of mesocosm predation rates, for example, based on average values can over- or underestimate the true rates occurring in the "patchy" natural system.

Movement patterns are also altered in mesocosms. For example, diurnal vertical migration of zooplankton species may be restricted in a spatially constrained mesocosm with realistic levels of turbulence. Movement is unrealistically restricted for nekton species that are maintained at nonnatural densities or confined to restricted spaces that preclude important dynamics such as schooling. In addition, the confined spaces of smaller mesocosms prevent or alter escape mechanisms normally employed by prey species, resulting in predator-prey contact rates between species that can exceed those under natural conditions (Heath and Houde, this volume).

Large-Scale Long-Term Dynamics

Many problems associated with mesocosm studies are a result of their truncated spatial and temporal scales. Some of these include small physical size, short-duration experiments, and a lack of spatial and temporal heterogeneity (Peters 1991; Carpenter et al. 1995). Experiments performed at too short of a time scale can fail to capture potentially important factors such as transient dynamics, environmental variability, indirect effects, and multiple equilibria (Tilman 1989). Because mesocosm experiments have such a short duration compared to natural marine processes, the application of results will be limited to similar temporal and spatial domains. For example, many marine species occupy a sequence of habitats during a life history that takes several years to complete. An example is cod (*Gadus morhua*) whose larvae are pelagic and adults are benthic (Bigelow and Schroeder 1953). These several life stages rely on different food sources and suffer different mortality rates. Simulating complex dynamics that take several generations to occur, especially for larger organisms, is usually beyond the scope of mesocosm experiments. Often researchers are forced to use high-turnover species with simple life histories in mesocosms, thereby limiting the applicability of the results to the natural environment.

In marine systems, physical and biological variability generally increase at broader temporal and spatial scales (Steele 1978). This means that patch-generating processes at these broad, long-term scales have relatively large effects on ecosystem-level dynamics. Examples of such processes include climate shifts, the North Atlantic oscillation, and the ENSO cycle in the Pacific. These events have profound effects on community dynamics and ecosystem functioning (Southward 1980). For instance, a common consequence of the cessation of upwelling off the Peruvian coast due to an El Niño event is the collapse of populations that depend on the high supply of nutrients (Barber and Chavez 1986). The Russell cycle is another important large-scale process (Southward 1980). It results in a dramatic shift in the dominant species of the North Atlantic, thus affecting recruitment and growth in important fisheries such as the cod fishery. Mesocosm experiments cannot address questions at these broad time-space scales, leaving them at a disadvantage to tackle current environmental issues (e.g., fish stock fluctuation, marine pollution, marine mammal-fishery interactions, effects of aquaculture, or establishment of marine protected areas).

LINKING EXPERIMENTAL ECOSYSTEMS TO MARINE SYSTEMS

Given that marine mesocosms have distorted physical and biological dynamics that alter patchiness, how do we use experimental ecosystems to address broad-scale questions? We will first examine several theories that address this issue. Then we will examine the problem from a practical standpoint and suggest several ways to tackle patchiness when going from small-scale studies to larger systems. We hope that these two complementary methods will aid in designing better experimental ecosystem experiments and allow for a greater awareness of how patchiness affects our ability to extrapolate among scales.

Theory

Several theoretical concepts bear on the problem of linking results from mesocosms to observations from ecosystems. These concepts include general systems theory (Bertalanffy 1968), hierarchy theory (Allen and Starr 1982; O'Neill et al. 1986), theory of similitude in engineering (Taylor 1974), universal scaling (Stanley et al. 1996) between nonsimilar systems (Barenblatt 1996), and geophysical fluid dynamics (Batchelor 1967; Pedlosky 1979). The following sections describe each approach, followed by a brief treatment of its relevance to the problem of linking mesocosm results to ecosystem observations. Details of application can be found in the references in each section.

SYSTEMS THEORY. General systems theory starts with the idea that a system cannot be understood by looking at a single component—one must consider both the system and its pieces, or elements. Consequently, one defines elements as well as the system and its boundaries, to develop computational expressions that apply to the system as a whole. Table 12.1 shows general systems theory as described by Bertalanffy (1968).

General systems theory has little to say directly about the problem of linking experimental research in mesocosms to research on ecosystems.

When mesocosms are treated as systems consisting of elements (Beyers and Odum 1993), several of the ideas in table 12.1 can be carried further. One could ask whether either mesocosms or ecosystems are closed, whether either mesocosms or ecosystems become dominated by any one subsystem, and which if any subsystems can be treated as systems. If an ecosystem component does prevail, this might then be studied in more detail in a mesocosm. Ideas (2) through (5) apply to elements of mesocosms that function differently, such as walls, top surface, bottom surface, and fluid center.

HIERARCHY THEORY. Hierarchy theory is also a conceptual cluster, one with a substantial literature. A Web site (www.science.mcmaster.ca/Biology/ faculty/Kolasa/Biology/faculty/Kolasa/refs.htm) maintained by J. Kolasa at McMaster University in Ontario lists more than 230 publications from 1945 through 1999, many outside the field of ecology. Within this literature, "hierarchy theory" has a variety of meanings. One recurring idea is that systems consist of "nearly decomposable elements" (Simon 1962). Another central idea is that these elements become organized

TABLE 12•1 *Summary of General Systems Theory*

1. Define the system, starting with a complete list of elements and mathematical relations among these elements.

2. The system is summative if the elements function independently.

3. The same equation describes the system and the elements if the functions are linear.

4. Some systems show progressive mechanization. That is, interaction strength decreases with time, with the result that (2) above applies in practice.

5. Some systems show progressive centralization. That is, some elements have greater influence, hence only a few elements contribute to the system-level behavior.

6. Some systems show hierarchical order. That is, the system becomes structured as elements become differentiated into semi-independent units that can themselves be treated as systems. The classic example is organisms, composed of organs, that in turn are composed of tissues consisting of cells.

7. Open systems behave differently from closed systems.

SOURCE From Bertalanffy (1968).

TABLE 12•2 *Concepts in Hierarchy Theory*

1. Hierarchical levels are a useful construct in ecosystem research.

2. These levels are recognized by having different rates at different levels.

3. The system of interest stands at a middle level; faster rates occur at low levels, slower rates occur at higher levels, and these "constrain" the middle-level system of interest.

4. Results or statistics computed at one spatial scale cannot be applied at another.

SOURCE From Parker and Pickett (1998: 174–176).

into "nearly decomposable levels" (Simon 1973). Hierarchy theory includes concepts (5) and (6) from Bertalanffy (table 12.1) and thus is a recognizable offspring of general systems theory.

Hierarchy theory in ecology was developed by Allen and Starr (1982), who extended the traditional hierarchy of organisms, patches, populations, and communities to a variety of ecological phenomena. O'Neill et al. (1986) developed the idea that hierarchical levels arise during the course of energy dissipation in ecosystems. Neither book contains a list of concepts such as that of Bertalanffy (1968) for general systems theory. The phrases "hierarchy theory" and "hierarchical structure" appear in 11 out of 22 chapters in a recent book on scale in ecology (Petersen and Parker 1998). The terms are defined implicitly in these chapters, except for Parker and Pickett (1998). Key concepts extracted from this chapter are listed in table 12.2.

Hierarchy theory, like general systems theory, has had little to say directly about how to link mesocosms to ecosystem research (but see Allen, this volume). If one considers an ecosystem to be hierarchically structured, then a mesocosm is likely to be taken as representing a subsystem, rather than taken as representing an entire ecosystem.

Concept (4) in table 12.2 is called "aggregation error" (cf. O'Neill and King 1998). Hierarchical approaches are widely considered to be useful for ordering concepts, but there is debate about whether it can be used to generate testable hypotheses or mathematical formulations of theory (Steele 1989). Because of aggregation error, one can expect that computations of rates in mesocosms will not scale directly to the ecosystem, based on the area or volume of the ecosystem relative to the area or volume of a mesocosm.

THEORY OF SIMILITUDE IN ENGINEERING. For a variety of reasons, mesocosms cannot be treated as miniature ecosystems. Mesocosms are dominated by the effects of vertical walls because of their small size and round shape compared to aquatic systems. A mesocosm would need to be constructed as a wide and very shallow puddle to be similar in shape to Chesapeake Bay, for example (Uhlmann 1985). One well-worked-out solution to the problem is to apply dimensional methods to calculate the distortions that occur in going from a model system that can be manipulated to a larger prototype that cannot. The method has been highly successful in the design of engineered structures.

The dimensional method, as applied to engineered structures, proceeds by identifying dimensionless ratios that apply to the behavior of both model and prototype. The method computes the ratio that applies to the prototype, then reproduces this for the (smaller) model. Treatments of the theory of similitude for engineering structures (Taylor 1974) tend to be algorithmic with little reference to the underlying theory. Table 12.3 lists a typical sequence of steps.

The great advantage of the method is that it has a history of success in scaling from small to large systems; it allows one to compute changes that occur in moving from small scale models to larger scale prototype, rather

TABLE 12•3 *Steps in Applying Dimensional Analysis*

1. Define the model system and prototype.

2. List the variables that apply to both. In engineering this is often accomplished with a conservation equation for forces acting on model and prototype systems.

3. List the "dimensions" or groups of similar units that apply. Typically these are mass, time, Euclidean lengths L, areas L^2, and volumes L^3.

4. Using any of several textbook methods, form the dimensionless ratios that apply to the behavior of the system. The number of such ratios will be the number of variables, less the number of dimensions.

5. Obtain estimates of these ratios throughout the range that includes both model and prototype.

6. Introduce distortions in the model system so as to mimic the dimensionless ratio of the prototype, allowing the dynamics of the prototype to be investigated with the model system.

SOURCE Adapted from Taylor (1974).

than having to judge the change by verbal or intuitive methods. One can produce numbers to show that typical values encountered in an ecosystem have been reproduced in a mesocosm (see Kemp et al., this volume). Dimensional methods are thus an indispensable component of the design of experiments in mesocosms (Sanford 1997).

The dimensional method has limits when applied to engineering structures, and has yet more limitations when applied to mesocosms. The method relies on dimensionless ratios (table 12.3), but only one ratio at a time can be distorted. It usually proves impossible to mimic the field values of more than one dimensionless ratio in a mesocosm. The dimensions used in engineering fail to capture the biology, although this can be remedied by introducing nontraditional dimensions (Schneider 1994).

Dimensional methods have yet more limitations when applied to natural ecosystems. Dimensional analysis uses Euclidean geometry, but the geometry of an ecosystem is fractal, with surfaces that are rough at all scales (e.g., lake bottoms, ocean floors, coral reefs, and kelp forests) and lines that are convoluted at all scales (e.g., shorelines). Dimensional analysis uses Newtonian dynamics for which rates of change are continuous, linear, and clocklike in regularity. In contrast the dynamics of ecosystems are often complex, proceeding by lurches (e.g., epidemics, landslides, and fisheries collapses). Consequently, the classical principles of similarity that apply to an engineered model and prototype are hard to apply to ecosystems.

UNIVERSAL SCALING IN COMPLEX SYSTEMS. Universal scaling (Stanley et al. 1996) applies to systems whose structure arises from the effects of two opposing exponential rates, giving rise to a power law relationship between two variables (see also Schneider, this volume). Biological examples of power laws attributable to opposing exponential rates include the diameter of bacterial colonies along a strip and the perimeter of an expanding muskrat population (Stanley et al. 1996). A key idea in universal scaling is the renormalization group (Wilson 1971), which is a set of transformations that leave the relations of variables unchanged in going from one scale to another. Barenblatt (1996) calls these relations "incomplete similarity," thus emphasizing that they are generalizations of the completely similar relations used in standard dimensional analysis. Barenblatt (1996) provides an algorithm for applying incomplete similarity (table 12.4).

Examples of the application of incomplete similarity to mesocosms and ecosystems can be found elsewhere in this volume (Schneider). Power laws based on incomplete similarity are becoming increasingly common in

TABLE 12•4 *Analysis of Incompletely Similar Systems*

1. List variables.

2. Define similarity groups (dimensions) such as mass, length, time.

3. Reduce the list of variables to a smaller list of dimensionless ratios, using the methods of dimensional analysis.

4. Obtain estimates of these new (dimensionless) variables and plot them to investigate their behavior.

5. Test for fit to simple power laws based on Euclidean geometry.

6. If this fails, estimate power law scalings, such as fractal dimensions.

SOURCE From Barenblatt (1996).

ecology. Examples include fractal landscapes, allometric relations of organism function to size, and species-area curves.

It is natural to suppose that if the same power laws appear for mesocosms and ecosystems, then a single scaling law might be developed to compute from mesocosm results to ecosystem behavior. Unfortunately, the same power law can appear when the underlying dynamics of mesocosms differ from that of ecosystems. Consequently, empirically derived power laws must be used with caution in any attempt to compute from very small systems (such as mesocosms) to very large systems (such as Chesapeake Bay).

Power laws are commonly encountered in both ecosystems and mesocosms (Schneider, this volume). Universal scaling theory provokes the question "What opposing exponential rates were responsible for the observed power law relation of two variables?" Mesocosms provide an experimental basis for addressing this question.

GEOPHYSICAL FLUID DYNAMICS. Rapid progress in understanding the ocean and the atmosphere occurred when principles of fluid dynamics, developed for flow in pipes, were applied in a geophysical grid (Batchelor 1967; Pedlosky 1979). The approach is applicable to biological components of marine ecosystems (Wroblewski and Hofmann 1989), but examples are few. This approach recognizes the limits of theory and the impossibility of manipulative experiments at the scale of ecosystems. Intense computation rather than mathematical analysis is used to apply theory to complex dynamics. Table 12.5 shows the typical sequence of development followed by cyclic testing of computational model against field data.

TABLE 12•5 *Model Development and Testing in Geophysical Fluid Dynamics*

1. Define the system of interest, its boundaries, appropriate spatial elements (typically in a grid), and an appropriate time scale at which to compute the dynamics.

2. Write the governing equations for the dynamics. In geophysical fluid dynamics, the equations for fluid transport among cells (spatial elements) are based on Newton's laws of motion, with constraints (such as incompressibility of water) introduced as appropriate.

3. Use dimensionless ratios to simplify the equations by eliminating terms that do not apply at the space and time scales of interest.

4. Use the governing equations to compute from one time step to the next.

5. Design observational programs to test predictions from the computational model.

6. Revise the model as appropriate.

These models require estimates of biological rates (Wroblewski and Hofmann 1989), which can be obtained under defined conditions from mesocosms. Models based on fluid dynamics could, in principle, be used to develop predictions of the behavior of mesocosms placed into the water column, as with large bag experiments in ponds, lakes, or oceans. The cycle of research by model prediction and testing might then be extended to yet larger and more natural systems, such as a cove or section of an ecosystem, eventually being applied to an entire ecosystem. Geophysical models are a potentially important but untried means of linking mesocosm results to ecosystem questions.

Practice

This section examines how to address patchiness explicitly when trying to link mesocosm studies to natural systems. Three issues will be addressed: how to create patchiness within and among experimental ecosystems, how to link patchiness seen at broad scales with results obtained from mesocosm studies, and how to use computer simulations to address patchiness over multiple scales.

PATCHINESS INTO THE MESOCOSM. In a well-mixed mesocosm, the reproduction of pelagic patchiness is a difficult task. The problem is exacerbated because, in the ocean, the characteristic length scales of physical and biological patches are very large. Except for very unusual environments and for severely under-mixed mesocosms, realistic patch sizes are typically orders of magnitude larger than most experimental ecosystems (Sanford 1997). Given these constraints, there are still many ways to examine the effects of patchiness in mesocosms and to gain a better understanding of how results can be extrapolated to larger scales.

One way is to incorporate patchiness directly into the experimental ecosystem itself. The Marine Ecosystem Research Laboratory (MERL) at the University of Rhode Island created a thermocline within a long cylindrical mesocosm through differential heating of the top and bottom, in effect creating a temperature gradient (Donaghay and Klos 1985; Klos 1988). By isolating the original well-mixed mesocosm into two distinct "patches," one can now investigate questions related to benthic-pelagic coupling or examine the rate of fluxes across a boundary. Caution needs to be exercised when incorporating physical patchiness into mesocosms because often the creation of the patch boundary (e.g., thermocline) may interact with the mesocosm boundary to distort ambient mixing (Sanford 1997).

Patchiness may also be incorporated by the use of pulse perturbation experiments. For example, nutrients can be added to a mesocosm at a frequency matching that of the natural environment, or added as a spike to simulate an isolated event such as fall turnover. Pulse experiments can also examine the effects of higher trophic levels. Fish can be added for a specified time and then removed to approximate patchiness created by the "pulsed" predation of a fish school. Alternatively, fish can be excluded altogether and predation effects simulated by removing a percentage of the biomass each day. In addition, physical environments containing different levels of patchiness can be approximated by varying the mixing and exchange rates of mesocosms. Artificial waves can be introduced to simulate a reef environment, mixing levels can be adjusted to simulate different turbulence and light levels (Petersen et al. 1998), and exchanges can be modified to reflect different residence times within the mesocosm.

Patchiness within experimental ecosystems can also be created by linking separate mesocosms into a single experimental unit. This method has two main advantages. First, it allows patchiness to be created between mesocosm units, thereby avoiding some of the problems mentioned above of trying to incorporate patchiness within a single mesocosm. The mixing within the separate mesocosms can be maintained at realistic levels, enabling experiments to be conducted where exchanges between mesocosms are emphasized. For example, by creating a low exchange (i.e., mixing) rate between different mesocosms, refugia can be created for prey species, or an environment with low exchanges with adjacent parcels of water can be simulated. In addition, multiple mesocosms allow for the examination of how patchiness in the form of multiple habitat types (e.g., submerged aquatic vegetation, pelagic, or marsh habitats, etc.) affects system response. Systems could be linked in different configurations with different exchange rates to approximate the residence time typical of each subsystem. The Smithsonian's Chesapeake Bay and Everglades mesocosms employed this method to link various habitat types together to create a functioning ecosystem model (Adey and Loveland 1998). A more complete picture should be possible by linking mesocosms because the interaction of the different subsystems can be addressed, as would be needed, for example, if one wanted to examine the role that marshes play in moderating the nutrient fluxes to the pelagic water column. These questions simply cannot be addressed in an integrated fashion with isolated mesocosms of each habitat type.

Still another approach to the inclusion of patches within experimental ecosystems is to simply use larger systems. With the expanded temporal and spatial scales, one is able to capture more of the biological and physical patchiness found in the natural system. Experiments in freshwater systems have used this method for many years by doing whole lake manipulations or performing the experiment in lakes of different size (Fee and Hecky 1992). In marine systems, due to their open nature, an appropriate analogue is either missing, or too cost prohibitive. One possible surrogate could be the use of warm/cold core rings from major current systems. These are isolated ecosystems with minimal exchange with the surrounding environment, a captive community at least at the phytoplankton-zooplankton level, with a life span of weeks to months and sometimes years (Tranter et al. 1980; The Ring Group 1981). These rings are common features in some oceanic environments and if one can find a range of sizes, then some scale-dependence comparisons may be possible. The ultimate extrapolation for natural mesocosms would be the use of mid-oceanic gyres as isolated experimental systems. They would encompass all the scales of physical and biological variability present in the pelagic environment. Unfortunately, at these larger scales one loses homogeneity, replicability, and the possibility of experimental manipulation. In addition, there are large costs associated with sampling such broad areas for the long periods of time necessary to capture the dynamics at this scale (e.g., climate cycles, multiple life stages, etc.). Recent advances in remote sensing technology may help to alleviate this problem or at least make it more manageable.

PATCHINESS FROM A MAP. Instead of putting patches into a mesocosm, one might design mesocosm experiments to sample across the range of variables obtained from a "map" of the ecosystem. This approach recognizes the difficulty of inclusion of a patch into an experimental ecosystem and, instead, relies on subsampling portions of a patch within replicate mesocosms. Most mesocosm experiments are designed to represent the mean or expected values. In this case, experiments will be designed to sample the range of ecosystem behavior. This mode of linking mesocosms to an ecosystem holds greatest promise when dynamics measurable in an experimental ecosystem possess a known relation to a variable that can be mapped in great detail in the field.

A map sufficient to the task can be constructed in one of several ways. One obvious way is to employ satellite imagery, or some other form of remote sensing. This produces a detailed map with a large scope (i.e., with high resolution of local variation over wide spatial extent). Patchiness can be characterized at fine and broad scales, allowing maximum and minimum values of the mapped variable to be identified. When the ecosystem variable cannot be mapped in detail, it can often be mapped from a survey. The methods for constructing a map from survey points are numerous (Cressie 1991). The resolution of such a map can be no finer than the separation between locations used in the survey, hence information will be less and patchiness can be characterized at just a few scales. Nevertheless, the map still holds considerable spatial information and so can be linked to mesocosm results to gain a better understanding of the spatial dynamics of an ecosystem.

At least as important as the map is a secure means of linking the mapped variable to the dynamic variable measured under controlled conditions in the mesocosm. An example of such a linkage is that between ocean color and primary production—a linkage that allows ocean color to be used to compute primary production via the relation of color to chlorophyll, and chlorophyll to production. Both relations can be investigated under the controlled conditions of a mesocosm; the relations so developed can be tested under field conditions. Once verified, the relations can be used to compute production from a mappable variable, ocean color.

When patchiness is contained in a map, rather than in a mesocosm, the range of variability found in the field becomes a central factor in the design of the experiment. This focus on the range of variability in the mapped variable differs from the more intuitive approach of designing mesocosm experiments relative to average values of the variable mapped in the ecosystem. A typical research sequence might be to identify the range of the mapped variable at some scale of interest, design a mesocosm experiment across this range of the mapped variable, identify the functional relation of dynamic variable to mapped variable in the mesocosm, test this relation in the field if possible, then compute the map of the dynamic variable from the map of the ecosystem variable. Conclusions are drawn by embedding mesocosm results into a "surface" represented by the mapped ecosystem rather than from the average or "typical" mesocosm result. This embedded experiment approach (Schneider 1978) was used recently to draw conclusions about ecological processes acting over an entire sandflat, from experiments necessarily limited to a few meters in extent (Thrush et al. 1997). A similar approach could be used to link mesocosm experiments to a mapped ecosystem variable, to draw conclusions at the scale of a marine ecosystem.

COMPUTATIONAL METHODS. Computer simulations are powerful tools for studying patch effects in marine systems as well as extrapolating mesocosm results to the natural environment. If a computer model adequately simulates the system of interest (including appropriate levels of physical and biological patchiness) then changes over a range of spatial, temporal, and complexity scales may be examined quantitatively. In addition, the scale dependence inherent in these systems may also be investigated. The "experiments" that may be performed using a computer simulation are often impossible within an isolated mesocosm or may be too cost prohibitive to conduct in the field.

Spatial effects can be incorporated into the design of a computer model by dividing the model domain into a gridded framework. This approach (Eulerian method) follows the bulk flow of constituents and organisms past a fixed point. Rules and equations are used to govern the exchange of material and movement of organisms between cells. Experiments can then be run under different grain sizes (i.e., cell size) and extents (i.e., total model domain) to characterize how the effects of patchiness within a given model simulation (i.e., in the physical and biological constituents) change with scale. In addition, this method also allows the simulation of experimental mesocosms with limited (or regulated) exchanges. Thus, a model adequate to explain the fine-grained dynamics of individual mesocosms can also be used to extrapolate results to the scales of the natural system.

A different yet complementary approach can be employed to investigate patchiness within a computer simulation framework. This technique, called the Lagrangian method, follows particles around as they move within a predefined model domain. The domain, or size of the system, can range from a small mesocosm up to a system as large as the Chesapeake Bay. The rules and equations used to move the particles from one place to the next are similar to those described above, but an advantage of this method is that the fate of individuals, not just biomass, can be tracked over the duration of a simulation. For example, in a recent paper (Hood et al. 1999), the technique was used to follow the flow of passive particles (i.e., phytoplankton) within a simulated Chesapeake Bay environment. The simulation showed how the particles oscillated back and forth with the tides, then identified regions where particles would accumulate and form patches. This simulation even identified an accumulation zone that moved around but still maintained its integrity over extended time periods. The existence of these "emergent" features would be hard to detect with typical field surveys. In addition, questions related to behavior and altered feeding patterns can be addressed with the Lagrangian method. Heath and Houde (this volume) used simulated fish within a simulated mesocosm to test the effects that container size and behavior have on feeding efficiency.

Both the Eulerian and Lagrangian methods can incorporate a wide variety of habitat types and physical conditions, thereby providing a valuable link between isolated mesocosm experiments and the natural system. A multitude of different habitat configurations, physical conditions, and biotic components can be explored within a modeling framework that is far less costly than the corresponding suite of experimental mesocosms. The models can be adjusted to give similar dynamics and then used as tools to investigate dynamics beyond the boundaries imposed by the restricted time, space, and organismal scales of typical experimental ecosystem experiments.

COMMENTS

Patchiness created by physical and biological processes acting over many scales is a critical component of marine ecosystems. Often, the effects of patch structures cannot be reproduced easily in mesocosm studies. Because of the wide range of physical variability that occurs in the ocean, pelagic length scales (kilometers) often exceed the size of a single mesocosm experiment (meters). Attempts to simulate larger-scale physical and biological phenomena in mesocosms inevitably distort smaller-scale processes (such as turbulence-dependent planktonic predator-prey interaction). The net result is that it is difficult, if not impossible, to simulate the full suite of marine biological-physical interactions in individual mesocosms.

Short of enclosing extremely large volumes of water, the inclusion of patch effects within mesocosm experiments remains a significant challenge. On potential solution is to use the strengths of mesocosms—that is, the homogeneity, control, and replication—to study within-patch dynamics. For instance, experimental manipulations can systematically examine mixing effects, disturbance, and predator-prey dynamics. Using various mixing rates to simulate different light and energy regimes and press- and pulse-experiments to examine both top-down and bottom-up effects on community dynamics are also possible. One recent experiment used mesocosms to scale primary productivity under light- and nutrient-limited conditions (Petersen et al. 1997). The search for additional scale-sensitive relationships has obvious implications.

Altered patch levels within mesocosm studies are a major factor limiting extrapolation. Experiments using mesocosms of different sizes can capture the change in process that occurs as a result of changes in the geometric dimension of the mesocosm, thus allowing the effects of artifacts due to enclosure to be quantified (Chen et al. 1997). Examples of this approach include the mesocosm experiments at the Multiscale Experimental Ecosystem Research Center (MEERC) at the University of Maryland. At the largest scales, use of naturally isolated systems, such as warm/cold core rings of various sizes, might be considered.

Unfortunately, employing a suite of experimental sizes to study the scale-dependent nature of a particular question is expensive. Alternatives are to use methods to extrapolate results beyond the confines of a particular experimental ecosystem size. A number of techniques are possible, including measurements of the ranges of variables of interest from field experiments or using larger-scale synoptic images and linking these to the results obtained in the mesocosm studies. Computer simulation models are also suggested for running virtual scale experiments as a way to cycle between mesocosm experiments and field experiments. The goal would be to incorporate the important biological and physical patch-generating mechanisms between the fine-scale study (i.e., experimental ecosystem) and its broad-scale application (i.e., environmental problem). This matching of the scale of the experiment with the scale of the question is not always done. In a recent literature review of mesocosm studies (Petersen et al. 1999), mismatches between the physical and biological processes and the temporal and spatial scales over which the experiments were conducted occurred often. Under these circumstances, the failure to account for longer-term broad-scale effects, and the corresponding patchiness that they can create, can drastically limit the applicability of results.

Patchiness will always affect experimental ecosystem results and their extrapolation. If these systems are to be useful tools for addressing broad-scale environmental problems, then we need ways to deal with patchiness, both within and outside the mesocosm. This chapter examined this pressing issue by reiterating the importance of patchiness in marine systems, listing some ways in which patchiness is inevitably excluded from mesocosm experiments, and outlining methods to link results from experimental ecosystems to descriptions (including models) of natural systems at larger scales. Failure to consider patchiness when conducting experimental ecosystem experiments will result in well-replicated and controlled results of artifacts that have little relevance to current marine environmental issues.

ACKNOWLEDGMENTS

D. Breitburg, C-C. Chen, S. Crawford, M. Heath, E. Houde, T. Malone, R. Mason, and E. Porter were active participants in the workshop discussions which provided the foundation for this chapter. Support was provided by the U.S. EPA STAR program as part of the Multiscale Experimental Ecosystem Research Center (MEERC) at the University of Maryland's Center for Environmental Science (R819640). The manuscript was improved by comments from Victor Kennedy, John Petersen, Robert Gardner, Steven Walters, and an anonymous reviewer.

LITERATURE CITED

Abbott, M. R. 1993. Phytoplankton patchiness: Ecological implications and observation methods. In S. A. Levin, T. M. Powell, and J. H. Steele, eds, *Patch Dynamics*, pp. 37–49. Lecture Notes in Biomathematics 96. New York: Springer-Verlag.
Adey, W. H., and K. Loveland. 1998. *Dynamic Aquaria: Building Living Ecosystems*. San Diego: Academic Press.

Allen, T. F. H. 1977. Scale in microscopic algal ecology: A neglected dimension. *Phycologia* 16:253–257.

Allen, T. F. H., and T. W. Hoekstra. 1992. *Toward a Unified Ecology.* New York: Columbia University Press.

Allen, T. F. H., and T. B. Starr. 1982. *Hierarchy: Perspectives for Ecological Complexity.* Chicago: University of Chicago Press.

Barber, R. T., and F. P. Chavez. 1986. Ocean variability in relation to living resources during the 1982–83 El Niño. *Nature* 319:279–285.

Barenblatt, G. I. 1996. *Scaling, Self-similarity, and Intermediate Asymptotics.* Cambridge: Cambridge University Press.

Batchelor, G. K. 1967. *An Introduction to Fluid Mechanics.* Cambridge: Cambridge University Press.

Bertalanffy, L. von. 1968. *General Systems Theory.* New York: Braziller.

Beyers R. J., and H. T. Odum. 1993. *Ecological Microcosms.* Berlin: Springer-Verlag.

Bigelow, H. B., and W. C. Schroeder. 1953. Fishes of the Gulf of Maine. *Fishery Bulletin* 74:1–577.

Bloesch, J., P. Bossard, H. Buhrer, H. R. Burgi, and U. Uehlinger. 1988. Can results from limnocorral experiments be transferred to in situ conditions? *Hydrobiologia* 159:297–308.

Brown, J. L. 1975. *The Evolution of Behavior.* New York: Norton.

Carpenter, S. R. 1996. Microcosm experiments have limited relevance for community and ecosystem ecology. *Ecology* 77:677–680.

Carpenter, S. R., S. W. Chisholm, C. J. Krebs, D. W. Schindler, and R. F. Wright. 1995. Ecosystem experiments. *Science* 269:324–327.

Chen, C,-C., J. E. Petersen, and W. M. Kemp. 1997. Spatial and temporal scaling of periphyton growth on walls of estuarine mesocosms. *Marine Ecology Progress Series* 155:1–15.

Cressie, N. A. C. 1991. *Statistics for Spatial Data.* New York: Wiley.

Denman, K. L. 1976. Covariability of chlorophyll and temperature in the sea. *Deep-Sea Research* 23:539–550.

Denman, K. L. 1992. Scale-determining biological-physical interactions in oceanic food webs. In P. S. Giller, A. G. Hildrew, and D. G. Raffaelli, eds., *Aquatic Ecology: Scale, Pattern and Process,* pp. 377–402. Oxford: Blackwell Science.

Denman, K. L., and T. Platt. 1976. The variance spectrum of phytoplankton in a turbulent ocean. *Journal of Marine Research* 34:593–601.

Denman, K. L. and T. M. Powell. 1984. Effects of physical processes on planktonic ecosystems in the coastal ocean. *Oceanography and Marine Biology: An Annual Review* 22:125–165.

Donaghay, P. L., and E. Klos. 1985. Physical, chemical and biological responses to simulated wind and tidal mixing in experimental marine ecosystems. *Marine Ecology Progress Series* 26:35–45.

Fee, E. J., and R. E. Hecky. 1992. Introduction to the Northwest Ontario Lake size series (NOLSS). *Canadian Journal of Fisheries and Aquatic Sciences* 49:2434–2444.

Franks, P. J. S. 1992. Phytoplankton blooms at fronts: Patterns, scales, and physical forcing mechanisms. *Reviews in Aquatic Sciences* 6:121–137.

Frost, T. M., D. L. DeAngelis, S. M. Bartell, D. J. Hall, and S. H. Hurlbert. 1988. Scale in the design and interpretation of aquatic community research. In S. R. Carpenter, ed., *Complex Interactions in Lake Communities*, pp. 229–258. New York: Springer-Verlag.

Gaskin, D. E. 1976. The evolution, zoogeography and ecology of Cetacea. *Oceanography and Marine Biology: An Annual Review* 43:163–178.

Giller P. S., A. G. Hildrew, and D. G. Raffaelli. 1992. *Aquatic Ecology: Scale, Pattern and Process*. Oxford: Blackwell Science.

Gower, J. F. R., K. L. Denman, and R. J. Holyer. 1980. Phytoplankton patchiness indicates the fluctuation spectrum of mesoscale oceanic structure. *Nature* 288:13–15.

Hamner, W. M., P. P. Hamner, S. W. Strand, and R. W. Gilmer. 1983. Behavior of Antarctic krill, *Euphausia superba*: Chemoreception, feeding, schooling, and molting. *Science* 220:433–435.

Hardy, A. C., and E. R. Gunther. 1935. The plankton of the South Georgia whaling grounds and adjacent waters, 1926–1927. *Discovery Reports* 11:1–456.

Harris, G. P. 1980. Temporal and spatial scales in phytoplankton ecology: Mechanisms, methods, models, and management. *Canadian Journal of Fisheries and Aquatic Sciences* 37:877–900.

Haury, L. R., J. A. McGowan, and P. H. Wiebe. 1978. Patterns and processes in the time-space scales of plankton distributions. In J. H. Steele, ed., *Spatial Pattern in Plankton Communities*, pp. 277–327. NATO Conference Series 4, vol. 3. New York: Plenum.

Hood, R. R., H. V. Wang, J. E. Purcell, E. D. Houde, and L. W. Harding. 1999. Modeling particles and pelagic organisms in Chesapeake Bay: Convergent features control plankton distributions. *Journal of Geophysical Research-Oceans* 104:1223–1243.

Horne, J. K., and D. C. Schneider. 1997. Spatial variance of mobile aquatic organisms: Capelin and cod in Newfoundland coastal waters. *Philosophical Transactions of the Royal Society of London, B* 352:633–642.

Inman, A. L., and J. Krebs. 1987. Predation and group living. *Trends in Ecology and Evolution* 2:31–32.

Kierstead, H., and L. B. Slobodkin. 1953. The size of water masses containing plankton blooms. *Journal of Marine Research* 12:141–147.

Klos, E. 1988. An experimental estuarine salinity gradient. In *Proceedings of Oceans 88 Conference*, pp. 1529–1534. Piscataway, NJ: Institute of Electrical and Electronics Engineers.

Kolmogorov, A. N. 1941. The local structure of turbulence in an incompressible viscous fluid for very large Reynolds number. *Akademiia Nauk SSSR Comptes Rendus* 30:299–303.

Lasker, R. 1975. Field criteria for survival of anchovy larvae: The relation between inshore chlorophyll maximum layers and successful first feeding. *Fishery Bulletin* 73:453–462.

LeBrasseur, R. J., W. E. Barraclough, O. D. Kennedy, and T. R. Parsons. 1969. Production studies in the strait of Georgia. Part III. Observations on the food of larval and juvenile fish in the Fraser River plume, February to May, 1967. *Journal of Experimental Marine Biology and Ecology* 3:39–50.

Levin, S. A. 1992. The problem of pattern and scale in ecology. *Ecology* 73:1943–1967.

Levin, S. A., A. Morin, and T. M. Powell. 1989. Patterns and processes in the distribution and dynamics of Antarctic krill. In Scientific Committee for the Conservation of Antarctic Marine Living Resources, *Selected Scientific Papers Part 1*, pp. 281–299. SC-CAMLR-SSP/5. Hobart: SC-CAMLR.

Mackas, D. L., K. L. Denman, and M. R. Abbott. 1985. Plankton patchiness: Biology in the physical vernacular. *Bulletin of Marine Science* 37:652–674.

Marquet, P. A., M.-J. Fortin, J. Pineda, D. O. Wallin, J. Clark, Y. Wu, S. Bollens, C. M. Jacobi, and R. D. Holt. 1993. Ecological and evolutionary consequences of patchiness: A marine-terrestrial perspective. In S. A. Levin, T. M. Powell, and J. H. Steele, eds., *Patch Dynamics*, pp. 277–304. New York: Springer-Verlag.

National Research Council (NRC). 1995. *Understanding Marine Biodiversity.* Washington, DC: National Academy Press.

Norris, K. S., and C. R. Schilt. 1987. Cooperative societies in three-dimensional space: On the origins of aggregations, flocks, and schools, with special reference to dolphins and fish. *Ethology and Sociobiology* 9:149–179.

Okubo, A. 1980. *Diffusion and Ecological Problems: Mathematical Models.* Biomathematics 10. New York: Springer-Verlag.

Okubo, A. 1986. Dynamical aspects of animal grouping: Swarms, schools, flocks, and herds. *Advanced Biophysics* 22:1–94.

O'Neill, R. V., D. L. DeAngelis, J. B. Waide, and T. F. H. Allen. 1986. *A Hierarchical Concept of Ecosystems.* Princeton: Princeton University Press.

O'Neill, R. V., and A. W. King. 1998. Homage to St. Michael: or, why are there so many books on scale? In D. L. Peterson and V. T. Parker, eds., *Ecological Scale*, pp. 3–15. New York: Columbia University Press.

Oschlies, A., and V. Garcon. 1998. Eddy-induced enhancement of primary production in a model of the North Atlantic Ocean. *Nature* 394:266–269.

Parker, V. T., and S. T. A. Pickett. 1998. Historical contingency and multiple scales of dynamics within plant communities. In D. L. Peterson and V. T. Parker, eds., *Ecological Scale: Theory and Applications*, pp. 171–192. New York: Columbia University Press.

Pedlosky, J. 1979. *Geophysical Fluid Dynamics.* New York: Springer-Verlag.

Peters, R. H. 1991. *A Critique for Ecology.* Cambridge: Cambridge University Press.

Petersen, J. E., C.-C. Chen, and W. M. Kemp. 1997. Scaling aquatic primary productivity: Experiments under nutrient- and light-limited conditions. *Ecology* 78:2326–2338.

Petersen, J. E., J. C. Cornwell, and W. M. Kemp. 1999. Implicit scaling in the design of experimental aquatic ecosystems. *Oikos* 85:3–18.

Petersen, J. E., L. P. Sanford, and W. M. Kemp. 1998. Coastal plankton responses to turbulent mixing in experimental ecosystems. *Marine Ecology Progress Series* 171:23–41.

Peterson, D. L., and V. T. Parker, eds. 1998. *Ecological Scale: Theory and Applications.* New York: Columbia University Press.

Platt, T. 1972. Local phytoplankton abundance and turbulence. *Deep-Sea Research* 19:183–187.

Powell, T. M. 1989. Physical and biological scales of variability in lakes, estuaries, and the coastal ocean. In J. Roughgarden, R. M. May, and S. A. Levin, eds., *Perspectives in Ecological Theory*, pp. 157–176. Princeton: Princeton University Press.

Raffaelli, D. G., A. G. Hildrew, and P. S. Giller. 1992. Scale, pattern and process in aquatic systems: Concluding remarks. In P. S. Giller, A. G. Hildrew, and D. G. Raffaelli, eds., *Aquatic Ecology: Scale, Pattern and Process*, pp. 377–402. Oxford: Blackwell Science.

Rothlisberg, P. C., J. A. Church, and A. M. G. Forbes. 1983. Modelling the advection of vertically migrating shrimp larvae. *Journal of Marine Research* 41:511–538.

Sanford, L. P. 1997. Turbulent mixing in experimental ecosystem studies. *Marine Ecology Progress Series* 161:265–293.

Saunders, P. M. 1992. Space and time variability of temperature in the upper ocean. *Deep-Sea Research* 19:467–480.

Schneider, D. 1978. Equalisation of prey numbers by migratory shorebirds. *Nature* 271:353–354.

Schneider, D. C. 1991. The role of fluid dynamics in the ecology of marine birds. *Oceanography and Marine Biology: An Annual Review* 29:487–521.

Schneider, D. C. 1992. Scale-dependent patterns and species interactions in marine nekton. In P. S. Giller, A. G. Hildrew and D. G. Raffaelli, eds., *Aquatic Ecology: Scale, Pattern and Process*, pp. 441–467. Oxford: Blackwell Science.

Schneider, D. C. 1994. *Quantitative Ecology: Spatial and Temporal Scaling*. San Diego: Academic Press.

Sheldon, R., W. A. Prakash, and W. H. Sutcliffe Jr. 1972. The size distribution of particles in the ocean. *Limnology and Oceanography* 17:327–340.

Simon, H. A. 1962. The architecture of complexity. *Proceedings of the American Philosophical Society* 106: 467–482.

Simon, H. A. 1973. The organization of complex systems. In H. H. Pattee, ed., *Hierarchy Theory*, pp. 3–27. New York: Braziller.

Sinclair, M. 1987. *Marine Populations*. Seattle: University of Washington Press.

Skellam, J. G. 1951. Random dispersal in theoretical populations. *Biometrika* 38:196–218.

Southward, A. J. 1980. The western English Channel: An inconstant ecosystem? *Nature* 285:361–366.

Stanley, H. E., L. A. N. Amaral, S. V. Buldyrev, A. L. Goldberger, S. Havlin, H. Leschhorn, P. Maass, H. A. Makse, C.-K. Peng, M. A. Salinger, M. H. R. Stanley, and G. M. Viswanathan. 1996. Scaling and universality in animate and inanimate systems. *Physica A* 231:20–48.

Staples, D. J. 1979. Seasonal migration patterns of postlarval and juvenile banana prawns, *Penaeus merguiensis de Man*, in the major rivers of the Gulf of Carpentaria, Australia. *Australian Journal of Marine and Freshwater Research* 30:143–157.

Steele, J. H. 1978. Some comments on plankton patches. In J. H. Steele, ed., *Spatial Pattern in Plankton Communities*, pp. 157–176. NATO Conference Series 4, vol. 3. New York: Plenum.

Steele, J. H. 1989. Discussion: Scale and coupling in ecological systems. In J. Roughgarden, R. M. May, and S. A. Levin, eds., *Perspectives in Ecological Theory*, pp. 177–180. Princeton: Princeton University Press.

Steele, J. H. 1991. Marine ecosystem dynamics: Comparison of scales. *Ecological Research* 6:175–183.

Stommel, H. 1963. Varieties of oceanographic experience. *Science* 139:572–576.

Strass, V. H. 1992. Chlorophyll patchiness caused by mesoscale upwelling at fronts. *Deep-Sea Research* 39:75–96.

Sund, P. N., M. Blackburn, and F. Williams. 1981. Tunas and their environment in the Pacific Ocean: A review. *Oceanography and Marine Biology: An Annual Review* 19:443–512.

Taylor, E. S. 1974. *Dimensional Analysis for Engineers*. Oxford: Clarendon Press.

The Ring Group. 1981. Gulf Stream cold-core rings: Their physics, chemistry, and biology. *Science* 212:1091–1100.

Thrush, S. F., V. J. Cummings, P. K. Dayton, R. Ford, J. Grant, J. E. Hewitt, A. H. Hines, S. M. Lawrie, R. D. Pridmore, P. Legendre, B. H. McArdle, D. C. Schneider, S. J. Turner, R. B. Whitlatch, and M. R. Wilkinson. 1997. Matching the outcome of small-scale density manipulation experiments with larger scale patterns: An example of adult-juvenile interactions. *Journal of Experimental Marine Biology and Ecology* 216:153–169.

Tilman, D. 1989. Ecological experimentation: Strengths and conceptual problems. In G. E. Likens, ed., *Long-Term Studies in Ecology*, pp. 136–157. New York: Springer-Verlag.

Tilman, D., and P. Kareiva. 1997. *Spatial Ecology: The Role of Space in Population Dynamics and Interspecific Interactions*. Princeton: Princeton University Press.

Tranter, D. J., R. R. Parker, and G. R. Cresswell. 1980. Are warm-core eddies unproductive? *Nature* 284:540–542.

Turner, G., and T. J. Pitcher. 1986. Attack dilution, search, and abatement. *American Naturalist* 128:228–240.

Uhlmann, D. 1985. Scaling of microcosms and the dimensional analysis of lakes. *Internationale Revue der Gesamten Hydrobiologie* 70:47–62.

Venrick, E. L. 1982. Phytoplankton in an oligotrophic ocean: Observations and questions. *Ecological Monographs* 52:129–154.

Weber, L. H., S. Z. El-Sayed, and I. Hampton. 1986. The variance spectra of phytoplankton, krill and water temperature in the Antarctic Ocean south of Africa. *Deep-Sea Research* 33:1327–1343.

Wiens, J. A. 1989. Spatial scaling in ecology. *Functional Ecology* 3:385–397.

Wilson, K. G. 1971. Renormalization group and critical phenomena: (I) Renormalization group and the Kadanoff scaling picture. *Physical Reviews* 4:3174–3183.

Wroblewski, J. J., and E. E. Hofmann. 1989. U.S. interdisciplinary modeling studies of coastal-offshore exchange processes: Past and future. *Progress in Oceanography* 23:65–99.

INDEX

A

Abies 70
Acacia aneura 68
acidification 262
aerial remote sensing, use in water sampling of 319
aggregation error 115–116, 343
alleles 226
allometric scaling 118
   use in ecophysiology 309
allometry 303
   scaling of 118
American shad (*Alosa sapidissima*) 306
anisotropy 74
ant communities 75–76
aquaculture, effects on marine ecosystems 342
aquatic ecosystems
   effects of gravity on 19
   effects of observational scale 6
   effects of sunlight on 19
   experimental designs 5
   minimum size of enclosures in 46
   regulation of water residence time 32
   time scale in 254
   See also ecosystems
archipelago 13
areas, use in scaling functions 116
arrow worms 184
artifacts 10, 313
   scale dependence of 17
   sources of 10
artificial ecosystems 157
artificial gradients 339
artificial waves 349
assumptions 91
   use in experimentation 97–100
Astagalus canadensis 226
Atlantic cod (*Gadus morhua*) 312, 341
Atlantic croaker (*Micropogonias undulatus*) 306
Atlantic puffins (*Fratercula arctica*) 311
autocorrelation 320
autotrophs 30, 308

B

bacterial production 165
bags, use as enclosures of 9
Bak sandpile 140
Baltic Sea 184
basidiocarps 64
bay anchovy (*Anchoa mitchilli*) 193, 212–213
bay scallop (*Argopecten irradians*) 186
Bayesian methods 73
BD-EF (biodiversity-ecosystem functioning) 234–243
   experimental studies 236–241
beetles 78
benthic algae 258
benthic macrofauna